Regenerative Urban Development, Climate Change and the Common Good

This volume focuses on the theory and practice of the regenerative development paradigm that is rapidly displacing sustainability as the most fertile ground for climate change adaptation research.

This book brings together key thinkers in this field to develop a meaningful synthesis between the existing practice of regenerative development and the input of scholars in the social sciences. It begins by providing an expert introduction to the history, principles, and practices of regenerative development before going on to present a thorough theoretical examination by known theorists from disciplines including sociology, geography, and ethics. A section on regenerative development practices illustrates the need to significantly advance our understanding of how urbanization, climate change, and inequality interact at every scale of development work. Finally, the book ends with a serious consideration of the ways in which integrated systems thinking in higher education could result in a curriculum for the next generation of regenerative development professionals.

Regenerative Urban Development, Climate Change and the Common Good will be of great interest to students, scholars, and practitioners of regenerative development, climate change, urban planning, and public policy.

Beth Schaefer Caniglia is Editor-in-Chief of *The Solutions Journal*, where she amplifies the voices of dreamers, innovators, change-makers, and risk-takers in service to building an equitable and sustainable world.

Beatrice Frank is the Social Science Specialist for the Capital Regional District Regional Parks and an adjunct professor at the University of Victoria, Canada. As a social scientist, she focuses on engaging people in decision-making around natural resources management and conservation. She has been engaged in discussion around climate change, sustainability, and environmental justice over the last decade, which has flourished in her contribution on regenerative development discourses.

John L. Knott, Jr., a third generation master builder/developer with over 50 years of experience, is an internationally recognized leader in green building, sustainable development, and urban regeneration. He is the creator and Founder of CityCraft®. Knott is recognized as an influential thought leader in molding the nation's sustainable development movement. He has served as an advisor to Department of Housing and Urban Development (HUD), Department of Energy (DOE), Environmental Protection Agency (EPA), and the National Park Service throughout his career, also serving as Chairman of the US Working Group for Urban-Suburban Indicators in compiling the landmark Heinz Center report, The State of the Nation's Ecosystems.

Kenneth S. Sagendorf is the Founding Director of the Innovation Center and Professor in the Anderson College of Business at Regis University. Working in a college focused on stewardship, he built the Innovation Center to challenge the status quo of business education to focus on creative and systems thinking and love and heroism. He deeply believes in the power of education to create the humans and the humanity needed for our future.

Eugene A. Wilkerson spent 15 years in the private sector working primarily in Human Resources and Education and Training. In 2008, his passion for learning led him to pursue PhD in Educational Leadership. Wilkerson is currently an Assistant Professor at Regis University and Assistant Dean, Anderson College of Business.

Routledge Advances in Climate Change Research

Contemplating Climate Change
Mental Models and Human Reasoning
Stephen M. Dark

Climate Change, Moral Panics and Civilization
Amanda Rohloff
Edited by André Saramago

Climate Change and Social Inequality
The Health and Social Costs of Global Warming
Merrill Singer

Cities Leading Climate Action
Urban Policy and Planning
Sabrina Dekker

Culture, Space and Climate Change
Vulnerability and Resilience in European Coastal Areas
Thorsten Heimann

Communication Strategies for Engaging Climate Skeptics
Religion and the Environment
Emma Frances Bloomfield

Regenerative Urban Development, Climate Change and the Common Good
Edited by Beth Schaefer Caniglia, Beatrice Frank, John L. Knott Jr., Kenneth S. Sagendorf, Eugene A. Wilkerson

For more information about this series, please visit: https://www.routledge.com/Routledge-Advances-in-Climate-Change-Research/book-series/RACCR

Regenerative Urban Development, Climate Change and the Common Good

Edited by Beth Schaefer Caniglia, Beatrice Frank, John L. Knott, Jr., Kenneth S. Sagendorf, and Eugene A. Wilkerson

Routledge
Taylor & Francis Group
LONDON AND NEW YORK

earthscan
from Routledge

First published 2020
by Routledge
2 Park Square, Milton Park, Abingdon, Oxon OX14 4RN

and by Routledge
52 Vanderbilt Avenue, New York, NY 10017

Routledge is an imprint of the Taylor & Francis Group, an informa business

British Library Cataloguing-in-Publication Data
A catalogue record for this book is available from the British Library

Library of Congress Cataloging-in-Publication Data
A catalog record has been requested for this book

ISBN: 978-1-138-55692-8 (hbk)
ISBN: 978-1-315-15050-5 (ebk)

Typeset in Goudy
by Cenveo® Publisher Services

Contents

Figures

Tables

Contributors

Tom W. Boyd, David Ross Boyd Professor Emeritus of Philosophy and Religious Studies, University of Oklahoma

Thomas J. Burns, Professor of Sociology, University of Oklahoma

Beth Schaefer Caniglia, Editor-in-Chief, *The Solutions Journal*

Jennifer Eileen Cross, Associate Professor in the Department of Sociology, Institute for the Built Environment, Colorado State University

Allison Dake, Affiliate Professor College of Business and Economics, Regis University

Thomas Dietz, Professor of Sociology and Environmental Science and Policy, Michigan State University

Beatrice Frank, Social Science Specialist, Capital Regional District of Victoria & Adjunct Professor, School of Environmental Studies, University of Victoria, Canada

Barbara J. Jackson, Director, Franklin L. Burns School of Real Estate and Construction Management, University of Denver

John L. Knott, Jr—CityCraft® Founder

Carrie M. Leslie, Doctoral Student in Environmental Sociology, University of Oklahoma

L. Hunter Lovins, President, Natural Capitalism Solutions

Nicholas S. Mang, Principal of Regenesis Group, Inc.

David N. Pellow, Dehlsen Chair and Professor of Environmental Studies and Director of the Global Environmental Justice Project, University of California

Josette M. Plaut, Executive Director, Institute for the Built Environment, Colorado State University

Kenneth S. Sagendorf, Professor and Director, Innovation Center, Anderson College of Business Regis University

Carol Sanford, Award-Winning Author, Contrarian Speaker, Exec. Educator

Rebecca Sheehan, Associate Professor in the Department of Geography, Oklahoma State University

Eugene A. Wilkerson, College of Business and Economics, Regis & University Assistant Dean, Anderson College of Business

Foreword

A quick search for all the business-related books and articles with the term "leader" or "leadership" in the title uncovers that 593,033 were written between 2006 and 2016.[1] That is more than 150 books or articles published every single day in the decade leading up to the election of Donald Trump as president of the United States.

Documented within the research, there are dozens of leadership styles described eloquently and in great detail. One would assume that the skills gleaned from the voluminous methods offered have inspired countless leaders to improve the quality of life for everyone in society. Unfortunately, business leadership's enduring commitment to a wealth maximization economic development model created a world where 90% of the wealth is owned by less than 10% of the population.[2]

Despite our obsession with leadership—or more accurately, our obsession with writing about leadership—the future still seems bleak on other fronts as well. Each year we destroy our forests at a rate roughly equivalent to the size of Ireland. Accelerating deforestation is largely the result of converting forests to arable land to grow more food, yet infant mortality rates in low income countries due to poverty-related health issues is 14 times greater than that of high income countries.[3] In the United States, the world's largest exporter of agricultural products, childhood obesity rates have more than tripled since the 1970s.[4] Business leaders have been aware of the diverging and incongruent trends, but actions to solve the problems would cost precious resources in pursuit of wealth maximization for the wealthiest.

In 2006, violence against citizens in the most volatile regions in the world led to 2,500 deaths. After a decade of prolific leadership penmanship, in 2017 those same regions killed over 44,000 citizens.[5] Refugees fleeing violence in their home countries seek asylum in the United States, a country where guns killed nearly as many people last year as the number of deaths from the most violent regions in the world combined. Leaders in the United States have seeded gun violence through perverse interpretations of the second amendment to the point where about a third of the 320 million US citizens own nearly 400 million guns. Worse than that, every single state allows concealed carry.[6]

Current leaders have delivered the world to an important crossroads. We can continue on our current trajectory of death, destruction, and poverty, or we can change course. Some believe the question is not which way will we turn, but is it too late to even decide? In working together to create a less dire future, it helps to reconsider time itself. The Greeks describe time with both quantitative and qualitative aspects. "Chronos" is the counting of time in a linear fashion, whereas "Kairos" considers the interactions of various forces in the world to determine the right time for certain actions.

In 2015, I brought this thinking to Regis University to become the Founding Dean of what was then called the College of Business and Economics. As the only Jesuit University in the Rocky Mountain region, I was charged with the task of leading a different business school. The 450-year plus history of the Jesuits and their role in transformational education provided guidelines for what this might look like. Our foundational value *Magis*, which translates to the more universal good, challenged us to seek for ways to accept this Kairos moment and how we would influence the world as we launched at nearly the same time of the writing of *Laudato Sí*.

I choose to believe in the revolutionary idea of regenerative development. At its core regenerative development is premised on business abandoning the old models of leadership and embracing stewardship. Unlike leadership, stewardship is much more intentional. Business, the most powerful force in society, must adapt a new categorical imperative: to protect and care for society and improve the quality of life on earth. Regenerative development might just be the most important enabler of that new vision for business stewardship. It applies an intentional process that views society as a system of integrated activities, each renewing rather than depleting resources. Every aspect of regenerative development aspires to improve the quality of life for all. This is the heart of our college's stewardship vision. With a universal commitment from business to embrace stewardship, regenerative development will be the tool that effectively changes society's current trajectory while ensuring a thriving planet in the future.

Tim Keane, Dean of the Anderson College of Business, Regis University

Notes

1. http://web.b.ebscohost.com/ehost.
2. https://inequality.org/facts/global-inequality/.
3. http://www.who.int/gho/child_health/mortality/mortality_under_five_text/en/.
4. https://www.cdc.gov/healthyschools/obesity/facts.htm.
5. http://www.acleddata.com—including the Middle East, Africa, and Southern and Southeastern Asia.
6. https://www.cnn.com/2018/02/15/politics/guns-dont-know-how-many-america/index.html.

1 Regenerative development: Urbanization, climate change, and the common good

Beth Schaefer Caniglia, Beatrice Frank, John L. Knott Jr., Kenneth S. Sagendorf and Eugene A. Wilkerson

Urbanization, climate change, and inequality are among the greatest challenges facing the world today. More importantly, these forces combine in ways that make addressing the impacts of one difficult without considering the others. Traditionally, the intertwined relationships among people, prosperity, and planet have been tackled through sustainability practices. However, in recent times, the most innovative responses to the integral nature of these factors have been found in the regenerative development community.

The regenerative development paradigm is catching on in cities such as Denver, Colorado; Portland, Oregon; and Pomona, California (Caniglia 2018, 2019). Small enclaves of architecture, design, and development practitioners have advocated this approach to development for decades, and the scholarly communities in these disciplines have made progress articulating the theory, principles, and practices surrounding regeneration. The theoretical underpinning of the systems approach espoused by scholars in traditional fields associated with regenerative development has been vetted by social scientists from a variety of fields such as environmental sociology, political ecology, and interdisciplinary practitioners of coupled human and natural systems (CHANS) approaches. However, the theories and practices of regenerative development have yet to be engaged in most social science disciplines. As regenerative development is grounded in a worldview where humans are deeply embedded in nature; where the emphasis is given to the complex, dynamic, and interconnected relations of all living organisms; and where change is constant and inevitable, the full potential of the regenerative approach cannot be reached until its principles, practices, and literature are fully explored by social scientists from a spectrum of disciplines (Benne & Mang 2015; Du Plessis & Brandon 2015). This book is dedicated to that effort. In this chapter, we lay out the complex ties between urbanization, climate change, and inequality; define regenerative development; and provide an overview of the book.

The intersectionality of urban problems

Cities present us with tremendous challenges and opportunities. Currently, the vast majority of citizens of industrialized nations live in or near urban centers, and by 2050, the majority of the world will be urbanized (United Nations, Department of Economic and Social Affairs, Population Division 2018). Until recently, urbanization was a challenge most faced by industrialized nations; however, urbanization is impacting the developing world at the most rapid rate (Worldwatch Institute 2016). In the industrialized world, urbanization is characterized by challenges such as waste management, gentrification, displacement of populations of origin, and lack of affordable transportation and housing. In the developing world, the challenges can be quite different, especially when basic services are not available, the informal economy is prominent, and shanty towns are abundant In both cases, the economies of nations are intimately tied to their cities, with the majority of production and consumption taking place in and near urban areas (Kilroy, Mukim & Negri 2015). Approximately 80% of all skilled jobs are located in cities (Talbot 2018). Relatedly, approximately 75% of all energy consumption takes place in cities, placing the role of cities in the amelioration of climate change center stage. Cities are also the most unequal places on the earth, and the larger a city grows, the larger the gap between the rich and the poor becomes (Gordon 2013).

These challenges—urbanization, climate change, and the common good—are intimately interwoven in urban systems. As populations grow, competition for housing increases, the poor are displaced at more rapid rates, and cost of living rises. The existing housing stock undergoes transformations that can be good for energy consumption, but the new homes often have larger ecological footprints overall. Public transportation often improves with urbanization, but the cost is frequently higher than what the poor and lower income families can afford; and those in the service sector are forced into communities that lie outside of public transportation zones. In order to create livable cities for all inhabitants, far more attention is needed in the intersectionality of urban socioecological problems (Caniglia, Vallee & Frank 2017).

The economies of cities are the cornerstones of national and state economies (Dobbs et al. 2011). In the industrialized world, cities house 80% of all skilled jobs (Talbot 2018). The healthcare sector is increasingly concentrated in cities, with approximately 90% of all hospitals and 65% of all specialty physicians located in urban centers. Not only is the education sector concentrated in cities, but the cost of higher education is 30–50% cheaper in urban centers (McKinsey Global Institute 2011). Not surprisingly, the rich and powerful are highly concentrated in cities (Burrows, 2018). For example, Kiplinger cites that most millionaires in the United States live in Los Angeles and New York City. Because the top 10% of all Americans own 84% of all the stock, the vast majority of economic power is concentrated in very few locations—all cities. This trend is repeated around the world (United Nations, Department of Economic and Social Affairs, Population Division 2018): Urban centers are economic, political, and population power houses.

Of course, the distribution of quality of life in the city is highly unequal. Whether we examine São Paulo, Moscow, Chicago, Nagoya, London, or Madrid, the data consistently show that cities exhibit the highest gaps between their rich and poor occupants (United Nations, Department of Economic and Social Affairs, Population Division 2018). The service economy in part structures this outcome, because the sector itself is bifurcated between highly skilled positions and lower skilled positions—which contributes to the disappearing middle class (Burgers & Musterd 2002; Sassen 1996). There is close coupling between the concentration of economic production, consumption, banking, stocks and development, urban inequality, and the overall impact of urban development on the production of the common good. Unchecked, these forces consistently lead to the development of a growing and nearly permanent and vulnerable underclass.

The advent of climate change enhances the challenges faced by those who reside in cities, rich and poor. Cities around the world are experiencing environmental fluctuations outside of normal ranges, such as particulate matter in the air, record high and low temperatures, flooding, droughts, wildfires, and other extreme weather events. Scholars have shown that a variety of factors predict which population groups will be most impacted by these changes (Caniglia, Vallee, & Frank 2017; Frank, Delano & Caniglia 2017). All populations are exposed to more air pollution, along with more water and airborne diseases as a result of climate change. However, higher exposure to these threats increases vulnerability, suggesting that the homeless and those living without adequate shelters are particularly at risk. Farmers, herders, and those whose livelihoods depend on subsistence lifestyles are also at higher risk. Local authorities and first responders around the world are challenged to protect their communities and to enact effective measures when disasters strike. The costs of climate change for citizens, businesses, and government officials are growing incrementally and are difficult to bear.

These challenges are consistently cited as a primary reason why cities have emerged as global leaders in climate-related innovation (C40Cities). Urban greenways, public transportation, green roof initiatives, and energy-efficient buildings are just a few examples of the innovations cities are implementing to ameliorate the costs and risks of their changing climates. Furthermore, urban leaders know that they are most at risk to the impacts of climate change, because 90% of all cities are located in coastal communities (see C40Cities) making their infrastructure, economies, and citizens the most vulnerable worldwide. As a result, cities lead the way in important innovations. Significant work has been done to improve measurement of urban climate impacts, to develop adaptation frameworks, and to create increasing resilience through urban climate action plans (C40Cities; Dobbs et al. 2011; UN 2018). Cities are also leading the way in technological innovations that are central to decarbonization, including new electrical grid technologies, transportation, waste management, and building efficiencies (Caniglia 2019).

The primary difficulty cities face is how to approach the impacts of rapid urbanization, climate change, and inequality *at the same time*. Systems thinking and complex-adaptive management practices offer the most promising approach to the tapestry of interactive challenges cities now face (see Chapter 3). While many cities are systematically addressing climate risks, urbanization, affordable housing, transportation needs, etc., most are not providing integrated solutions that account for the feedback loops that intimately tie these challenges together. Moving the needle in a positive direction in one of these categories can have important negative impacts on the others. For example, transportation improvements in low-income communities make those communities more desirable to people with higher incomes. This often causes gentrification to the extent that the residents of origin can no longer afford to remain in their homes. Similar processes take place with neighborhood greening initiatives (Gould & Lewis 2017). Providing improved access to green space in low-income communities leads to "green gentrification," causing displacement of populations of origin. Rather opposite challenges are experienced in the urban South, where deep preferences for self-governance in informal settlements confound attempts of local authorities to provide improved housing conditions in systematic ways. Cities are making impressive improvements in managing their ecological, economic, and infrastructure capital. However, urban governance systems can only succeed in creating equitable and sustainable cities if they incorporate cultural capital into their frameworks. This is precisely the motivation for writing this book. Regenerative development offers a framework that provides a systems approach to complex-adaptive management that incorporates cultural, ecological, and economic capital with the need to preserve and improve existing infrastructure.

What is regenerative development?

Regenerative development is rapidly displacing sustainability as the most fertile ground for climate change adaptation research. The term "regeneration" is flooding the literature as a refreshing substitute for the concept of sustainability. From its original description within the broader context of sustainability, scholars and practitioners have increasingly articulated the meaning of regenerative development, providing a holistic and bio-regionally based framework that stands in contrast to the mechanistic, more static, and anthropocentric worldview embedded in current approaches to sustainable development (Zhang et al. 2015). This distinction has taken root with the changes the traditional sustainability paradigm has undergone over time and by stepping up the quality of thinking and engaging around the interactions between the human-made and the natural environment (Du Plessis & Brandon 2015). Said differently, the unique approaches to systems thinking that evolve from regenerative development contemplate feedback loops that emerge from non-environmental systems.

Research and practice have increasingly shifted toward applying holistic processes to create feedback loops between physical, natural, economic, and social capitals (Caniglia et al. 2014; Lovins et al. 2018;). This idea of focusing on the whole and on integrated complex systems has propelled the contemporary notions of sustainability forward and allowed for the development of socioecological systems and CHANS thinking (Caniglia, Vallee & Frank 2017; Frank, Delano & Caniglia 2017). This push toward "the reconnection of human aspirations and activities with the evolution of natural systems, essentially coevolution" (Mang & Reed 2012, p. 6), has advanced its full potential in the regenerative paradigm. The evolution of this new thinking affirms the idea that holistic processes and feedback loops are not only mutually supportive but also contain the capacity to restore equitable, healthy, and prosperous relationships among the different forms of capital (Caniglia 2018; Du Plessis 2012).

As nicely put by Du Plessis and Brandon (2015), regenerative development—or the regenerative sustainability paradigm—"has the potential to create a future where the damage done to the biosphere and to our social systems has been restored, and people can live in mutually supportive symbiosis with their social and biophysical environment (their whole ecological system)—the one nurturing and growing the potential of the other" (p. 56). Regenerative development is grounded in a worldview where humans are deeply embedded in nature; where the emphasis is given to the complex, dynamic, and interconnected relations of all living organisms; and where change is constant and inevitable (Benne & Mang 2015; Du Plessis & Brandon 2015).

The idea behind regenerative development is to offer an approach that fosters a strong adaptive capacity and evolutionary potential for the whole system, thus allowing for a thriving future for all through restoration, regeneration, and the commitment to creating ongoing positive impacts on the health of ecosystems and the entire biosphere (Benne & Mang 2015; Hes & Du Plessis 2015). To summarize it in a simple way, regenerative development is about humanity and its coevolutionary role with nature in building sustained social and natural capital—or to "make sustainability real" (Zhang et al. 2015, p. 2).

As with any growing, building, and evolving field of study and exploration, the terms used by different authors in this book will vary. Regenerative Development, Regenerative Design, Just Sustainability, and Regenerative Sustainability are just some of the conceptual frameworks used to describe this holistic system perspective that has attracted increased interest from different disciplines and practices (Agyeman, Bullard & Evans 2003; Zhang et al. 2015). Because of its novelty and evolving nature, regenerative development definitions, applications, and expected outcomes remain to be clearly articulated. However, most authors agree that regenerative processes have three primary goals:

- Catalyzing increased prosperity and health of human and natural environments through holistic design and meaningful community participation.
- Fostering positive feedback loops that create mutually beneficial relationships between natural, human, economic, and physical capital that self-replicate to build abundance in all four categories.
- Respecting and having a deep consideration of local contexts, whether economic, cultural, or ecological, so that development is properly adapted to the local ecosystem, cultural, and economic circumstances.

The principles of regenerative development are reflected in global institutions and evolving international environmental law. The Sustainable Development Goals (SDGs) recommend that countries prioritize ending poverty, increasing equal access to education and opportunity, building sustainable and resilient cities, and improving quality of life for all in harmony with the land, sea, and air (https://www.un.org/sustainabledevelopment/sustainable-development-goals/). These regenerative development principles also interlink quite closely to the following three principles of the New Urban Agenda document adopted at the United Nations Habitat III meeting held in Quito, Ecuador, in September 2016[1]:

- **Leave no one behind:** This principle is about ending poverty in all its forms and dimensions, including the eradication of extreme poverty, ensuring equal rights and opportunities, providing socioeconomic and cultural diversity, enhancing integration in the urban space and livability, and propelling education, food security, health, and well-being forward. It includes ending the epidemics of AIDS, tuberculosis, and malaria, promoting safety and eliminating discrimination and all forms of violence. It focuses on ensuring public participation through safe and equal access for all to physical and social infrastructure, basic services, and adequate and affordable housing.
- **Sustainable and inclusive urban economies:** This principle is about leveraging the agglomeration benefits of well-planned urbanization, high productivity, competitiveness, and innovation. It aims at promoting full and productive employment and decent work for all, ensuring decent job creation and equal access for all to economic and productive resources and opportunities, and preventing land speculation. It promotes secure land tenure and the proper management of urban shrinking where appropriate.
- **Environmental sustainability:** This principle is about promoting clean energy, sustainable use of land and resources in urban development, as well as protecting ecosystems and biodiversity. The key focus of this principle is adopting a healthy lifestyle in harmony with nature and promoting sustainable consumption and production patterns. Such an approach is key to build urban resilience, reduce disaster risks, and adapt to climate change.

A multitude of concepts and approaches have been implemented to achieve sustainable urbanization as clearly highlighted in the special issue "Toward a Regenerative Sustainability Paradigm for the Built Environment: From Vision to Reality" (see https://www.sciencedirect.com/journal/journal-of-cleaner-production/vol/109), among others. Common traits to the range of regenerative approaches found here are to put the holistic and coevolving core ideas of regeneration into practice in cities (i.e., sustainable built environment, green technologies) and to enhance life conditions and ecosystem health at a global scale. Such a shift is not only key to building resilience and adaptive capacity toward climate change, but it is also necessary to help to shift societal worldviews toward the common good—or toward citizenship, collective action, and active participation that strives for a thriving and abundant future for all.

The book aims at offering a common platform to examine the ways the regenerative paradigm can benefit from the empirical work conducted by the former and the theoretical frameworks developed by the latter in their disciplinary domains of expertise. To move beyond some current literature that conflates regenerative development with sustainable development, the authors were provided with a standard definition of regenerative development that served as a guide for their thinking. This a priori approach provides a sense of rigor to the process and serves as connective tissue for the different perspectives in the text.

Regenerative development is a development paradigm designed to push beyond sustainability. While sustainable development focuses on development today that protects the ability of future generations to develop, the priority of regenerative development is to apply holistic processes in a bioregional context to create feedback loops between physical, natural, economic, and social capital that are mutually supportive and contain the capacity to restore equitable, healthy, and prosperous relationships among these forms of capital.

Why this book is different

We have challenged authors, and invited the broader social science community, to interrogate their existing contributions and insights into urbanization, climate change, and the common good by using the lens of regeneration, which we believe will provide mutual benefits and advancements in practices and theories. As a construct, regenerative development's inclusion of all four types of capitals—physical, natural, economic, and social—makes it important that this book is written for a variety of audiences. As such, it only makes sense to include contributions from a multitude of perspectives. You will read chapters from traditional scholars who are turning over the truths as they currently know them to find the new truths that regenerative development births and that coupled and human natural systems need. Chapters are written by consultants who spend time encouraging others to see regenerative development as a new possibility. Experienced practitioners and appliers of regenerative development who are busy blazing new paths and partnerships share their experiences in this book, bringing the wisdom of experience and a lens where regeneration may have been an influence across previous generations. And educators who are pushing the boundaries of disciplinary and, especially, academic traditions contribute chapters examining if what we do can contribute to developing the field of professionals we need and, if not, how do we educate for a regenerative and common good-focused future. This book is intended to reach widely to those dreaming of a regenerative future.

Through this book, we hope to begin to break down some of the silos that exist, but most importantly, we seek what the Jesuits call the *Magis*—the greater good—that comes from encouraging and including broad examination and application of a topic. In each chapter you read, you will encounter work aimed at honoring the dignity of the human spirit and the earth. You will wrestle with concepts such as cultural lag and institutional mismatch as well as social and environmental vulnerability. You will consider why regenerative development stems from an understanding that one is developing an interconnected habitat and a connected and reliant set of systems, not merely structures. And we hope that you, as a reader, will clearly see how the capacity for a regenerative future can evolve beyond individual champions.

How this book is put together

The book opens with a contribution of the editors (Chapter 1) that engages with existing literature regarding sustainability, resilience, and regenerative development and highlights the centrality of systems thinking as a pathway to fruitfully combine and enhance this disciplinary approach through a more holistic and comprehensive vision. This introductory chapter is followed by Chapter 2 contributed by Carol Sanford, in which she discusses the most recently emerging paradigms in regenerative development. Based on her extensive knowledge in this domain, the author explains the evolve capacity, do good, arrest disorder, and extract value paradigm and how these concepts form a way of organizing, acting, and pursuing regenerative development by generating a series of worldviews (i.e., living systems, human potential, behavioral, machine, and aristocracy). She continues by highlighting the Seven First Principles of Regeneration, which offer guidelines to promote living systems understanding of people, watersheds, businesses, communities, and most other systems on the earth. This contribution is followed by Beth Caniglia's discussion (Chapter 3) about the centrality of systems thinking in regenerative development. In this chapter the author argues that regenerative development holds significantly more promise for systems thinking in ways that allow multiple forms of capital to coevolve to improved states than current applications of resiliency and sustainability, especially in the face of rapid urbanization and increasing climate variability. In Chapter 4, Nicholas Mang continues by highlighting the need for regenerative psychology perspective. The way we define ourselves is key to develop regenerative mindsets and frameworks that influence the way we relate to and work on the development of places such as cities. Nicholas Mang discusses regenerative psychology with a focus on four interrelated psychological paradigms: Psychologies of Adjustments, Psychologies of Human Potential, Psychologies of Living Systems, and Psychologies of Spiritualization. These paradigms help in better understanding how humans can become integral members and contributors to the ecological systems in which they exist, and thus practice systems thinking holistically. An examination of the concept and practice of regenerative development through the lens of environmental justice scholarship and politics follows. As advocated by David Pellow in Chapter 5, regenerative development holds a great deal of promise for moving scholars, planners, and community leaders beyond the limitations of sustainability discourses and development practices; the promise of this concept may be more fully realized if it can successfully be applied to contexts and cases where populations face extreme social, economic, political, and environmental harm and marginalization (i.e., environmental injustice). The regenerative development paradigm can indeed be pushed further by better addressing the impacts of discrimination to human populations and ecosystems. The first step to do so will be recognizing that environmental justice is not only about broad economic social issues—such as class—but also about racism, patriarchy, nativism, colonialism, and other systems of power discrimination. The second step will be addressing engrained social and environmental inequalities and power imbalances at the root of unsustainable social behaviors, so to build regenerative, coevolutionary, and *just* human and environmental systems well into the future.

Sustainability emerged as the fusion of thinking about development and conservation, human well-being, and the biosphere. Improving human well-being while allowing the biosphere to flourish—as argued by Thomas Dietz (Chapter 6)—requires governance that supports complex coupled human-environmental system. Governance requires norms consistent with the goals of regenerative development and decision-making processes that instantiate those norms effectively. And since change is ongoing and the future uncertain, analytic deliberative governance is a key approach to facilitate social learning and adaptive risk management—key themes discussed in Thomas Dietz contribution. To close this more theoretical section, Thomas Burns, Tom Boyd, and Carrie Leslie discuss the gravity of ecological overshoot from an ethical standpoint in Chapter 7. Key topics are the mismatch between societal institutions and the alienated individualism of modern humanity, problems of cultural lag particularly as they have come to dominate the ethics around environmental issues, and the inability to making the environment a central organizing principle of regeneration. The authors offer practical steps to best prepare society to live sustainably, concentrating on an emergent ethic of ecological peace. An important point brought forward by these contributions is that regenerative development should not be limited to the restoration of socioecological systems. Many of the issues faced today, such as climate change, are grounded in the social and environmental injustices, unethical behaviors, and weak governance embedded in past and present socioecological systems. Regenerative development, with the integration of environmental justice, deliberative governance, and ethics discourses, has the potential to make a real change, and thus establish new and coevolving socioecological systems that are mutually beneficial for the society and the environment.

The more theoretical contributions of this book are followed by chapters on regenerative development practices, which illustrate the need to significantly advance the approaches to the ways urbanization, climate change, and inequality are addressed at every scale. In Chapter 8, Hunter Lovins discusses regenerative economy and the need to change the current economic model to create an economy in service to life, not consumption. Being nature sustainable because regenerative, she describes the principles and means needed to align economy with the laws of natural systems, and thus create a regenerative economy. She suggests that decarbonization of the economy and the shift to regenerative agriculture are key transformations needed to achieve regenerative economy. Regenerative businesses and economies are implemented when holistic thinking and management are applied. This is especially evident in Chapter 9 in which John Knott, the founder of CityCraft®, explains how regeneration is implemented in light of regenerative building and design. In this chapter, the author discusses building with regeneration in mind. He describes the evolution of the principles and processes of regenerative planning and presents their application in two case study areaswho describes the evolution of the principles and processes of regenerative planning and presents their application in two case study areas: Dewees Island

and Noisette. The future potential of regenerative planning is discussed in this chapter in the context of a third case study, the West Denver community in Denver, Colorado. Applying regenerative thinking is pushed a step further in Chapter 10, in which Rebecca Sheehan talks about the regenerative sociocultural capital of memorialization. In public spaces, social memory is often portrayed by static monument that tributes dominant powers, leaving little room for change, process, and different interpretations. Rethinking memorialization with regenerative principles in mind is central to address these social spatial justices. As a matter of fact, when sociocultural landscapes are memorialized through ongoing dynamic processes focused on holistic systems thinking, inclusive and just public spaces emerge, significantly improving the regenerative potential of cities and space memorialization for current and future human generations. The on the ground experiences drawn from these authors' works in business, economics, and building offer a unique insight to the regenerative development discourse and its potential to make sustainability work in the real world.

The last part of this book starts with a discussion around the importance of including social science knowledge in regenerative discourses. In Chapter 11, Jennifer Cross and Josette Plaut first discuss the principles of regenerative development from a social science standpoint and proceed by describing the five core capabilities needed to be a regenerative practitioner—system actualizing, framework thinking, self-actualizing, developmental facilitating, and living systems understanding. These core capabilities help regenerative practitioners to become competent in Developmental Facilitating, thus fostering positive psychology principles and social networks through facilitation. Cultivating human thriving and connections between groups are indeed key in informing and deepening regenerative development thinking and processes. This chapter is followed by a regenerative definition of workforce development by Eugene Wilkerson and Allison Dake (Chapter 12). Traditionally this term was understood as the link between economic development and business competitiveness with the mission of addressing economic disenfranchisement. This definition lends itself toward a supply and demand view of workforce development. The authors of this chapter highlight how the workforce in a developed country is not a capital asset that can be consumed and discarded based on economic changes. Instead, there is a need to develop regenerative workforce solutions that benefit all within society by effectively adjusting for innovative disruption, developing people through business to have a societal impact, creating work environments that have meaning, and honoring the uniqueness of all within a workforce. The last contribution of this book (Chapter 13) by Kenneth Sagendorf and Barbara Jackson is a reflection about how integrated systems thinking can influence the curriculum for the next generation of regenerative development professionals. The authors define what is necessary to educate for regeneration and explore differences in education when the outcomes of learning become one of systems thinking rather than critical thinking. As the authors of this chapter outline, systems thinking is a necessary precursor and not a substitute of critical thinking when educating for regeneration. The authors go a step further by overlaying the principles of heroic leadership to systems thinking, where heroic leadership is defined as a blend of self-awareness, ingenuity, love, and heroism toward the path of developmental integration for regenerative education. In a complex, uncertain, and ever-changing world, leadership is the key building block for fostering regeneration and achieving the greater good. Finally, Chapter 14 by Beth Caniglia, John Knott and Beatrice Frank concludes the book.

Conclusion

If we can solve the challenges faced by cities, we can significantly improve quality of life on our planet. This book represents a concerted effort to bring together scholars and practitioners to examine the potential for the regenerative development paradigm to provide a framework that enables improved management of the complex challenges facing cities in the new climate era. The scholars in this text represent a sophisticated set of theoretical, ethical, practical, and innovative examinations that will catapult regenerative thinking into the center of social science conversations. We are confident that this book will bring social science voices directly to bear in the practice of regenerative development—a necessity if it hopes to achieve its full promise to ameliorate climate change while significantly uplifting vulnerable populations in the context of rapid urbanization. It will also advance social science research and the relevance of social science work by engaging with an alternative to sustainability that is expanding in application. The ultimate vision is a book that seamlessly unites practice and scholarship for the planet and the common good.

Note

1. You can read the full version of the New Urban Agenda at this URL: <https://www2.habitat3.org/bitcache/97ced11dcecef85d41f74043195e5472836f6291?vid=588897&disposition=inline&op=view>.

References

Agyeman, J, Bullard, RD & Evans B 2003, *Just sustainabilities: Development in an unequal world*, The MIT Press.

Benne, B & Mang, P 2015, "Working regeneratively across scales insights from nature applied to the built environment", *Journal of Cleaner Production*, vol. 109, pp. 42–52.

Burgers, J & Musterd, S 2008, "Understanding urban inequality: A model based on existing theories and an empirical illustration", *International Journal of Urban and Regional Research*, vol. 26, no. 2, pp. 403–413.

Burrows, D 2018, "Where millionaires live in America," viewed 3 March 2019, <https://www.kiplinger.com/slideshow/investing/T064-S001-where-millionaires-live-in-america-2018/index.html>.

Caniglia, BS, Frank B, Delano D & Kerner B 2014. *Enhancing Environmental Justice Research and Praxis: The Inclusion of Human Security, Resilience & Vulnerability*. International Journal of Innovation and Sustainable Development vol. 8 (4) pp. 409–426.

Caniglia, BS, Vallee M & Frank B 2017, *Resilience, environmental justice and the city*, Routledge.

Caniglia, BS 2018, "The path to a regenerative future: The importance of local networks and bioregional contexts", *The Solutions Journal*, vol. 9, no. 2.
Caniglia, BS 2019, "Why Denver? Sustainability innovation in the Mile High City", *The Solutions Journal*, vol. 10, no. 1.
Dobbs, R, Sit, S, Remes, J, Manyika, J, Roxburgh, C & Restrepo, A 2011, *Urban world: Mapping the economic power of cities*, viewed 10 March 2019, <https://www.mckinsey.com/~/media/McKinsey/Featured%20Insights/Urbanization/Urban%20world/MGI_urban_world_mapping_economic_power_of_cities_full_report.ashx>.
Du Plessis, C & Brandon, P 2015, "An ecological worldview as basis for a regenerative sustainability paradigm for the built environment", *Journal of Cleaner Production*, vol. 109, pp. 53–61.
Du Plessis, C 2012, "Towards a regenerative paradigm for the built environment", *Building Research & Information*, vol. 40, no. 1, pp. 7–22.
Frank, B, Delano, D & Caniglia, BS 2017, "Urban systems: A socio-ecological system perspective", *Sociology International Journal*, vol. 1, no. 1.
Gordon, L 2013, The world's largest cities are the most unequal, viewed 19 February 2019, <https://blog.euromonitor.com/the-worlds-largest-cities-are-the-most-unequal/>.
Gould, K & Lewis, T 2017, *Green gentrification*, Routledge.
Hes, D & Du Plessis, C 2015, *Designing for hope: Pathways to regenerative sustainability*, Routledge, Earthscan, Oxon, UK.
Kilroy, AFL, Mukim, M & Negri, S 2015, *Competitive cities for jobs and growth: What, who, and how (English). Competitive cities for jobs and growth*, World Bank Group, Washington, D.C., viewed 10 March 2019 <http://documents.worldbank.org/curated/en/902411467990995484/Competitive-cities-for-jobs-and-growth-what-who-and-how>.
Lovins, LH, Wallis, S, Wijkman, A & Fullerton, J 2018, *A finer future: Creating an economy in service to life*, New Society.
Mang, P & Reed, B 2012, "Designing from place: A regenerative framework and methodology", *Building Research & Information*, vol. 40, no. 1, pp. 23–38.
McKinsey Global Institute 2011, *Urban world: Mapping the economic power of cities*, McKinsey&Company, New York, <https://www.mckinsey.com/~/media/McKinsey/Featured%20Insights/Urbanization/Urban%20world/MGI_urban_world_mapping_economic_power_of_cities_full_report.ashx>.
Musterd, S & Burgers, J 2002, "Understanding urban inequality: A model based on existing theories and an empirical illustration", *International Journal of Urban and Regional Research*, vol. 26, no. 2, pp. 403–413.
Sassen, S 1996, "Service employment regimes and the new inequality", In M Enzo (ed.), *Urban poverty and the underclass: A reader*, Wiley.
State of the World Forum 2016, *Can a city be sustainable?*, The Worldwatch Institute.
Talbot, D 2018, *Why cities get the best jobs*, viewed 19 February 2019, <https://www.forbes.com/sites/deborahtalbot/2018/09/12/why-cities-get-the-best-jobs/#722848ac1492>.
United Nations, Department of Economic and Social Affairs, Population Division 2018, *The world's cities in 2018—Data Booklet* (ST/ESA/SER.A/417).
Zhang, X, Skitmore, M, De Jong, M, Huisingh, D & Gray, M 2015, "Regenerative sustainability for the built environment from vision to reality: An introductory chapter", *Journal of Cleaner Production*, vol. 109, pp. 1–10.

2 The regenerative paradigm: Discerning how we make sense of the world

Carol Sanford

I have often wondered how fast we could create powerful shifts toward vital, viable ecosystems and societies, if we gathered the really big players in business, philanthropy, and education around the table to work on making a better world. I was hopeful that I might get some sort of answer when I was invited recently to participate in a dinner with 20 leaders, from a huge retail business, a respected technology company, and a major foundation driving innovation on global challenges. The purpose was to discover ways to slowdown and even to reverse ecological destruction. I was sorely disappointed not in their good faith and aspiration, but in the thinking they brought to the subject. Of course, they had come up with new content and ideas, but they were thinking about ecosystems and restoration using outdated paradigms and ways of exploring opportunity. They spent two hours mired in archaic ways of working on change, only repeating the thinking that created racism, inequity in social systems, and climate change in the first place. In the following chapter, I hope to provide a basis for individuals and groups in regard to regenerative development, especially for those with the grandest intentions, who wish to explore the potential for change. I will offer a way to gain a new perspective engaging with the thinking itself that undermines discovering a new path. It is a way of thinking based on the actual living systems (humans, planet) under consideration.

Discerning paradigms

I was a sophomore in college in 1962 when Thomas Kuhn released *The Structure of the Scientific Revolution* (1970), introducing us all to the idea of paradigm shift. Kuhn taught at the University of California, Berkeley, in both the philosophy and history departments. I was questioning everything at that stage of my life, and Kuhn, who invited us as a species to consider the evolution of our interpretation of the world, gave me the framework for an appropriate and productive reasoning process. He described how we got stuck in a single prescribed view, which limited our ability to discern how nature and society work, and he offered as an alternative a more whole and complete perspective. He made us aware that the world is made up of alive and dynamic processes. Kuhn defined a paradigm as an era-based, normalized protocol, prescribed by the scientific community for discovering answers to puzzles. A paradigm codifies a set of concepts and practices that *define* a discipline in the quest for truth. Most researchers in a given era ascribe to its dominant paradigms, and thus they become the right and only way to create new knowledge. These mental boundaries tend to blind people to other ways of considering, and they become like the water that fish swim in—invisible and normal. Like fish, we humans have no idea that we are experiencing the limitations of unexamined paradigms.

I vividly remember the shock with which I reacted to Kuhn's claims: "You mean that there is no *agreed-on, absolute truth*?" I had believed that everything I was being taught was proven science, coupled with settled religious facts. In the middle of the free speech movement and the war in Vietnam, when paradigms were shifting and new ways of thinking were still unnamed, I began to examine and give names to the paradigms that were driving our governing, educational, and economic processes. Now, after decades of exploration, I am able to discern four major paradigms through which we observe and attempt to make sense of our universe. These paradigms arose in different historical eras and are associated with distinct *worldviews*. Taken together, paradigms and worldviews guide our thinking, often without any conscious awareness on our part.

Here is a brief clarification of the difference between a paradigm and a worldview based on my personal understanding. A paradigm has to do with how we pursue knowledge: what we are able to perceive, the ways we acquire knowledge, especially in science, and what we consider to be reliable knowledge (the study of which is formally known as epistemology). Paradigms set the boundaries for the questions we pursue and the answers we are able to find. A worldview, on the other hand, is a cosmological framing of how things work. It is based on societal values and beliefs and has mainly to do with how we ought to live. We are willing to live in accord with worldviews because they help us make sense of how the world works. They vary among cultural groups, from atheist to Christian, for example, and they define the possible range of answers within disciplines, such as sociology, history, musicology, and aesthetics. They also shape agreements between disciplines, framing them so that they align with one another and work together to describe how the world operates. Within disciplines or fields of endeavor, it is worldviews that describe origins and provide coherence. In the remaining of the chapter, I will explore both paradigms and worldviews, and describe ways in which our interpretations of unfolding events are informed by paradigms. In particular, I will examine the most recently emerging global paradigm, *Evolve Capacity*—or knowing by examining the dynamics of living systems—and the *living systems worldview* that it informs and is framed by. I will also describe some ways to bring the Evolve Capacity paradigm's perspective into any kind of work.

I believe that the Evolve Capacity and the living systems worldview are the basis of regenerative practice. Since Einstein advocated for a different mind than the one that created the mess, I am offering a powerful set of frameworks for that discernment. I think it is the only way to generate a new perspective and not meet the definition of insanity, which is continuing to do things in the same way over and over and expecting a new outcome. The best approach to see a new path is to understand how we see and work on it which is the real problem we are facing, not new ideas. We will start with a look at four paradigms that affect how we think as a species. And notice how different people have different paradigms. We will layer onto that the eras of history that have created different worldviews (ways of defining the working of things). Then, we finish with a system of Seven First Principles that define how we can bring such ideas into practices in the world around us. And do so with consciousness of our how thinking is driving our efforts and effects.

Seeing paradigms and their effects

We often sense the paradigms that people operate from when we observe their language, stories, or behavior, and we use this information to determine whether or not they are in our tribe. But much more important is to learn to clearly see our own adopted paradigm and the way it continually shapes and limits, or perhaps has the potential to expand our ways of thinking, relating, and working. To make it easy to grasp how paradigms show up, individually and collectively, I will look at examples from home and work, following a framework that anyone can use to develop a more encompassing perspective on how they are making sense of their immediate world. As a frame of reference for this exploration, I will use human-to-human interactions and the ways we navigate with worldviews and disciplines to make sense of the social world that we inhabit, coupled with a few brief examples of our engagements with the natural world.

Most people work or have worked in places where choices are made about managing people, based on one or another prescribed management system. Each of these management systems is invisibly based on a paradigm included in the human-human framework, and thus each offers different methods, results, and measure of success. The effects of the systems on work and employee well-being are what first become visible, not only to the people being managed and led, but also to the people doing the leading, although they may not at first seem to be in line with the paradigm that sourced the thinking. Some paradigms degrade commitment up and down the business and across all work systems. Some give extraordinary results to the systems and each person in them. But how can you know in advance what the results will be?

Each of us has unknowingly chosen, by luck of birth and family or education or other experience, a paradigm that shapes our perception and interpretation of the world and affects the choices we make and the understanding that we are able to develop in all of the activities of our lives, from parenting and education to business and governance. At the most fundamental levels, our chosen paradigms control our emotions and appetites and impose lenses between our eyes and the world. They dictate what we are allowed to know, what we accept as worth examining and embracing, where we put our energy and resources, and what we see as possible or plausible. They frame and are in turn framed by the cosmologies of religions and tribes that shape our collective ways of living. They direct how we raise and educate our children and frame how we as leaders engage our workforces and even our customers. Left unexamined, they blind us far more than they inform us.

The four modern paradigms

I have addressed the four governing paradigms of modern living, which I named, *Extract Value*, *Arrest Disorder*, *Do Good*, and *Evolve Capacity* (Figure 2.1). They run from the most pervasive to the least influential and understood, and from the oldest historically to the most recently articulated (although they have existed in parallel for decades, generations, and in some cases, centuries). Most of us today think and behave inconsistently because, although we are in general governed by one of the four, we are constantly influenced by all. We do not reflect on them and do not perceive them ruling us, and therefore we are not able to sort out and order ourselves.

Extract Value

In business settings, the Extract Value paradigm assumes some people—such as powerful civic and religious leaders—know more than others. We accept what they focus us on, which is primarily getting the most out of people by employing their human skills for the benefit of the powerful, with insufficient thought to the return workers receive beyond pay or professional experience. In physical terms, we get the most from materials by using them efficiently and from resources by not worrying about how they will be replenished. In businesses ruled by this paradigm, workers may see managers as the benevolent sources of rewards, while managers tend to see workers as interchangeable cogs in a machine. Management may offer skills training, but with the intention of improving the performance of a machine, rather than benefiting human lives. This paradigm is built on the assumption that powerful businesses own the labor of their employees through processes of transactional

Evolve Capacity—Know by examining the dynamics of living systems

---

Do Good—Know by moral teachings and metaphors

Arrest Disorder—Know by scientific method with sensorial inputs

---

Extract Value—Know by accepting the authority of those with power over us

Figure 2.1 The four modern paradigms framework

exchange—human knowledge, effort, and energy are purchased; employment contracts are drawn up; employee manuals are signed onto. "What is good for General Motors is good for the rest of us." Managers have the right to control the behaviors and attitudes of workers because their time and skills are bought and paid for. As concerns materials and resources, the assumption is that the earth belongs to humans who have paid for its lands and waters and thus have the right or obligation to dominate their nonhuman assets and use them for human benefit.

Arrest Disorder

The Arrest Disorder paradigm limits knowledge to what can be learned by following the scientific method. In any study, a question is asked that includes only a single variable, a control group not subject to the variation is set up for comparison, and thus, exploration is limited to a narrow focus on a single, isolated aspect of the subject under examination. This way of seeing only the limbs of trees in a vast forest results in narrow understanding and a bias toward problem-solving that neglects the larger, dynamic context of living wholes. More importantly for our understanding here, it leads to work on making problems less bad, rather than making whole systems more fully alive. Limited by the scientific method, we have no way to know if we have selected the right variable or if our work has been effective in making the whole work better. The only framing allowed by this paradigm is to seek change by starting from the definition of a problem. This allows for no other way forward than to solve the problem. Not until we get beyond a problem view of the world, or criticizing variations from the ideal, can we see the way to determine what we can. and must, work on, will we ever get beyond arresting disorder and begin to make systemic, beneficial changes.

The shift from Extract Value to Arrest Disorder is the shift from the apparent assumption that "only experts and other authorities know the truth," to the belief that "anyone can define an opportunity and pursue it." Arresting disorder activates us to work primarily on imperfections or variances from targets or ideals. It is based on standards and best practices defined by the organization, which everyone involved is expected to pursue and achieve. The focus is on set tasks and measurable behaviors, which gives a sense of concrete reality. Leaders operating from this paradigm tend to regard people as having fixed personalities and unalterable intelligence levels, and thus manageable only by external interventions targeting changes in behavior. Emphasis is placed on solving personnel problems, reducing shortfalls, and setting workers on right performance paths. In ecosystems, the Arresting Disorder paradigm focuses on reducing harm by limiting negative impacts on living aspects of the landscape. It drives the sustainability, circular economy, prevention, restoration, and resilience movements.

Do Good

The Do Good paradigm calls on us to use culturally accepted ideals (e.g., competencies) as the basis for defining what is worth promoting and contributing to.

Its epistemology acknowledges *moral* boundaries to what can be known and held as true. Individual communities determine their own moral good and work to convince others that it alone is the true good. Doing good invites us to hold the belief that people are able to change and become both more responsible and more skilled over time, and it is the source of social rules and "good management guidelines." These are seen as universal and applicable to all persons and situations and are developed in line with the intention to make meaningful contributions to something valuable and to benefit persons we care about.

The Do Good paradigm is the source of the feeling that philanthropy and volunteering are rewarding and worthy of approbation. It is a heart or feeling orientation that generates work assessed in terms of its effect on the organization and its stakeholders. We become passionate about our causes. When we give an employee a review and observe their response or when we see an employee we have been coaching achieve a goal, we are most likely engaging from the Do Good paradigm, the essence of which is achieving standardized, generic ideals of good. In ecosystems, this paradigm focuses on restoration—for example, resforestation, revivifying riparian systems, putting private lands into conservation trusts, and legislating to preserve public lands as wilderness preserves. The shift from Arrest Disorder to Do Good is from defining a problem as the motivation for creating good in the world to acknowledging that change for the better is the effect of the capacity of every person to be their own researcher and searcher for truth. The Doing Good paradigm dictates decisions based on moral choices that affect whole communities. Its source of knowing is the self-examination of individuals and communities.

Evolve Capacity

In business, the Evolve Capacity paradigm fosters commitment to the development of capacity in every employee and company team, focusing on their potential to evolve themselves and contribute to the living systems in which they are nested. Its methodology is continual regeneration. Managers focus on the *essences* of the persons whose work they oversee, the individuals who are evolving *right in front of us, now, today*, seeking to bring forward and develop each person's unique potential. They support the growth and development of employees in ways that allow their essences to be increasingly expressed. This entails becoming fully present with one person or group and acting in the specific situation, with the intention of enabling them to act from personal agency, to become more uniquely themselves, and to achieve more. When an employee discovers something that might be called "her true self," and then reveals new capacity and takes on ever-bigger challenges, she is likely to be working in an organization committed to evolving capacity.

The shift from Do Good to Evolve Capacity is from individual moral compass to the working of living systems. In the entire process of our human collective development, we are required to move our boundaries from the authority of power to starting from problems to moral choices by individuals and collectives to imaging the working of whole, living systems. The Evolve Capacity paradigm requires a much greater ableness than the other paradigms to see the potential effects of our actions. Most often, this is a capability that we have to consciously develop in ourselves in order to exercise it more often and more completely in all of our activities. It also requires moving away from standardized ideals and projections of our personal or cultural standards on others. Self-directed responsibility arises in people when we connect them to external effects and their potential to contribute to them.

The Evolve Capacity paradigm is especially concerned with the consideration of communities and ecosystems. Every community and every watershed—or really, every *lifeshed*—is unique, with an essence and distinctive potential of its own. When working regeneratively, there can be no standardized management practice. Like every other living entity, each watershed demands that we approach it individually, using the First Principles of Regeneration to guide an exploration of *this specific place* to reveal and support its unique potential. The overarching intention of this paradigm is to consider all actions in terms of the ability of evolving life in all its forms. This is seen through the ableness of each form to express its unique potential in its own right, as well as aspect of the larger living system within which it is nested. To diminish the life of any form by extraction of its life energy is avoided. The intention is to go beyond only restoring or sustaining it by arresting harm, and to the tendency of imposing a uniform set of ideals. These older paradigms are at odds with the unique essence and the underlying wholeness of every particular place, community, and person. Evolving Capacity is pursuing that essence expressing its role in a place, community, or family.

Five worldviews from the perspective of work settings

Living inside a paradigm, we form a way of organizing actions and pursuing them—a worldview. Worldviews, the cosmologies for coherence in living, are the practical day-to-day coherence we create to enable social communities to function. It is possible to see immediate correlations between worldviews and paradigms, but they are not perfectly aligned. I will relate the formations and characteristic experiences of worldviews in terms of work settings and offer a bit of a history as I go along. We will consider the worldviews in terms of five eras in history in which they arose.

Before you read further, select a work practice that seems completely up-to-date or that is among the most popular or, in your experience, the most useful. Spend a moment or two assessing it. I would like you to challenge yourself to just see how open and honest you can be. Come up with a program, a practice, or a process that you would stand up for, one that would inspire you to say, "I will stand for this. I will fight for it. I will claim it as the best we can do on work design." Now, think for a moment about when and how this practice originated. In which century do you think it was created? Then as you read through the following pages, examine it going back in time through the eras in which each of the five worldviews originated. Pause along the way to note how your understanding of it is developing, and what might be changing in your attitude toward it.

Aristocracy worldview

The wisest and usually most powerful individuals have control over ownership of assets.

The first worldview in the framework originated in the era when city-states, nations, and religions were governed by aristocracies, when in particular the Catholic Church was influencing the development of cultures throughout the Western world. Work design today is based on hierarchies because religion based its work design on hierarchies that were adopted from royal courts. The belief here is that a select few people, both through birth or ordination, were smarter than everyone else and endowed with higher authority.

We forget how much of business management today derives from the churches and courts of old, some of which still exist. In the 16th through 19th centuries, churches were governed as firmly entrenched hierarchies as were governments led by royal families. Over time these were challenged by the rebels and revolutionaries who left the Catholic Church and/or fled from Europe to the New World, rejecting aristocracies of all kinds and government by unelected authority. The archetype of the rugged individual at odds with external authority was brought into being in the 16th and 17th centuries by those who challenged aristocratic tyranny and by the Protestants sects, which rejected priests and popes as intermediaries between themselves and their God. The rejection of hierarchy birthed the era of the craftsman, and for a period of time, the prevailing commercial practice was to make a living person-to-person. Today we call this "small business." If a man made a leather coat, he made it for someone specific. He did not sell it off the rack or make it for the shelf. He made it for an individual, as a birthday or wedding gift, to meet a specific need, or as part of an arranged trade. He raised the cattle, and slaughtered them, tanned the hides and tended the pond where the tanning solution was dumped (or did not tend it, which was common). Then he cut the coat to a common pattern, seamed it, finished its edges, and delivered it with great pride to be worn for decades. He probably may also have repaired it from time to time.

Handicraft was common to all enterprise in the crafts era. There were silversmiths and blacksmiths, spinners and dyers, weavers and seamstresses, cheesemakers, butchers, brewers, basket and barrel makers, carvers, joiners, farmers, and herders—all mostly learning a variety of skills from older relatives and relying on extended family members for whatever their own household could not produce. One person or one family unit saw each product through, from raw material to polished object, and as a result a much higher level of quality was the norm. Exquisite handiwork was common because craftspeople cared about what they made. They knew they were going to see their work walking around on the backs or in the hands of people familiar and sometimes dear to them, and so they wanted the coat or the basket or the ax handle to be extraordinary. They lived with their work and saw it at market or at church. In this era people grew up seeing the wholes of things. The aristocratic worldview was in suspension, and for a short time the notion of the self-reliant, robust, courageous, and nimble-fingered pathbreaker took its place. Today this short-lived but unforgotten archetype is associated with the entrepreneurial worldview, which evolved alongside the larger, still predominate worldviews that came into being in later eras.

In the craft era, authority was based on land ownership, which also determined which families received the right to vote as democracies emerged. Land and control over assets, the ability to produce wealth, was the basis of governing power. Land and the beings it supported, including human tenants, were perceived to be inert or soulless, subject to the will of the landowner. This notion is still inherent in the practices of industrial agriculture, mining, and chemical refining, which are based on the view that soil nutrients, minerals, and fuels are present on the earth solely to be extracted and exchanged.

Machine worldview

People are cogs in mechanical processes.

The English economic historian Arnold Toynbee announced in the 1850s that in the past hundred years Europe had experienced an Industrial Revolution. Technological changes—such as the invention of new machines and mechanical systems, advances in transportation, and the shift of work from workshops to factories—had brought about mass production of manufactured goods. These changes also resulted in an economic and management revolution. In Britain and the United States, people without the means or skills to create their own craft workshops joined assembly lines.

In a short time, the shift in work design from handcraft to routinized assembly led to loss of connection with the whole—worker, customer, economy, and community. Although much good came from the Industrial Revolution, most importantly the creation of the middle class, in general working people came to be regarded as machinery to be managed and manipulated. The dominate worldview shifted from makers as controllers of land and resources to individuals as cogs in machines. The human relationship to ecosystems under this view remained the same as it had been in the era of the aristocracy worldview, with the exception that now more people could own and extract value from land.

The machine worldview was shaped by the Extract Value paradigm and infused by its way of knowing and understanding. Its architect was Frederick Taylor, who created what we now call "scientific management." He proposed that work could be done more efficiently and at less expense, if the production process was broken into small pieces and assigned, one each, to workers who could learn them easily and repeat them uniformly over and over again. Taylor was a fan of the economist Adam Smith's treatise on capitalism, which described this way of working as a narrowing of focus or fine tuning. Smith imagined the human mind as a kind of clockwork and the universe as an infinitely complicated machine, set in motion by God and left to run on its own. Living systems and processes could be fully understood and mastered through the sciences of physics, chemistry, and mechanics. The machine worldview that grew out of this paradigm reduces workers to interchangeable parts in machines connected in linear manufacturing processes. Work becomes rote, and because workers are replaceable, their safety and well-being are disregarded. Google cofounder Larry Page tells a story that illustrates the effects of the machine worldview. His grandfather always carried with him to work an old metal rod soldered to an iron ball because, in the factory where he worked, supervisors regularly beat men for not doing their jobs as management saw fit. Page's grandfather and his fellow employees were seen as machine parts to be hammered into line. After years of abuse, they seized control of the factory and locked out management in a successful strike that led ultimately to the creation of the United Automobile Workers Union. This was a major accomplishment and a step in the progression beyond the Extract Value paradigm and the machine worldview. But still today, these ways of knowing and living prevail around the world and are the primary sources of authority that we think of escaping when we speak of making a paradigm shift.

Behavioral worldview

Human behavior is controlled by external conditioning.

In the machine worldview, humans were interchangeable parts and so the management and the idea of management had to do with production and how quickly it could proceed. The mechanical worldview started giving way in the 1920s to a behavior modification worldview. John Watson, a psychologist at Johns Hopkins University, was mindful of the numbers of unhappy industrialists who were having a hard time getting their workers to do what they were told to do. Workers were rebelling. They were unionizing. They were considered lazy and ungrateful for the opportunities offered to them. Watson proposed that if industrialists funded a research laboratory for him, he would give them control over labor. He would provide ways to manipulate employees that would consistently result in their doing everything management asked for, on demand. Money poured in. Watson built his laboratory and spent a few years determining what really motivates people. The workplace, and eventually schools and families, were flooded with ideas derived from the "ABC theory"—antecedent, behavior, consequence. Provide an enticement and a behavior will result in response, which can then be reenforced or extinguished with an appropriate consequence. Pretty soon businesses and the military, and not long after that the educational system, and eventually parents were all trying to figure out what actions and re-enforcement were needed to control people's behavior. Whole generations were treated with rewards and punishments in attempts to motivate them to comply with the demands of managers. But, one little flaw in Watson's research was that all of it was based on the study of rats and how they moved through mazes. He never questioned whether his results translated to human minds and behavior. Recent research is making it clear that they do not transfer well and only in some instances, and that as regards the development of human capacity, they are often simply irrelevant. However, this has been ignored, and the behavioral worldview continues to claim that humans are like rats to be manipulated with incentives, rewards, and punishments.

Human potential worldview

All people have free will and can develop the capability to motivate and develop themselves.

By the 1960s a large group of people, led by humanists and a few psychologists, could see that the assembly line and rats-in-mazes metaphors for humans were incomplete. This new understanding led to what we now call the Human Potential Movement. This movement was concerned with self-mastery; it posited that people can be self-aware and self-directed, and experience intrinsic motivation. They researched ways to get people to take on their own behavior modification, such as goal setting and personal rewards. They used such processes as affirmations—for example, the repetition of personal mantras based on aspirations—as a way to affect different behaviors and outcomes. Their big shift, in human relationships and organization management, was away from the ABC theory of conditioning to the position that personal agency makes intrinsic motivation practicable and delivers a much better result.

Behaviorists believed that humans could not motivate themselves because they had no self-control or self-determination. But the early humanist psychologists—Virginia Satir, Carl Rogers, and Abraham Maslow—and eventually hundreds of others in the field posited that this was not true. They developed their own research methods (they are also paradigm dependent) and demonstrated that people can observe themselves, make conscious choices, and become self-directed. They argued that all people have free will, not just the aristocracy or the well-educated or those who have succeeded in becoming wealthy. They claimed that all individuals could bring about change in themselves, and this became the starting point of the self-help movement and the industry that grew out of it.

Living systems worldview

Regeneration is the core characteristic of living systems, including human systems.

There has always been a worldview more immediately connected to nature and natural systems than the aristocracy, machine, behavioral, and human potential views. It has persisted for centuries in limited communities, and mostly in the spiritual practices of people who believe that they can change themselves through everything from communing with nature to meditation to psychedelic substances. These were early expressions of what has come to be understood as a living systems worldview. Now, with science and technology making it possible to study life, a completely modern paradigm and a worldview based on it have developed.

In the first years of the 21st century, a leap in the understanding of living systems has occurred. For example, we learned not so long ago that if a forest is attacked on one side by a disease or an infestation of insects, changes in structure begin to occur on its other sides, in trees that can be hundreds of miles away, as they begin to prepare themselves for what is coming. The forest uses biological patterns to re-express its whole being. It remains itself, and at the same time it has regenerated and become more resilient. Likewise, when a person regenerates understanding from reflecting on their experience with a restraint they face success. One becomes better able to experience being fully alive and able to take on the future. Regeneration, connecting to and reexpressing the unique pattern of the whole, is the core characteristic of a living system. In the living systems that are individual humans—and in the individuals of many other animal species—it extends beyond physical DNA and patterning into the unique essence of a self. The humanists gave us this understanding, but they failed to grasp that it is true for nature as well. We are so conditioned by the machine and behavioral worldviews—which dominate our parenting, schools, and organizations—that we look at life as fixed, predetermined, and fragmented, and often fail to develop the capability to see the living wholes in life processes.

Regeneration is the innate ability of a living system to bring itself to a new level of organization and expression after it has been destabilized or disrupted. This kind of renewal requires that the living system reconnect to the core of its life—what it is in its essence. It is possible for two reasons. First, every living system has a unique core, each natural system—such as a human body, a forest, or an ecosystem—at a physical level and each person at a psychological level. When it regenerates, it is not creating a brand new, unrelated part or aspect of itself; rather, it is actualizing the potential that was always there, bringing it forward into a new moment or context. Some works of art, music, and literature, for example, are manifestations of regeneration—produced in response to transformative events (destabilizations) experienced by their creators. Second, every individual is always nested within a larger system that it contributes to and is nurtured by in return. This reciprocity *invites* the expression of the individual's essence. Regeneration is the expression of essence in service to the whole systems that it depends on to thrive. This is why understanding regeneration requires understanding the working of living systems and the systemic reciprocity they offer to each of the living entities within them.

The regenerative way of understanding is a particular philosophy, which differs considerably from the human potential movement and radically from behaviorism, and mechanism. Each of these forms the basis for different ways of working, and out of each a comprehensive worldview gets formed and human characters get shaped. A worldview gives people a way to agree on the truth. They live by that truth whether it is substantiated or not; they agree on it, and often it becomes the only lens through which they are able to see the world.

Worldviews summary

We have looked briefly at the four dominant paradigms that determine how we study and make sense of the world (epistemologies), and we have seen how they shape cultural and religious worldviews (cosmologies). We have looked at five worldviews that give us a practical way to organize life on a daily basis and live in communities of shared understanding. Both paradigms and worldviews are implicit to our thinking and acting, but paradigms are less visible than worldviews, shaping how and what we can use to form truth and providing the basis for agreements on how to live together within a shared understanding. Looking from one nation to another, we can quickly see how much more or less freely some people are allowed to accept teachings and ideas or reject them based on their sources of knowledge. We can also see that in current years both paradigms and worldviews are becoming more fluid, as media in all of its forms exposes us to foreign ideas and the Internet becomes a shared way to discover and validate truths.

The primary worldview of current work design

Amelioration—this is the prevailing worldview or mindset that we live with now. The focus is to ameliorate what is seen as harmful human behavior by managing people behaviorally. And if the new management program has side effects, they create yet another program to manage that. For example, until recently feedback was used in most businesses as a tool to change unwanted behavior. But it had a negative impact on spirit, and so people added their own comments to their reviews. And then feedback was invited from all directions, 360 degrees, as a way to avoid bias. What resulted when human resource departments managed work design by building off the paradigms underlying the behavioral worldview was a never-ending effort to fix people or manipulate them to unlock their human potential. Those efforts left us with a need to ameliorate all kinds of problems. One of the reasons why my favorite work is to engage directly within companies is that I get to help bring human beings back to wholeness. We expose paradigms and the resulting worldviews that perceive and engage people as machines or rats. We undo the practices that cause negative side effects that demand continual amelioration. You cannot have good businesses or a good educational system or a democracy that works—you cannot even really have happy families if people cannot think critically and be self-managing, if they are always reactive.

Extraordinary outcomes of the living systems worldview

Every company that I have worked with to build regenerative work systems from the living systems world view has been seeking extraordinary, seemingly impossible outcomes—ones beyond the usual notions of continuous improvement. For example, I worked with Kingsford Charcoal that owned Hidden Valley Ranch Dressing. When they bought the business, it was pretty average in the food industry. That meant it was taking two to five years to get a great product idea into the market. We changed that, moving to a six-month cycle from ideation to execution. In our change process, we moved Hidden Valley dressings to numbers one, two, and three within a year and a half. These illustrate the kind of returns I am talking about, when I propose that we shift to the living systems world view and start thinking regeneratively. And it does not end with a bestselling salad dressing. Whatever you are working on, you need to be ready to take what Google's leaders call a "moonshot" and go for it. Once you have set up your regenerative work system, the next thing you want to do is to shift to an *urgent and compelling expectation for changes in social and planetary imperatives*—which is not negotiable if society and the earth are to work as they need to. What I mean by social imperatives are the nonnegotiable changes required if our society is ever truly going to work well. For example, we must create the conditions that will make it possible for all people to become self-managing and develop critical thinking skills—the basic conditions required in order for democracies to work well. Planetary imperatives are the changes that will keep us in right relationship with the earth, many of which are absolutely necessary, such as stopping extracting resources at rates that exceed their ability to replenish themselves. We must stop creating toxins and dumping them where they can do no good. You want to acknowledge ambitious imperatives like these and you want your business to go after some of them.

Regeneration takes an organization or community beyond the ideas of "being less bad" and "doing good." You want to pursue what makes your people and your business whole and healthy. Let go of outmoded practices such as hiring A-level talent. Business leaders often ask me, so what does the hiring process look like from a living systems perspective. I reply that the regenerative path is to build, not buy, talent. Human beings all have innate talents waiting to be developed. Each one of us is unique and capable of growing, and has still more potential to be realized over time. Understanding this is foundational to regenerative business practices built from the living systems worldview and the Evolve Capacity paradigm. Developing people enables them to exceed all the goal markers thought possible, and this inspires their loyalty. And there are even more extraordinary effects; when you develop an entire organization, you create a revolution. When you begin to see all people and activities through the developmental lens, not only your own people, but also your suppliers and distributors begin to grow beyond whatever they thought they could. Your own and your employees' families, your children's schools, and your local government may as well. With an Evolve Capacity approach, everyone in the market and community has the idea that they are a force, a united team that can grow together in order to produce extraordinary results. Life appears as it actually is, dynamic rather than fixed. Everything becomes alive to you.

Regeneration as a practice

Humans are nested in living systems, which are not separated into two realms, natural and human. We create false ideas when we try to mimic nature. There is a wonderful quote on this subject by Buckminster Fuller "Work against 'Life Centered' principles, and you will find yourself thwarted at every turn. Work with 'Life Centered' principles, and the Universe itself pitches in to help." To work with life-centered principles, we must understand how life works in a dynamic, engaged way as nested systems.

I will now spend a moment discussing the First Principles of Regeneration, which I suggest for use in any situation in which you wish to create integrity and universal meaning. First Principles serve for sourcing creation or an examination or evaluation of material or thinking. First Principles are basic, foundational, self-evident propositions, or assumptions that cannot be deduced from any other propositions or assumptions. In philosophy, First Principles were initially formalized by the Aristotelians. First Principles thinking, which is sometimes called reasoning from First Principles, is one of the most effective strategies you can employ for conceptualizing and understanding complex situations and then generating original solutions.

The Seven First Principles of Regeneration

Several years ago, working with a cadre of scientists, ecologists, and naturalists, I articulated and began working with the Seven First Principles of Regeneration. These principles can be used to enliven and promote living systems understanding of people, watersheds, businesses, communities, and most other systems on the earth, not including computer or machine systems. Next are described the Seven First Principles of Regeneration (I made them up from watching living systems for four decades).

Image a whole at work

A living system is a *whole*, defined by natural boundaries. It cannot continue to live if it is broken into parts or fragments. To see or know the working of a living system, it is necessary to image it engaged in being alive. For example, a frog is a living system when it is living in a pond nested in a specific lifeshed or ecosystem, able to jump around, feed itself, and reproduce. It is not a whole when it is vivisected and then dissected in a laboratory. We can explore any system at work, from markets to economies, citizens to nations, and neighborhoods to ecosystems.

We can understand it firsthand by using First Principles to examine it directly, imaging its work as it engages in transactions with others, transitions through time, even transforms from one state of being to another. This understanding will far surpass any insight we might gain by explaining it within the context of static, preestablished thought structures (e.g., anatomy books, psychological treatises, engineering schematics, statistical analyses). We now have whole groups of scientists who study rivers as if they existed independently of watersheds and lifesheds,[1] and trees as though they lived apart from the forests and ecosystems they grow in. We can do the same with all living systems, from smallest to largest, and this can be reflected in how we speak about them.

Work toward potential not ideals

The second principle is to see the uniqueness of the whole and its distinctive potential. The opposite is to think in terms of ideals or problems, which leads us to ask if a person or city matches our ideal. Comparisons immediately drop out when we think in terms of potential because potential can only be conceived of concretely, inherent within one life form, one person, or community. We have created a mess by imposing ideals on children, First People, people of other faiths than our own. We establish ideals as a way to certify and license them. This can only diminish living systems, as there is no practical generic ideal in any area of life.

First Principles are guides rather than ideals, ways to test our ideas about living entities not ways to control them. When a new person or community comes into play, there is no transferability of knowledge, except in the process of examining one's own ideas and actions towards them in light of the First Principles. A second way we undermine potential is by defining everything in terms of problems, without awareness that the problem-solution concept derives from the machine worldview and the Arrest Disorder paradigm. Defining problems and working with standards and ideals inevitably means starting from current existence and limiting ourselves to improving rather than regenerating. My favorite illustration of this limitation is parents who look at their own children as if they were little bundles of problems. They talk back, make messes, do not do their homework; they fall short in a score of other ways. When the parents try to fix these problems and teach children to overcome their failings, endless battles ensue. To work on problems splinters children's lives and denies their unique potential; it casts them into the generic role of troublemaker.

Problem-solving and problems are always defined in terms of variation from an ideal. In order to get past the trap of the problem perspective, in order to really bring it into view, I might point out that, throughout history, ideals have led to colonization and often to genocide and the eradication of entire cultures. In the past we have collapsed the cultures of many indigenous peoples by imposing ideals of language and behavior. If people accept that they need other people's approval, adopting their ideals and the lists that go with them, then they stop reflecting and finding out who they are and what living systems they wish to serve. They lose their authenticity. This is also true of certification programs. I have been asked why I do not offer certification for the work that I do. My answer is that the only reason to certify is to make money. Certification entails the creation of ideals and the selection of a list of fragmenting items by which to judge people as a way to convince them that you know the best ways to do what they want to do. This shuts out all questions and denies the potential of unique individuals.

The second principle of regeneration guides us away from problems and ideals to the discovery of potential in each of the people and all of the other living systems that we encounter in our lives, both at work and home.

Reveal and express essence

Everything that is alive has an essence. Essence is that which, if taken away from us, we would no longer be who we are. It is sometimes called "our being DNA." If we do not start from essence, we move immediately to categorizing things and putting them into boxes based on patterns that we invent. An example is personality tests. You have probably taken one that defines you as one of four types. We all want to know what makes us who we are, and we are raised to expect to see our essence in the feedback and observations of others. You may also have taken a test in some job or other to certify your strengths—even though your strengths are mostly socially conditioned and can change easily when situations change. The drive seems to be to contain people in boxes that make them easier to manage.

If you define an entity by its type, then you have missed its life, its uniqueness, and reduced it to a thing among similar things. The third principle of regeneration is core, and it is one that is dropped out completely in the Doing Good and Arrest Disorder paradigms. That is, it has no place in sustainability practices. Learning to do essence thinking is what discernment is really about.

Engage with living systems developmentally

Developmental work makes essence expression possible for anyone, it reveals patterns and removes all the veils, particularly personality traits, that otherwise inhibit it. We have been examining how regenerative practices affect humans, especially in work settings, which are the places where most people are able to make regular, beneficial contributions to the lives of others, and help make them more effective in all of their activities. In a business, there are two arenas of developmental practice that serve to benefit the organization and have the potential to improve societies and democracies. The first of these arenas is the development of critical thinking skills. This includes primarily building the capacity for discernment and complex systems thinking. When engaged routinely by an organization, in the course or regular work events and with reflection on how they were handled, they increase the ability of participants to see how they are thinking and to know whether they are whole and complete in their decision-making and execution.

The second arena for development is that of personal mastery and the ability to manage one's state of being in difficult situations. Learning to see our own mental and emotional processing and to monitor interaction with others is core to working together effectively. With education and reflection, we can learn to recognize what upsets us, what energizes us, and how to manage our reactions—core skills of well-functioning individuals. These skills are developed only haphazardly throughout our lives. We rarely have opportunities to learn them intentionally. In an organization committed to working regeneratively, personal development can be integrated into daily work. When development becomes a focus, it establishes the foundational abilities necessary to practice the other six principles.

By practicing critical thinking and self-mastery, we learn to know ourselves and to see the effects of our actions on other people and the outcomes that result. In particular, we develop the ability to experience these in real time and to change our choices of words and actions, and our decisions, in order to purposefully alter the effects that we see unfolding in the moment. In my experience, building such skills in organizations has made them resilient and produced innovations at much higher rates than would have been possible had employee education been limited to functional skills only.

Design from a mindfulness of nestedness

Living wholes are always nested within and among one another. Nothing exists independently. A family is nested in a community, which is nested in a society, which is nested in a culture. The only way to really know a living system is by imaging it within nested systems—the frog within the pond within the meadow within the lifeshed. Nesting is defined as the interrelationships and interdependencies among smaller and larger wholes. It is the opposite of ranking, in that it does not discriminate based on size and position, but understands the interbeing or interweaving of whole systems. You have probably had a job at an organization where you were ranked within a hierarchy. Yet, a business organization is a living system of people and processes at work with one another to improve the lives of customers and other stakeholders. In the businesses I work with, employees are always organized within teams, and these teams are always nested within stakeholder groups—customers, suppliers, distributors, investors, local communities, and lifesheds. They are not nested in the company, as you might expect, because in fact they are integral to the whole that is the company—they are the company. The next ring out from any team is its stakeholders. The quality of reciprocal relationships with them is what determines how well the business does. From the perspective of the living systems worldview, there is no desire to break wholes apart or to disengage one from another. Instead, there is always an imperative to understand the work of systems in terms of reciprocal relationships. We human as individuals are nested in families, and those families have unique essences and look very different from one another depending on how individuals connect within them and how they move about in the world. But all families exist in communities and nations and within a global whole.

Intervene at systemic nodes

The idea of working nodally is challenging as it requires imaging a system that is alive and dynamic and engaged in reciprocal relationships with other living systems. Life moves and changes for us, based on activity at nodal points, not in scattered parts. Nodal interventions are the foundation of acupuncture, for example. You do not place needles in all the places that hurt or are off balance. You put them where energies cross, intersections through which the system as a whole can be affected. Our senses can connect us only with small aspects of a whole working system. When we are considering negotiations between nations, looking for an intervention that is likely to shift all conversations going forward, or when we are working on the health of a human body, where core changes can reverse illness and promote wellness, we are looking for the confluence of energy where an intervention can effect a change.

Nodal intervention is not about priorities or leverage because living systems have no priorities. It is working in a way that is whole and systemic all the time, looking for nodes where relatively small actions can affect large, whole-system changes. We have borrowed this idea as a metaphor for the way technological networks function. In nature it works less as a template and is far more complex. When you consider a lifeshed and its health, for example, you have only to examine the wetlands, which can tell you everything you need to know about the state of the whole. They, in their total living complexity, constitute a nodal point. We do not fully understand nodes in ecosystems or in terms of how human communities or the entire planet work. We do not understand them in business terms, either. When I work with Europeans, one of the things that fascinates me is their understanding of nodes. They speak about them directly. The interventions that attract many Europeans are free health care, to make society's foundation strong, and free education, to develop the capacity for democracy. For example, most Europeans agree that free education, when it is done right, is a nodal intervention with great potential to improve the health of societies and economies. Free education is a node where individuals can make contributions with the potential to change society for the better. Education is core, foundational to life, and therefore it is a node. Those are the keystones, the *nodal* interventions with the potential to produce a vital and viable nation.

Innovate for systemic reciprocity

This principle is the opposite of "every man for himself." The nested nature of life makes us all interdependent, and this is the source of living systems thinking: systemic reciprocity, rather than transactional engagements with one or more entities for our exclusive benefit. The core to systemic reciprocity and the regeneration of whole systems is to evolve capacity to be fully capable participants in the systems they are nested in. This is the principle through which regenerative interventions do the work of the Evolve Capacity paradigm. Innovating for systemic reciprocity focuses on evolving capacity of the system and all beings in the system. Stepping in for others, making improvements for them, is the practice of other paradigms—Doing Good, Arresting Disorder, and even sometimes Extracting Value. And although we may sometimes benefit people as a way to further the aims of these paradigms, we are not truly evolving capacity in living systems until everyone involved is making their own, self-directed contribution to regeneration. This is how potential is expressed, in the collective, regenerative endeavors of unique individuals expressing their essences and becoming more able through the effort. And this is also how we stop the depletion of—or the extraction of value from—people who are only giving.

The Seven First Principles as capabilities

Working from principles is a unique capacity that is currently not being developed in most people. Instead, we are educated to follow specific sets of rules based on someone else's idea about what is right for the groups they considered. We are not using systemic mental frameworks to build the capacity for reflection and discernment that is required to think regeneratively. For living systems, there can never be one right answer for every situation. Working with principles requires consciousness, the ability to separate from a situation while being in it, to discern how things are working, use judgment to locate nodes, and to predict the likely effects of an action or decision. We must learn as individuals how to do these things in real time in the midst of real work, and then we must make it possible for whole organizations to learn how to do them. The highest aim of regenerative practice is to evolve capacity in the beneficiaries of our offerings so that they can contribute from their essence to the vitality, viability, and evolution of the systems that are being actualized.

The Seven First Principles as a system at work

Each of the Seven First Principles is an opportunity to develop capability (Figure 2.2). We have been so thoroughly educated and conditioned to work from the older, less complete, paradigms, and worldviews that we must now develop almost from scratch the full complement of regenerative capabilities. The way our uneducated brains work is by seeking and confirming what is familiar. Preservation is the goal of the undeveloped brain and it either does not see anything new or worth exploring or it rejects it in order to preserve the safety of adopted ideas and practices.

This is why a developmental culture and community is needed. Businesses and organizations are ethically called to create growth environments, rather than ones that reaffirm and deliver on unexamined ideas. The Seven First Principles of

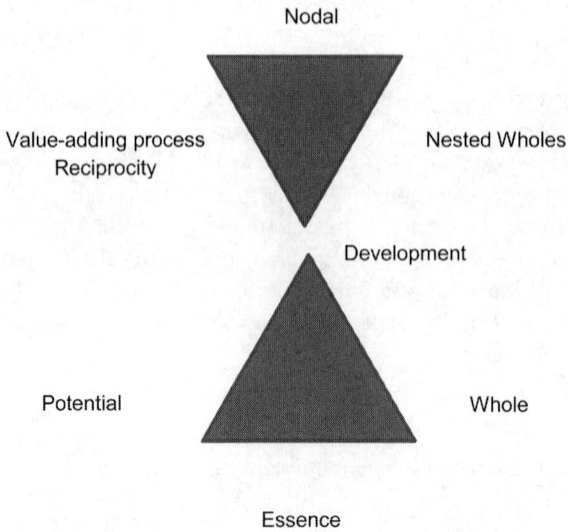

Figure 2.2 The Seven First Principles of Regeneration at work

Regeneration cannot be brought into a community unless they are coupled with intensive personal and organizational development processes. They simply will not be useful with the minds, infrastructure, and work designs that are mostly in place now. A conversion to regenerative practices will require deep questioning and rethinking, a process to evoke in people a consciousness of the effects of and outcomes from all of their actions. The process must be strategically embedded in the *way* work is done and in *how* everything is carried out. This is why most of my life's work has been about designing and engaging organizations in a developmental way of working and doing business, a process that develops individuals' critical thinking skills and personal mastery through education and reflection, and pushes across restraints by connecting every person in the organization directly to external effects.

The new way of working to regenerate requires setting up systems for each person to find a way to contribute, and that creates daily destabilization of people's certainty, comfort, and ideation process. It takes them off automatic and hands them the controls. This is what is required to bring about real change. For videos, books, and articles on regenerative business practices, visit www.CarolSanford.com.

Note

1. I use the word "lifeshed" instead of "watershed" because watershed is an anthropocentric term that cuts waters away from their living ecosystems, conceiving of them as resources. "My water!" Sometimes people say "airshed" or "foodshed." But no—water, air, food.

Reference

Kuhn, T 1970, *The structure of the scientific revolution* (2nd ed.), University of Chicago Press, Chicago, IL.

3 The centrality of the systems approach: Regenerative development, resilience, and sustainability

Beth Schaefer Caniglia

The sustainability approach to harmonizing environment, equity, and economies has come under strong critique in recent years. It has been over 30 years since the publication of *Our Common Future* (1987) and 25 years since the adoption of Agenda 21 at the Earth Summit in Rio de Janeiro, Brazil in 1992. Yet, environmental degradation continues to threaten livelihoods across the globe; climate change is driving more and more extreme weather events; and inequality between the haves and the have-nots is greater than ever (Caniglia, Vallee & Frank 2017). Many scholars and practitioners have abandoned sustainability as a framework and instead prefer to discuss resilience or regenerative development. How are these frameworks to be differentiated? As a scholar who works with practitioners in sustainability, resilience, and regeneration, I often hear explanations of sustainability and regenerative development that confound their goals and processes. The purpose of this chapter is to untangle these concepts from each other in ways that illustrate how they differ, what they have in common, and the specific barriers that hold us back from achieving the fundamental goals. I argue that the greatest challenges for achievement of sustainability, resilience, and regenerative development are essentially the same: A failure to understand and commit to systems approaches. I begin by defining terms and reviewing relevant literature and defining key dimensions of sustainability, resilience, and regeneration. Next, I define systems thinking and, finally, compare the ways in which each of these approaches to development incorporate systems orientations.

Sustainability

The most common definition of sustainable development was put forth by the World Commission on Environment and Development in *Our Common Future*, and states that sustainable development is "*development which meets the needs of the present without compromising the ability of future generations to meet their own needs*" (1987, p. 8). Multiple extensions of this definition have been offered in the years since the Brundtland Commission's initial description (Costanza, Daly & Bartholomew 1991; Costanza & Patton 1995; Dresner 2002; Gladwin, Kennelly & Krause 1995; Holden, Linnerud & Banister 2016; Pezzey 1992).

Much of the debate surrounding definitions of sustainability evolve around these questions:

- Sustainability of what?
- Sustainability for whom?
- Sustainability over what time frame?
- Sustainability at what scale?

Underpinning most definitions of sustainability is the desire to harmonize what are known as the three Ps: People, Prosperity, and Planet or the three Es: Environment, Economy, and Equity. There is an assumption underpinning the sustainability framework that these three pillars are interdependent—three legs of a common stool. Culture and the economy depend on the natural world to thrive. To quote the Preamble to Agenda 21: "*…integration of environment and development concerns and greater attention to them will lead to the fulfilment of basic needs, improved living standards for all, better protected and managed ecosystems and a safer, more prosperous future*" (1992). From the beginning, the assumption was that humans and nature are part of a common and interdependent system. Costanza and Patten (1995, p. 195) presented a definition of system sustainability: "*A system is sustainable if and only if it persists in nominal behavioral states as long as or longer than its expected natural longevity or existence time.*" They also argued that "*[w]ithin the socioeconomic subsystem, a social consensus on desired characteristics … must be arrived at. These characteristics also function as predictors of what kind of system will actually be sustainable*" (p. 196). This is a critical point to which we will return toward the end of this chapter.

A second concept that guides the sustainability approach is the *precautionary principle*. Some interpret the precautionary principle as an approach to sustainability that allows using business-as-usual techniques as long as we keep an eye out for indicators that we are nearing thresholds that might impinge upon the rights of future generations or endanger contemporary development goals. Dasgupta (1993) has argued that a balance should be struck between current levels of well-being, future levels of well-being, and optimal development scenarios. He advocates for a *laisse fare* approach where "*if you get the intertemporal optimization right, and use the correct shadow prices, the correct policies will simply fall out of the analysis*" (Dresner 2002, p. 79). Multiple scholars have argued, however, that it is very difficult, if not impossible, to estimate these calculations accurately (Costanza & Patten 1995; Meadows 2008). We too easily miss the importance of lag-times, for example, and we often completely leave out essential variables in our models.

According to Simon Dresner (2002, p. 64), when asked for his definition of sustainable development, Nitin Desai said "*definitions are useful only for the clue that they give you for the premises on which somebody works.*" The sustainable development paradigm was designed to address trade-offs with a common denominator phrase that acknowledged environmental limits and risk, while fostering continual growth. The word "sustainable" is a moderator to the primary concept of "development," which is prioritized as the solution to existing inequity. Holden, Linnerud and Banister argue, in contrast, that the concept of sustainability implies "*…a set of constraints on human behavior, including constraints on economic activity*" (2017, p. 213). In fact, they argue that "*economic growth cannot be one of the sustainable development goals*" (2017, p. 213). So, there is clearly a disagreement regarding the ultimate goal of sustainable development when it comes to equity.

Another central component of the sustainable development concept is "non-declining capital" (Pearce, Markandya & Barbier 1989). This dimension of sustainability is interpreted through the lens of either "weak sustainability" or "strong sustainability," where a strong sustainability lens assumes limited substitutability of various forms of capital and weak sustainability assumes quite a bit of flexibility and substitutability across the forms of capital. Of course, the weaker model of sustainability is predicated on a global economic model, where trade around the world provides access to the materials needed to sustain production processes. Faith in progress is a related dimension of the sustainability paradigm. There is an assumption that technology can produce solutions to environmental limits and the perils of resource scarcity. When taken together, non-declining capital, substitutability, increased efficiency and the ability of technology to address scarcity point to a concept environmental sociologists call dematerialization, or increased efficiency (Herman et al. 1989).

Understanding sustainable development as a clear set of practices with measurable goals and indicators becomes more complicated by its political context. At the UN Conference on the Environment in 1972, a global consensus emerged that the earth's ecosystems were in great distress, as indicated by increased species extinction, collapse of ecosystem services, expansion of waterborne diseases, among other ecological crises. Conflicts emerged at future UN meetings, however, over whose responsibility it was to address these crises. Developing nations argued that the industrialized nations were to blame, since they were responsible for large-scale resource extraction and pollution during their "dirty" development stages. The advanced industrialized nations, however, advocated for a shared approach, where developing nations were responsible to implement cleaner technologies in their development strategies (gained through technology transfer), and industrialized nations would set up financial mechanisms to help clean up the damages caused by their historical pollutive technologies. This debate between developing and industrialized nations intensified over time and resulted in a series of compromises that weakened the implementation of sustainability policies worldwide.

Those advocating for sustainability in international policy arenas were hoping to offer a happy medium for those who prioritized expansion of the economy and those concerned about rapid environmental depletion in an unequal world (Ciplet, Roberts & Khan 2015; Roberts & Parks 2007). Best known as *common but differentiated responsibility*, a framework emerged that emphasized sustainability as "*a participatory process that creates and pursues a vision of community that respects and makes prudent use of all its resources—natural, human, human-created, social, cultural, scientific, etc.*" (Gladwin, Kennelly & Krause 1995; Viederman 1994, p. 4).

In this way, the integrated perspective of sustainability as an approach that recognizes the interdependence of economic, environmental, and human prosperity devolved into a practice of indicators and measurements that allowed policy makers, practitioners, local authorities, and business professionals to choose the dimensions of "sustainability" they were most committed to implement (e.g., LEED Certification; see Cole 2012; du Plessis 2012; Mang & Reed 2012).

Resilience in urban socioecological contexts

The complications of the international policy arena and the difficulty in adapting an expanding barrage of indicators to local contexts, led many practitioners and local authorities to turn to the resilience framework as an alternative to sustainable development (Caniglia et al. 2014; Caniglia, Vallee & Frank 2017; Frank, Delano & Caniglia 2017). Existing approaches to resilience have their roots in ecology (Folke 2006; Folke et al. 2010; Lake 2013; Walker & Salt 2006). Ecologists in the 1970s developed this framework to address the ways ecological systems, as well as individual species within those systems, adapt to and recover from external disturbances. In particular, the term resilience refers to the ability of a system to absorb external shocks without altering the existing relationships between species populations and other ecosystem characteristics (Barr & Devine-Wright 2012; Brand & Jax 2007; Holling 1973; Lopez et al. 2013; Webb 2007). In this tradition, resilience can be measured by how long it takes to restore stability within the system or as the extent of disturbance a system can take before it crosses characteristic thresholds into a new stability regime (Caniglia et al. 2014; Caniglia & Frank 2017).

Sustainability and resilience are closely related in that they are both concerned with system functioning and the ability of relevant systems to maintain themselves. Nevertheless, the challenges of rapid urbanization and increasingly extreme weather events due to climate change have created an urgency that the sustainable development discourses lack, especially in regard to addressing environmental changes and natural calamities. Disasters and hazards scholars have strong relationships with military leaders and disaster recovery personnel, because they ask questions of direct relevance to how people respond to high-stress situations (Quarantelli 1987; Tierney 2007). Their research has brought systematic attention to issues such as how to avoid disasters; how to mitigate their impacts ecologically, socially, and economically; and how to most rapidly and adequately respond when disasters strike, both in practice and in the research realm (Walker & Salt 2006).

The resilience framework has been applied to social systems as well, and efforts have been made to understand the ways the resilience paradigm applies to urban centers (Caniglia & Frank 2017; Frank, Delano & Caniglia 2017; Pellow 2017). Findings of the work by these researchers suggest that the resilience scholarship tends to overlook existing inequalities in urban centers, while they emphasize the conditions under which existing institutions (e.g., government and economic activities) and infrastructure (e.g., roads, bridges, dams, etc.) are vulnerable during external shocks, such as floods, heatwaves, riots, etc. From this point of view, city planners and emergency managers often conclude that a city can be classified as resilient even when large groups of its residents never recover from a hurricane or financial crisis (Caniglia & Frank 2017). The need to incorporate social justice into analyses of resilience and vulnerability in urban systems is highlighted extensively by sociologists. For example, Roberts and Parks (2007) analyzed the impacts of natural disasters on industrialized and developing nations and found that the impacts of disasters were more profound in the developing world, particularly among the poor. Ciplet, Roberts and Khan (2015) argue that there is "*heightened and disproportionate vulnerability to climate-related harm by disadvantaged social groups*" (p. 5). They illustrate the ways in which the poor are disproportionately exposed to waste chemicals and other toxins, and that exposure to drought, floods, and disease are consistently correlated with being poor, elderly, and marginalized.

The challenges of vulnerability in urban environments require us to consider population dynamics, climate change impacts, and existing inequalities in complex ways. Gentrification, renters versus owners, immigration status, single families and youth, and language proficiency, among others, influence the ability of individuals and communities to access resources when disasters strike. Urban centers are particularly complex, requiring examination of subsystems within the overarching urban context, along with the feedback loops that foster vulnerabilities. Understanding the characteristics of subsystems (e.g., vulnerable communities or particular demographic groups) that predict their ability to withstand shocks is a critical dimension of understanding urban resilience (Frank, Delano & Caniglia 2017). More work is needed to determine what it means to "fail safely" (Biello 2012) when each disaster produces higher death rates and streams of climate refugees.

Regenerative development

Regenerative development is the newest framework for achieving shared prosperity on a healthy planet. It is designed to explicitly link human and natural systems at scale and to plan for a longer time horizon than sustainability or resilience. Furthermore, it is a paradigm designed to push beyond sustainability and resilience. While sustainability focuses on development today that protects the ability of future generations to develop and resilience is focused on recovering system functions after external shocks, the priority of regenerative development is to heal existing damages in communities and ecosystems in ways that create abundance for the people, the economy, and the planet. The framework advocates applying holistic processes to create feedback loops between physical, natural, economic, and social capital that are mutually supportive and contain the capacity to restore healthy and prosperous relationships among these forms of capital (Caniglia 2018). Practitioners in the planning, architecture, and design communities were among the first to advocate for this conceptualization (Caniglia 2018; Cole 2012), in part as a means of translating global sustainability indicators into practice in local, bioregional urban contexts (Cole 2012; du Plessis 2012). Many argued that the sustainability framework was decidedly mechanistic—focused on replacing parts of the system and tinkering with business-as-usual, while regenerative development recognized the interdependent dimensions and feedback loops that linked the fate of the entire system to human behavior and systems design (Cole 2012; du Plessis 2012). Although socioecological systems were part of the sustainability equation from the beginning, advocates for the regenerative development framework argue that the systems approach was missing in the implementation of sustainable community design. Without attention to systems dynamics, they argue that an inferior, mechanistic approach to human-environment interactions will remain the inevitable outcome. Thus, the bioregional context of coupled human and natural systems (CHANS) is central to this approach (see Knott in this volume; Knott 2018).

Mang and Reed (2012), for example, argue that the regenerative development paradigm suggests that human culture and nature should coevolve. A similar sentiment is represented in McDonough and Braungart's book *Cradle to Cradle* (2002), in which the authors argue that humans should equally serve as tools in nature's evolutionary processes as we have used natural resources to serve human needs. Du Plessis (2012, p. 17) argues that regenerative development departs from sustainability in four key ways:

- Humans and their artefacts and cultural constructs are an inherent part of ecosystems.
- Their actions should contribute positively to the functioning and evolution of ecosystems and biogeological cycles, enabling the self-healing processes of nature.
- Their endeavors should be rooted in the aspirations of the context.
- Development and design represent an ongoing participatory and reflexive process.

Clearly, it is debatable whether these are, in fact, points of departure from the original conceptualization of sustainability. However, the consistent critique found in the literature suggests that sustainability as a concept has failed to translate into a systems approach among practitioners in urban development—a failure that some explicitly attribute to the US Leadership in Energy and Environment (LEED) rating system (Clegg 2012; Cole 2012; du Plessis 2012; Mang & Reed 2012). This critique is understandable, given that a World Bank study in 2007 revealed that over 1,000 sets of indicators existed to measure the sustainability of cities (Troy 2017). Summarizing a variety of contributions to the regenerative building and design literature, Clegg (2012, p. 366) argues that "*regenerative development has the capacity not only to reverse the negative ecological impacts created by human development, but also should have the capacity to increase social and natural capital.*"

Many regenerative economics practitioners have adopted the eight principles of a regenerative economy put forth by John Fullerton of the Capital Institute (2015),

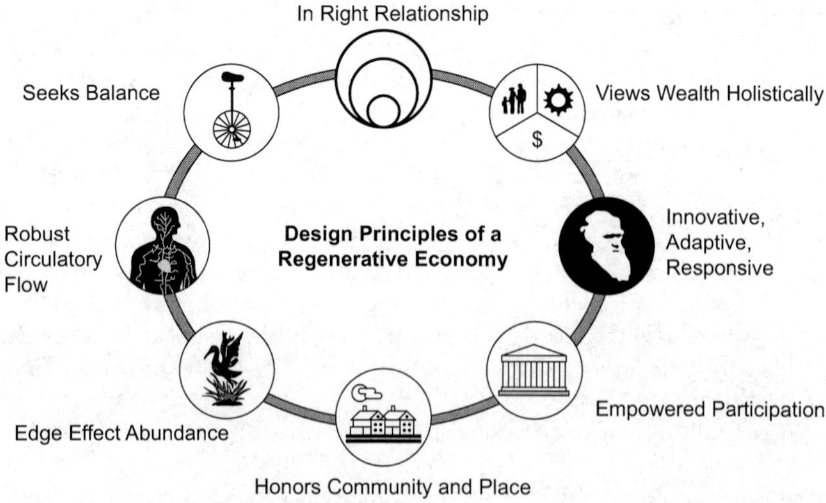

Figure 3.1 The eight principles of a regenerative economy taken from John Fullerton Capital Institute (2015)

as illustrated in Figure 3.1. Fullerton argues that a regenerative system relies upon putting the economy, human prosperity, and the ecological system into right relationship with one another. Rather than supporting a system where natural resources and human labor are put into service to an economic system that benefits the elites, we must put the economy into service of humans and the natural system upon which they depend for survival—an argument central to the discipline of ecological economics (Lovins et al. 2018). A multiple-capital perspective helps to motivate this right relationship by explicitly targeting the constant improvement of human, physical, ecological, and economic capital in all actions by government, educational institutions, businesses, and community organizations. Such systems also need to explicitly build in resilience at multiple scales through a commitment to constantly learning, increasing their flexibility, and thereby finding innovative responses that increase adaptability. Inclusivity and mutual respect for all communities within bioregional contexts are powerful ways to increase diversity of thinking and innovation and for revealing blind spots that perpetuate existing inequalities. These first four principles differentiate a regenerative approach from sustainability and resilience models, because they explicitly draw our attention to patterns of behavior, structural biases, and feedback loops in current bioregional contexts that unduly favor unequal prioritization of economic capital over human, physical, and ecological capital.

The remaining Fullerton principles adopt frameworks that have been advanced through understanding ecological systems and extend their application in ways that recognize the interdependence of human constructs and the ecological systems in which they exist. Place-based solutions recognize the unique configurations of climate, ecosystem services, industrial niches, and population dynamics that point to specific opportunities to develop "right relationships." Drawing boundaries around bioregional contexts is a contested mix between science and art. Some argue that using watersheds as a guiding geographical boundary is one of the best ways to capture the complex interactions that link ecosystem services to urban centers. Most ecologists recognize that the boundaries between particular clusters of species (e.g., fields and forests, cities and their hinterlands) contain some of the highest densities of diverse species, which makes those places informative regarding ecological resilience for specific places. We can look to those places for insights regarding what combination of species—flora and fauna—best serves the multiple ecological niches within our shared ecosystem. The final two principles—circulatory flow and balance—also stem from ecology but have significant implications for human systems as well. They point to the need for good, services, opportunities, and other benefits of a place to be accessible to all members. Where systems are blocked, parts of them become diseased, and fail to function for the good of the system. Therefore, a flowing system will also produce a more balanced system—one where talents and skills can be developed across sectors and subsystems in ways that support the overall positive function of the system as a whole.

Centrality of the systems approach

Underpinning contemporary work on sustainability, resilience, and regenerative development is the assumption that a new way of thinking is required to produce a new outcome (Meadows 2008; Senge 2008; Williams 2012). If we can only shift the focus of our developers, natural resource managers, business practitioners, and government officials to see in systems, we can create a more sustainable and equitable future. There are, however, many barriers to making such a shift in perspectives and practices. Systems thinking is very different from traditional ways of seeing the world. Cause and effect are complicated; the cause may be separated by a lot of time from the effect, for example, and truly identifying causes involves many more processes, feedback loops, delays, inputs, and outputs than we traditionally include in our mental models (Arnold and Wade 2015; Meadows 2008; Motesharrei et al. 2016; Seibert 2018). To achieve sustainability, resilience, or regenerative development, we have to integrate earth system models with human system models. Each of these is typically based on extensive knowledge of interactions among diverse elements at a variety of scales. Putting them together is no easy business. There are subsystems within each system that reinforce and/or create barriers to the achievement of particular system goals. Some may even argue that subsystems within larger systems have competing goals that work against harmonization for the common good.

I'm going to begin by laying out the basic framework for systems and systems thinking. A wonderful review of definitions of systems and systems thinking was presented by Arnold and Wade (2015) in which they ultimately define systems as "*groups or combinations of interrelated, interdependent, or interacting elements forming collective entities*" (p. 7). They further define systems thinking as follows: "*Systems thinking is a set of synergistic analytic skills used to improve the capability of identifying and understanding systems, predicting their behavior, and devising modifications to them in order to produce desired effects*" (2015, p. 7). I particularly value Arnold and Wade's discernment of eight essential components of systems thinking, which I illustrate in Table 3.1. This set of systems thinking requirements is consistent with the writings of numerous systems scholars (for example, Arnold 2015; Meadows 2008; Motesharrei et al. 2016; Seibert 2018; Walker & Salt 2006).

Traditional thinking emphasizes individual events, rather than patterns of behavior repeated through time. It focuses on the trees, rather than the entire forest. It stops at listing factors that combine to produce an effect at a given point in time, rather than examining feedback loops and interaction effects that impact patterns over periods of time. Traditional perspectives look at current resources, rather than the flow of those resources over time. So, let's examine the ways in which resilience, sustainability, and regeneration incorporate dimensions of these forms of thinking.

Table 3.1 Basic components of systems thinking

Systems thinking requirement	*Definition*
Recognizing interconnections	The ability to identify key connections between parts of a system.
Identifying and understanding feedback	Some interconnections combine to form cause-effect feedback loops. Systems thinking requires identifying those feedback loops and understanding how they impact system behavior.
Understanding system structure	System structure consists of elements and interconnections between these elements.
Differentiating stocks, flows, and variables	Stocks refer to pools of resources in the system. Flows are changes in stock levels. Variables are changeable parts of the system that impact stocks and flows, such as flow rates or limits to growth.
Identifying and understanding nonlinear relationships	Limits to growth, flow rates, inputs, and outflows often impact system behavior in nonlinear ways.
Understanding dynamic behavior	The behavior of systems varies in surprising and unanticipated ways. Delayed responses, exponential responses, as well as hard limits result in emergent behavior or the disappearance of behaviors.
Reducing complexity through conceptual modeling of systems	Systems are often too complex to model fully. Thus, we have to discern the primary drivers of system behaviors and understand their component parts in order to predict behavior or leverage the components in ways that produce desired outcomes.
Understanding systems at different scales	In truth, most systems do not have clear boundaries. It is necessary to consider the subsystems within systems and the ways our system of interest may be influenced by even larger, more complex systems.

Comparative summary of resilience, sustainability, and regenerative development in relation to systems approaches

The previous discussion of sustainability, resilience, and regenerative development suggest that all three approaches are grounded in a systems orientation. The extent to which and ways that each of these approaches explicitly carries the systems framework into practice differ in numerous ways, so let's take a look at these three concepts side by side. I will flesh out the ways in which each approach articulates with a systems orientation.

- **Sustainable development:** The most common definition of sustainable development was put forth by the World Commission on Environment and Development in *Our Common Future*, and states that sustainable development is "*development which meets the needs of the present without compromising the ability of future generations to meet their own needs*" (1987, p. 8).
- **Resilience:** The ability of a system to absorb external shocks without altering the existing relationships between species populations and other ecosystem characteristics (Barr & Devine-Wright 2012; Brand & Jax 2007; Holling 1973; Lopez et al. 2013; Webb 2007). In this tradition, resilience can be measured by how long it takes to restore stability within the system or as the extent of disturbance a system can take before it crosses characteristic thresholds into a new stability regime (Caniglia et al. 2014; Caniglia & Frank 2016; Walker & Salt 2006).
- **Regenerative development:** The priority of regenerative development is to heal existing damages in communities and ecosystems, which are connected in bioregional contexts, in ways that create abundance for the people, the economy, and the planet. The framework advocates applying holistic processes to create feedback loops between physical, natural, economic, and social capital that are mutually supportive and contain the capacity to restore healthy and prosperous relationships among these forms of capital.

I provide Figure 3.2 as a means of visually representing the relationship between resilience, sustainability, and regeneration. Resilience is concerned with maintaining the functions of existing systems in the face of external shocks. The "system" of concern is typically the human system, and the "external shock" may come from the natural system or other components of the human system. When examining ecological systems, the question of resilience pertains to the ability of a forest to recover from a fire, for instance, or the same forest to adapt to beetle invasions when trees are weakened by drought due to climate change. External shocks take the form of short-term crises, such as floods or hurricanes, or longer-term variations from normal ecological patterns, such as drought or species extinction; however, even the longer-term variations tend to last only a few years. In social systems, when examining resilience we concern ourselves with the ability of displaced populations to return home after floods or hurricanes or with the financial ability of households to afford repairs. Similarly, how insurance

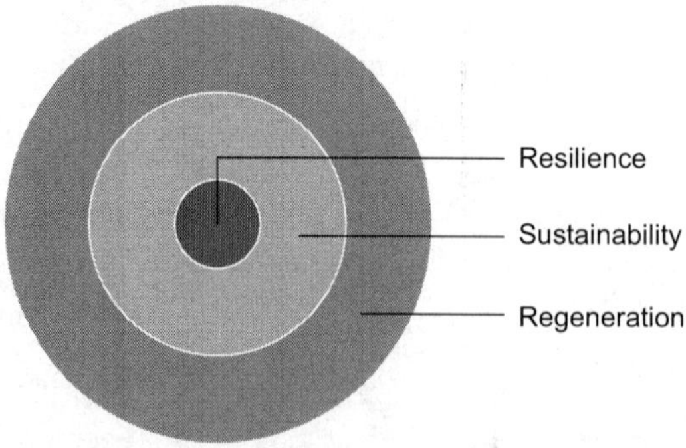

Figure 3.2 Conceptual comparison of sustainability, resilience, and regeneration

companies are affected by the losses they bear after severe storms and how long will it take for the oil refineries to get back on line after they are damaged by flooding? As David Biello so aptly highlights in his *Slate* article (2012), a system can be resilient without being sustainable. Resilience is about the ability of a system to "fail safely" when it is faced with extreme events, which is important for hazards preparedness or in the face of large market fluctuations. The time frame for resilience is much shorter than the duration of interest for sustainable or regenerative development approaches. Longer-term feedback loops and delayed responses are not generally concerns of resilience thinkers. This approach tends not to consider inequality within the "human system" of concern.

In contrast, sustainability is about striking a medium-term balance (one generation or 50 years) between desirable human, economic outcomes, and environmental limits. The goal is to meet current needs without compromising the ability of the next generation to pursue their own goals. This requires a change from linear and single-point-in-time focus to a reflexive and more adaptive orientation that considers ecological cycles, feedback loops, and interdependence between human and natural systems. Creating an approach to development that transitions away from insatiable resource consumption toward lifestyles based on consuming experiences and producing memories is at the center of this framework (Gardner 2016). Adopting a precautionary approach is central—an approach that incorporates an understanding that unforeseen consequences of technological advancements can have significant and sometimes irreversible outcomes that infringe on the rights of other countries, vulnerable communities, or future generations. In a currently unequal world, the sustainability paradigm accounts for the need of the poor countries to continue their development trajectories to improve livelihoods for their citizens, particularly in the least developed countries, where so many live on less than $1US per day. However, continual growth remains the central goal of the sustainable development paradigm with the desired outcome being to balance desirable outcomes for the people, the economy, and the natural environment.

Regenerative development requires practices that restore the capacity of systems to regenerate multiple forms of capital in a coevolutionary process. In this way, it is the most holistic approach—looking backward at dysfunctional patterns; examining the current balance of natural, physical, human, and economic capital; and focusing development activities in ways that replace dysfunctional patterns with institutions that create surpluses across the board. The central purpose of regenerative development is to put CHANS in right relationship with one another, as determined by bioregional contexts. As Girardet argues (2015, p. 11): "*To find solutions to the damage we have done to the world's ecosystems, we need to start thinking about regenerative rather than sustainable urban development.*" Regenerative development also emphasizes creating an economy in service of life (Lovins et al. 2018). Rather than a gentle nudge against the dominant economic paradigm in hopes that we can sustain it a little longer, regenerative development places the achievement of balance between equity, economy, and ecology as a central tenant. Sustainability provides lip service to this goal, but by making continued growth its central goal, equity and environment take a back seat to development in practice. The coupled human-natural systems (CHANS) model is most realized in this paradigm.

Conclusion

In this chapter, I have focused on disentangling the concepts of sustainability, resilience, and regenerative development. Resilience, sustainability, and regeneration are distinct concepts designed to guide development practice in ways that produce related but distinct outcomes. The primary goal of resilience thinking is that it leads practitioners to develop infrastructure and community practices that decrease the vulnerabilities of human systems in the face of external shocks (Caniglia et al. 2014; Mayer 2017; Pellow 2017; Walker & Salt 2006). Sustainability is concerned with modifying continual growth with assurances that future generations will have access to the resources needed for their growth. And regenerative development looks backward and forward to find ways to build systems that coevolve in ways that produce shared prosperity on a healthy planet.

I also highlight that resilience, sustainability, and regenerative approaches each build upon a foundation of systems thinking. Resilience does so over the shortest time frame, concerning itself primarily with the ability of the primary functions of the current system to "fail safely" in the face of external shocks to the human system. Sustainability extends its horizons to 50 years, while regeneration stretches in both directions—backward and forward in time—in an attempt to articulate the CHANS in ways that allow multiple forms of capital to coevolve to constantly improved states. As we evolve both scholarship and development practice toward a regenerative paradigm, it is essential that we carefully articulate the goals of our work and the outcomes we intend to create.

References

Arnold, RD & Wade, JP 2015, "A definition of systems thinking: A systems approach", *Procedia Computer Science*, vol. 44, pp. 669–678.

Barr, S & Devine-Wright, P 2012, "Resilient communities: Sustainability in transition", *Local Environment: The International Journal of Justice and Sustainability*, vol. 17, no. 5, pp. 525–532.

Biello, D 2012, "Can cities be both 'resilient' and 'sustainable' at the same time?", *Slate*, viewed 3 December 2018, <https://blogs.scientificamerican.com/observations/can-cities-be-both-resilient-and-sustainable/>

Brand, FS & Jax, K 2007, "Focusing the meaning(s) of resilience: Resilience as a descriptive concept and a boundary object", *Ecology and Society*, vol. 12, no. 1, p. 23.

Caniglia, BS, Vallee M & Frank B 2017, *Resilience, environmental justice & the city*, Routledge.

Caniglia, BS 2018, "The path to a regenerative future: The importance of local networks and bioregional contexts", *The Solutions Journal*, vol. 9, no. 2. <https://www.thesolutionsjournal.com/article/path-regenerative-future-importance-local-networks-bioregional-contexts/>

Caniglia, BS & Frank B 2017, "Revealing the resilience infrastructure of cities: Preventing environmental injustices-in-waiting", In BS Caniglia, M Vallee & B Frank (eds.), *Resilience, environmental justice & the city*, Routledge, pp. 57–76.

Caniglia, BS, Frank, B, Dalano, D & Kerner, B 2014, "Enhancing environmental justice research and praxis: The inclusion of human security, resilience & vulnerabilities literature", *International Journal of Innovation and Sustainable Development*, vol. 8, no. 4, pp. 409–426.

Ciplet, D, Roberts, TJ & Khan, MR 2015, *Power in a warming world: The new global politics of climate change and the remaking of environmental inequality*, MIT Press, Cambridge, MS.

Clegg, P 2012, "A practitioner's view of the 'regenerative paradigm", *Building Research & Design*, vol. 40, no. 3, pp. 365–368.

Cole, R 2012, "Regenerative design and development: Current theory and practice", *Building Research & Design*, vol. 40, no. 1, pp. 1–6.

Costanza, R, Daly, HE & Bartholomew, JA 1991, "Goals, agenda, and policy recommendations for ecological economics", In R Costanza (ed.), *Ecological economics: The science and management of sustainability*, Columbia University Press, New York, pp. 1–20, pp. 525 (Also reprinted as pp. 50–64 in: Costanza, R 1997, *Frontiers in ecological economics: Transdisciplinary essays of Robert Costanza*, Edward Elgar, Cheltenham, pp. 491).

Costanza, R & Patten BC 1995, "Defining and predicting sustainability", *Ecological Economics*, vol. 15, pp. 193–196.

Dasgupta, P 1993, "Optimal versus sustainable development", In I Serageldin & A Steer (eds.), *Valuing the environment*, World Bank, Washington DC.

Dresner, S 2002, *The priniples of sustainability*, Earthscan Publications.

du Plessis, C 2012, "Towards a regenerative paradigm for the built environment", *Building Research & Information*, vol. 40, no. 1, pp. 7–22.

Folke, C 2006, "Resilience: The emergence of a perspective for social-ecological system analyses", *Global Environmental Change*, vol. 16, no. 3, pp. 253–267.

Folke, C, Carpenter, SR, Walker, B, Scheffer, M, Chapin T & Rockstrom J 2010, "Resilience thinking: Integrating resilience, adaptability, and transformability", *Ecology and Society*, vol. 15, no. 4, pp. 20.

Frank, B, Delano, D & Caniglia, BS 2017, "Urban systems: A socio-ecological system perspective", *Sociology International Journal*, vol 1, no. 1, pp. 1–9.

Fullerton, J 2015, "Regenerative capitalism: How universal principles and patters will shape our new economy", viewed 3 December 2018, <http://capitalinstitute.org/wp-content/uploads/2015/04/2015-Regenerative-Capitalism-4-20-15-final.pdf>

Gardner, G 2016, *Imagining a sustainable city. Can a city be sustainable?*, Worldwatch Institute, Washington, DC.

Girardet, H 2015, *Creating regenerative cities*, Routledge, New York.

Gladwin, T, Kennelly, J & Krause, T 1995, "Shifting paradigms for sustainable development: Implications for management theory and research", *The Academy of Management Review*, vol. 20, no. 4, pp. 874–907, <http://www.jstor.org/stable/258959>.

Herman, R, Ardekani, SA & Ausubel, JH 1989, "Dematerialization", In JH Ausubel & HE Sladovich (eds.), *Technology and environment*, National Academy Press, Washington, DC, pp. 50–69.

Holden, E, Linnerud, K & Banister, D 2017, "The imperatives of sustainable development", *Sustainable Development*, vol. 25, no. 3, pp. 213–226.

Holling, CS 1973, "Resilience and stability of ecological systems", *Annual Review of Ecology and Systematics*, vol. 4, pp. 1–23.

Knott, J 2018, "Why regenerative development?", viewed 3 December 2018, https://regisseedinstitute.com/2017/10/24/why-regenerative-development/

Lake, PS 2013, "Resistance, resilience and restoration", *Ecological Management & Restoration*, vol. 14, no. 1, pp. 20–24.

Lopez, DR, Brizuela, MA, Willems, P, Aguiar, MR, Siffredi, G & Brand, D 2013, "Linking ecosystem resistance, resilience, and stability in steppes of North Patagonia", *Ecological Indicators*, vol. 24, no. 3, pp. 1–11.

Lovins, HL, Wallis, S, Wijkman, A & Fullerton, J 2018, *Creating an economy in service to life*, New Society Publishers, Gabriola Island, BC.

Mang, P & Reed, B 2012, "Designing from place: A regenerative framework and methodology", *Building Research & Information*, vol. 40, no. 1, pp. 23–38.

Mayer, B 2017, "A framework for improving resilience: Adaptation in urban contexts", In BS Caniglia, V Manuel & F Beatrice (eds.), *Resilience, environmental justice & the city*, Routledge, London.

McDonough, W & Braungart, M 2002, *Cradle to cradle*, North Point, New York, NY.

Meadows, DH 2008, *Thinking in systems*, Chelsea Green Publishing, White River Junction, VT.

Motesharrei, S, Rivas, J, Kalnay, E, Asrar, GR, Busalacchi, AJ, Cahalan, RF, Cane, MA, Colwell, RR, Feng, K, Franklin, RS, Hubacek, K, Miralles-Wilhelm, F, Miyoshi, T, Ruth, M, Sagdeev, R, Shirmohammadi, A, Shukla, J, Srebric, J, Yakovenko, VM & Zeng, N 2016, "Modeling sustainability: Population, inequality, consumption, and bidriectional coupling of the earth and human systems", *National Science Review*, no. 3, pp. 470–494.

Pearce, D, Markandya, A & Barbier, EB 1989, *Blueprint for a green economy*, Earthscan, London.

Pellow, DN 2017, "Critical environmental justice studies", In BS Caniglia, M Vallee & B Frank (eds.), *Resilience, environmental justice & the city*, Routledge, pp. 17–36.

Pezzey, J 1992, "Sustainability: An interdisciplinary guide", *Environmental Values*, vol. 1, no. 4, 321–362.

Quarantelli, EL 1987, "Disaster studies: An analysis of the social historical factors affecting the development of research in the area", *International Journal of Mass Emergency and Disasters*, vol. 5, pp. 285–310.

Redclift, M 2005, "Sustainable development (1987–2005): An oxymoron comes of age", *Sustainable Development*, vol. 13, pp. 212–227.

Roberts, JT & Parks, B 2007, *A climate of injustice: Global inequality, north-south politics, and climate policy*, Cambridge University Press, Cambridge.

Seibert, M 2018, "Systems thinking and how it can help build a sustainable world: A beginning conversation", *The Solutions Journal*, vol. 8, no. 3.

Senge, P 2008, *The necessary revolution*, Doubleday, New York.

Tierney, K 2007, "From the margins to the mainstream? Disaster research at the crossroads", *Annual Review of Sociology*, no. 33, pp. 503–525.

Troy, A 2017, "Datafying the city: The CityCraft integrated research center for West Denver", *Presentation at Mile High Data Day*, Denver, Colorado.

Viederman, S 1994, "Knowledge for sustainable development: What do we need to know?", In *A sustainable world*, California Institute of Public Affairs and Earthscan for IUCN, Sacramento and London.

Walker, B & Salt, D 2006, *Resilience thinking*, Island Press, Washington.

Webb, CT 2007, "What is the role of ecology in understanding ecosystem resilience?", *BioScience*, vol. 57, no. 6, pp. 470–471.

Williams, K 2012, "Regenerative design as a force for change: Thoughtful, optimistic, and evolving ideas", *Building Research & Information*, vol. 40, no, 3, pp. 361–364.

World Commission on Environment and Development 1987, *Our common future*, Oxford University Press, Oxford.

4 Toward a regenerative psychology of place

Nicholas S. Mang

How do you regenerate the collective psyche of a city? A city's structure can be rebuilt through legislative policy, economic funding, and building construction. But what gives a city life? Why do some cities feel more alive than others? Why in some cities do we feel our spirits lifted and rejuvenated, while in others we can feel more depressed, isolated, and alone? People often talk about the soul of a city, but what does this mean? When does a city have soul? How is it grown? And when is it lost? All of these questions grapple with issues of being and of the collective psychic being of a city. What gives a city life and what can regenerate this life should be central to any urban planner or designer's work. Yet, quite often, the work scope of an urban planner is reduced down to the development of solutions for strictly functional and material issues; for example, does the new stoplight reduces traffic congestion on street A? (Beatley & Manning 1997). To successfully work on regenerating the collective psyche of a city requires more than the introduction of new stoplights, new transportation systems, and new civic buildings. It requires of civic planners and designers that they develop a regenerative psychology, that is, a mind that can realize and actualize the potential of living systems to become increasingly vital, viable, and evolutionary (Krone 2005).

Regenerative Development requires much more of its practitioners than a just a newly developed skill set (i.e. a shift in "what we do"). Critical to regenerative development is also a shift in our internal practice (i.e. "what we take into consideration" and "how we think"). There is a common human tendency to try and assimilate unfamiliar concepts like "regenerative development" into our existing psychological paradigms and mental schemas, and in so doing, we run the risk of delimiting the potential that these concepts hold for truly transforming our practice and lives. Understanding the, often unconscious, frames of mind that we bring to this new field of development and the implications for how they influence and possibly delimit the potential of our practice is critical. Toward this end, this chapter explores four distinct psychological paradigms that, subtly and not so subtly, influence the way we relate to and work on the development of a place.

Toward a regenerative psychology

What we call the science of psychology is not a new or novel concept. Psychology, in fact, is one of the most ancient sciences known to humans (Goleman 1993; Wilber 1993). As long as humans have been able to observe themselves and their feelings, thoughts, and actions, psychology has been around. As psychologist James Baldwin states, "A history of psychology is nothing more nor less than a history of the different ways in which men [and women] have looked upon the mind." (Baldwin 1913, p. 1). Every culture form, to some extent, has a psychology, an understanding of mind, of motivation, and of cultural and individual sanity. Ancient civilizations, like the Egyptians and the Babylonians, developed very sophisticated psychologies (Frankfort et al. 1949; Schwaller de Lubicz 1957). This is also true for different indigenous peoples across the globe (Kim, Yang & Hwang 2006). Likewise, Taoist, Buddhist, Vedic, and Sufi psychologies, to name a few, have all greatly influenced Eastern cultures in their understanding of the mind (Fadiman & Frager 2002). Even in what we call Western (i.e., European) civilization, psychology existed far before Freud was ever born. Western psychology can be traced back at least to the times of ancient Greece with the works of such thinkers as Hippocrates, Socrates, Plato, Aristotle, and Protagoras (Daniels 1997). The life span of modern western psychology, therefore, is extremely short when viewed from this global and historical perspective. In this sense, modern psychology is less a process of invention as it is a process of rediscovery and regeneration of very ancient wisdoms and their evolutionary translation into modern times.

When viewed from this much broader cultural and temporal perspective, the science of psychology (i.e., the study of mind and its development) can be seen as comprising a very large and diverse set of schools and bodies of thought, each developed through particular cultural and contextual lenses and in pursuit of, sometimes, very different societal goals and aims. Yet behind this vast pluralism, when studied altogether, one can begin to discern universal patterns toward which the study and practice of psychology has been applied across cultures and through different eras of time (Goleman 1993; Wilber 1993). Based on a review of cross-cultural models of psychology (Mang 2009), and inspired by Krone's (1997) *levels of work framework*, this chapter posits four distinct yet nested orders of psychological development that are each in their turn critical to the development of a whole systems, regenerative psychology. These interrelated orders of psychological development can be depicted holarchically (Figure 4.1).

This chapter explores each of these four psychological orders in turn; their implications for the way we perceive, relate to, and work with the development of place; and how when taken altogether they help to inform a whole systems, regenerative psychology framework.

Psychologies of adjustment: Operational development and societal functioning

At one level, psychology can work toward enabling individual entities (i.e., individuals, families, and even societal organizations) to better function within society. This order of psychology focuses on cases of maladjustment to the norms of

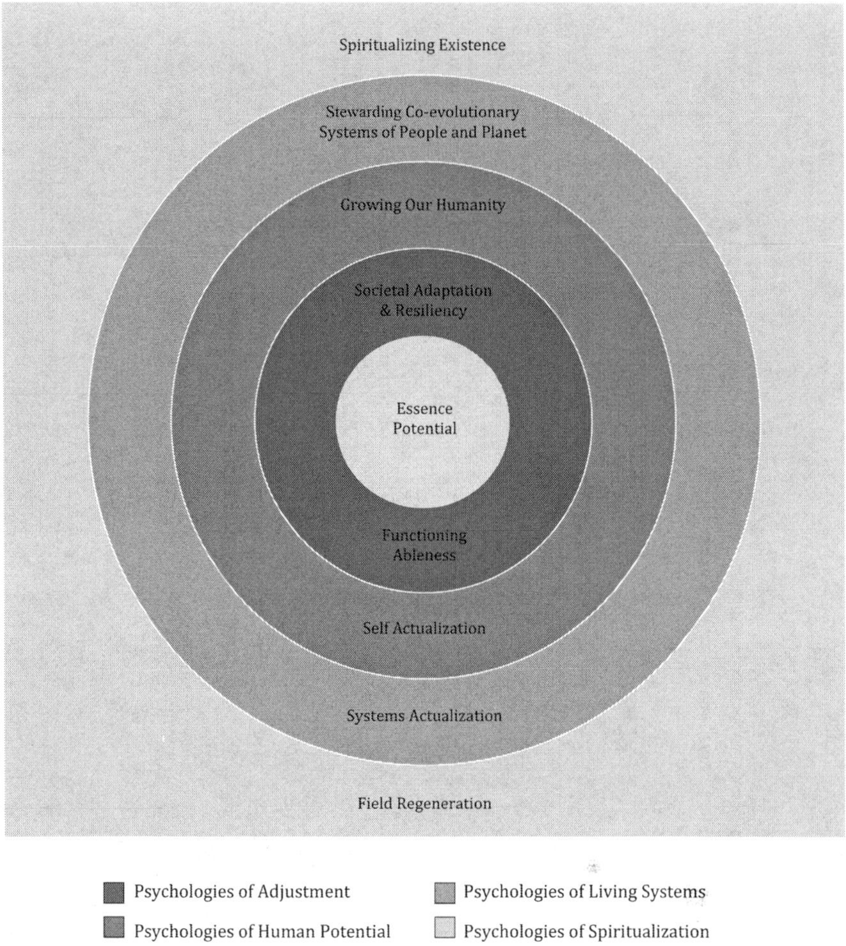

Figure 4.1 Regenerative psychology as a nested holarchy (Mang 2009)

society and remediation techniques for readjustment. According to Freud (1961), "In an individual neurosis we take as our starting point the contrast that distinguishes the patient from his environment, which is assumed to be 'normal'." Health, at this order of psychology, therefore, is defined in terms of one's ability to adaptively function and fully participate in societal life and illness as that which impairs one's ability to do this. As the *Diagnostic and Statistical Manual of Mental Disorders* (American Psychiatric Association 1994) states:

> Each of the mental disorders is conceptualized as a clinically significant behavioral or psychological syndrome or pattern that occurs in an individual and that is associated with present distress (e.g., a painful symptom) or disability (i.e., impairment in one or more important areas of functioning)....(p. xxi)

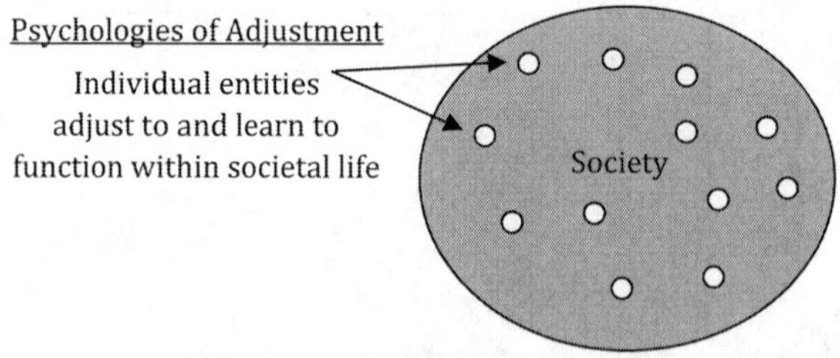

Figure 4.2 Psychologies of adjustment (Mang 2009)

In this chapter, this order of psychology will be referred to as a *Psychology of Adjustment*. Its designated scope of aim is depicted graphically in Figure 4.2.

At one level, adjustment psychologies are beneficial in that they help people learn to cope with the society in which they live. Ideally, they help a person to develop a strong and healthy ego with which to manage and engage in their social environment (Freud 1961). At another level, however, adjustment psychologies can be detrimental to individual, societal, and planetary health. Adjustment psychologies, because of the very nature of their focus, tend to be corrective rather than developmental (Mack 1993; Maslow 1968; Ouspensky 1981). One of the considerable dangers of adjustment psychologies is that they can help reinforce current societal norms and even label alternative (or outside of the box) cultural modes of behavior as *dysfunctional* (Anthony 1995). A psychology of adjustment, therefore, when adopted alone as a closed-system paradigm, runs the risk of reinforcing a maladjusted society (i.e., a society that is not harmonious with the natural health and development of its people and environment). According to Erich Fromm (1955), "mental health cannot be defined in terms of the 'adjustment' of the individual to his society, but, on the contrary, that it must be defined in terms of the adjustment of society to the needs of (hu)man(ity)." (p. 71)

How adjustment psychologies influence our relationship to a place

From an adjustment psychology perspective, humans, as individuals, communities, and collective societies learn to manage themselves and their environments in ways that enable and support their functioning capability. In this sense, we seek to create functional spaces for ourselves to live within. If a family moves into a house, for instance, they will first decide which rooms will serve as bedrooms, which room will be the living room, where they will dine, and so on and so forth. In this way they begin to organize the space in a way that serves their functioning lifestyle (Baldursson 2009). Toward this end, certain adjustments may also need to be made, either to the house itself (i.e., through remodeling) or to their established patterns of functioning together within their household. The same is true with a society that inhabits a particular ecological region. The organization and supply of food, water, shelter, transportation, and fuel are all operational needs of any human settlement. In this, a society will either adjust to the patterns and constraints of the existent ecosystems that it inhabits, and/or it will seek to adjust these patterns to better meet its societal demands and needs. Just as a healthy ego learns to adapt to a change, societies, when faced with changes (e.g., global climate change), need to learn to adjust their patterns and technologies to adapt to these changes in order to protect and sustain themselves.

Under a strictly operational mind frame, sense of place does not really exist, only space and the functioning use and organization of what fills a space. It is what fills a space that we pay attention to and evaluate land and region on this basis. Land is evaluated in terms of its utilizable resources and in terms of how it can be adjusted to best serve our functional societal needs and wants (Kaiser et al. 1995). Likewise, in seeking to adjust environments to best suit societal needs and functions, there is a tendency to eliminate perturbations that are seen as superfluous or even contradictory to our pursuits. Natural landscape and environment, therefore, is viewed in terms of its malleability to our pursuits, and forces within that environment are judged negatively to the degree to which they become destabilizing forces to our functioning capability. What we call wild, therefore, is seen as a destabilizing force that needs to be controlled for, either by fencing it off, domesticating it, or eradicating it (Nabhan 1997). In this, we seek to create controlled environments that are nondisruptive and conducive to normal functioning (Thompson, Sorvig & Farnsworth 2000). Thus, we create manicured lawns, sterilized landscapes and buildings, monoculture crops, chlorinated swimming pools, paved roads, and the list goes on.

The danger in this way of operating is that it leads to the creation of environments that are increasingly lifeless and artificial (McHarg 1971; Thompson et al. 2000). In open system terms, we increasingly cut ourselves off from the natural energies that feed us bodily, mentally, and spiritually. Famed urban historian Lewis Mumford (cited in Kunstler 1993) described post World War II development as leading to an "end product (that) is an encapsulated life, spent more and more either in a motor car or within the cabin of darkness before a television set" (p. 10). By seeking to create controlled environments that eliminate perturbations or variances to a norm and that are designed to maximize functional efficiency, there is also a tendency to create closed-system models for development that can be replicated anywhere in any place. We create standardized zoning and building practices and materials that produce increasingly homogenized developments and communities. We create strip malls and parking lots that look the same wherever you go. We create housing developments that are stamped replicas from place to place. In so doing, we end up creating the same developed environment everywhere, thus destroying any sense of a unique place (Beatley 2004; Kunstler 1993). This leads to the monotonization of a place, in which biological and cultural diversity is destroyed. In Relph's (1976) words, "There is a widespread and familiar sentiment that the localism and variety of the places and landscapes that characterized preindustrial societies and unselfconscious, handicraft cultures are being diminished and perhaps eradicated" (p. 79).

While there are many dangers that the operational mind, when left to its own devices, can fall into, there is also a value and therefore opportunity in the appropriate growth and development of this mind. When guided by the higher and more encompassing levels of mind, the operational mind can be employed to create and design functional space that works with and supports rather than against the patterns and dynamics of a given regional place. The opportunity, therefore, is to use the operational mind and its technologies to develop increasing discernment of the unique operational patterns that support life in a given socioecological place. This requires an interactive process of coadaptation and codevelopment with the natural place. It also requires the development of a comprehensive ecological knowledge of the place within which one lives and works.

Psychologies of human potential—Self-actualization and the development of being

Societies themselves can be evaluated in terms of the degree to which they help enable their constituents to become more fully human, more compassionate, more reflective, and of greater integrity as human beings. This is what Maslow (1993) referred to as a psychology that focuses on development toward the farther reaches of human nature. This second order of psychology works not toward the development of normalcy, but rather toward the excelling of individuals toward greater self-actualization and individuation (Rogers 1980). According to Maslow (1968), a person is healthy to the degree to which this actualizing force is enabled to flourish and unfold in his or her life. As he states, "perhaps we shall soon be able to use as our guide and model the fully growing and self-fulfilling human being, the one in whom all his potentialities are coming to full development, the one whose inner nature expresses itself freely" (p. 5).

As people self-actualize, they become both increasingly unique and individuated, while at the same time increasingly universal in the human values and virtues for which they live. This is what Maslow (1993) referred to as the meta-values that perennial philosophy speaks to through different eras, cultures, and religions. Self-actualizing people become increasingly compassionate to their fellow brothers and sisters and to other forms of sentient life. They develop increasing faith in humankind and in the inherent benevolence of people. They develop hope for their brothers and sisters and for a world that allows all people to self-actualize and live truly from their inner selves. As such, self-actualization stands in service of human actualization, of actualizing our humanity, our human being-ness in the world. A person like Gandhi, for instance, is beautiful not only because he lived so much from his true self; but also because his life speaks to the true self that lies within each of us, he speaks to who we are and can be as human beings (Easwaran 1978). I have entitled this level of psychological aim, a *Psychology of Human Potential* (Figure 4.3).

Psychologies of human potential help people to become fully actualized individuals. As such, they are highly value adding to the world in which we live. However, just like with adjustment psychologies, a danger lies in encapsulating this form of psychology in a closed-system paradigm. Such a danger occurs when the focus on self-actualization becomes isolated from an understanding of the

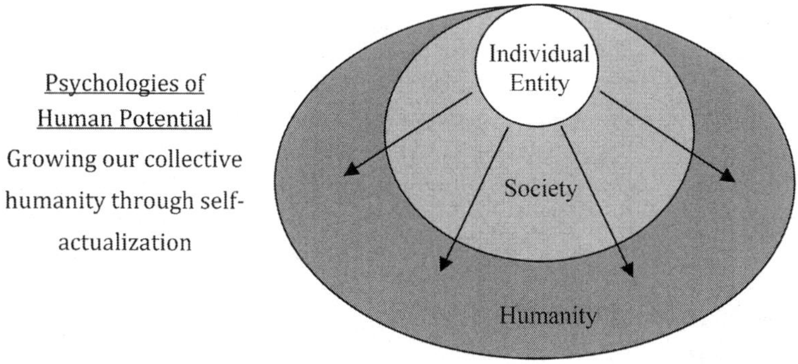

Figure 4.3 Psychologies of human potential (Mang 2009)

larger living systems we are a part of and have a value-adding role to play within (Metzner 1999). Without this larger systems awareness, we act and perceive as independent entities without an inner appreciation or caring for the collective living systems that source us and bind us together (Lovelock 1979; Orr 1994). As James Hillman and Ventura (1992) state,

> We still locate the psyche inside the skin. You go inside to locate the psyche, you examine your dreams, they belong to you. Or it's interrelations, interspyche, between your psyche and mine...—but the psyche, the soul, is still only within and between people... (L)ook what's left out of that. What's left out is a deteriorating world. (p. 3)

At such times we should ask ourselves: What are the fields and systems that support the self-actualization of life (both ours and others, both human and other-than-human) and how are we, as self-actualizing individuals, called to reciprocally serve and support their sustenance, growth, and ongoing evolution? Without working at this order of questioning, we as humans are perpetually in danger of falling into a narcissistic dance of self and human actualization that is not holistic or reciprocally caring for the planetary and cosmic systems that feed our growth and well-being.

How psychologies of human potential influence our relationship to a place

Beyond the development of a functional space within which to operate, humans also work with space at a second level. At this level, humans as individuals, communities, and societies seek to create environments that nurture and reinforce particular states of being and particular qualities of interrelationship between people. We create human places that are conducive to particular experiences and ways of living. This goes beyond the work of creating a functional space to that of creating the ambiance and mood of a place. In architecture, for instance, the way in which a structure is designed and built can produce very different effects on a person's psyche and their experience of moving through a given space. According to architect Christopher Day (1990), "to create nice and, more importantly, meaningful, appropriate atmospheres we need to focus our attention not on the quantities but on the qualities" (p. 24). A tall cathedral-like ceiling affects us differently than a cozy, low-ceiling room. In addition to the structure itself, the elements that make up these different rooms also affect the experience of a place. For instance, the way in which a room is lighted can create a very different ambiance and mood (Day 1990). The use of natural light, via skylights and windows, also creates a different affect than artificial lighting. As Day puts it, "all materials have individual qualities.... It is hard to make a cold-feeling room out of unpainted wood, hard to make a warm, soft, approachable room out of concrete" (p. 113). Colors also affect the experience and mood of a place (Bayes 1970). Sound is another element that affects the quality of experience in a place. The sound of falling water from a fountain, for instance, may have a soothing effect in a room. Likewise, urban restaurants often design their dining spaces in ways that amplify the level of conversational noise, thus creating the buzz-like effect of being a popular, bustling, and happening place.

As humans, we create different places to support different modalities and states of being. In a house, for instance, a family may choose to decorate certain rooms to feel more public, while others are more intimate and private. Some rooms may be designed to feel more lively and upbeat, while others may be designed to feel more soothing and as a place for quiet reflection. All together, these different qualitative spaces help to make up and balance out a place called home. In this sense, it is no longer just a functional household, it is now a place with a distinct quality of being and distinct elements within that together make up this holistic phenomenon called home. According to Bennett (1966), "the unity in diversity that characterizes it is the 'reality' of the home" (p. 15). What holds the different experiential spaces of a house together and helps them cohere into an integral sense of home is the core or essence of the family who resides in it. A place at this level is not just about the creation of atmosphere and mood for its own sake, but rather it is the creation of mood and atmosphere in service to the development and expression of one's unfolding inner life and essence.

At the societal level we, as humans, also create different places to feed and support different states of being and qualities of interaction. In any given city, for instance, there are places of worship that are built in ways that are conducive to particular qualities of experience and mentation (Cresswell 2004; Day 1990). Many Christian churches, for example, tend to have spires and cathedral ceilings that lift one's gaze toward the celestial heavens and offer us a sense of ascendance. Likewise, in any given city, one will most likely find cafes and plazas in which to sit and chat with one's fellow citizens. These cafes tend to create a very different ambiance than those of the churches and are conducive to a different quality of interaction with others and a different state of mind. In addition to religious places and cafes, there are also garden spots, marketplaces, nightclubs, and the list goes on, each helping to feed a different qualitative aspect of our being and soul. As an interrelating whole, each of these distinct places help to make up a distinct, experiential landscape that people move through and live within. As such, each cultural landscape, to the degree to which it is integral, helps to support a particular way of life and a particular way of experiencing one's lived world. This is a step beyond merely focusing on functional use of space to that of seeing the effects that a structured space has on one's psyche and well-being (Day 1990; Seamon 2000).

Each place tells a story of people's strivings, hopes, sufferings, and dreams. As Beatley (2004) puts it, "landscapes and places are embedded with memories, and the nature of these memories affect how we value and treat places" (p. 33). As we walk through the projects of an inner city ghetto, we feel something if we are at all human. If we walk through the landscape of a family farm that has been passed down through generations, we may feel something very different. As our modern, industrial society has begun to reawaken to this level of experiencing, we have realized that a mere functional space does not feed the heart, it does not feed the human spirit, and it does not feed meaningful interrelationships. Just providing low-income housing is not enough (Fuerst 2004). Just designing office space that is functionally efficient but pays little attention to people's subjective experience is not enough (Day 1990). What are needed are places that feed the human spirit, that support and encourage the growth of human community and well-being (Beatley 2004). A healthy place at this level can be defined as that which creates cultural landscapes that are conducive to people's self-actualization (Maslow 1998).

New Urbanism is one of the current movements in our society that is seeking to reclaim and reestablish life at a human scale of development (NewUrbanism.org 2006). Largely in reaction to the monotonous developments and suburban sprawl that spread in the United States with the advent of the automobile and its freeway infrastructure (Kunstler 1993; Relph 1976), New Urbanism seeks to recreate communities and neighborhoods that are human-scaled. The guiding principles of the New Urbanism design include walkability, connectivity, mixed-use and diversity, mixed housing, quality and aesthetically beautiful architecture and urban design, traditional neighborhood structure, increased density, smart transportation, and sustainability (NewUrbanism.org 2006).

Another movement in response to the modern sprawl of human development is Conservationism (Van Putten 1996). This movement seeks to protect and maintain natural wildlife areas from the incursion of human settlement. Wild, in this romantic sense, is viewed as beautiful, robust and untamed by humans, and therefore worthy of protection (Anderson 2005). Our human bodies, psyches, and spirit are fed by this pristine and wild nature (Metzner 1999). We go for walks in the woods, climb mountains, and swim in lakes and rivers, all to feed our growth, well-being, and overall self-actualization (Roszak 1993). Wilderness and nature, therefore, become identified as a place that we go to and visit to help feed our souls. Like with the first modality of human relationship to place, there are inherent dangers and opportunities in this second modality. The danger lies in artificially separating in our minds the idea of *human-made* places from *natural* places (McHarg 1971). We see human settlements as places that we create and can innovatively design however we see fit, while *wilderness* areas are places that should be largely left alone by humans so that nature can continue to create and maintain them. According to Anderson (2005),

> setting aside wilderness is only a reaction to the plundering of natural resources, and both spring from a mind-set of alienation from nature. Moreover, the wilderness concept tends to compartmentalize nature and culture, giving humans the illusion that activities done outside of protected areas will not affect what is within. (p. 120)

This approach leads to the development of human places, not socioecological places. We isolate a place, creating closed systems within the environment (McHarg 1971). Even the idealized goal of sustainable developments in this modality is to create a separate island of human habitat that no longer impacts on or answers to the ecological environment it is a part of (Dernbach 2002). Our urban settlements disrupt environmental flow and natural patterns of interrelationship in ways that are harmful and even degenerative to the natural systems that regulate and support life within the place and for the planet as a whole (Leopold 1949). In all of this, there is a blindness to our patterns of living in a place and on the planet and how these patterns are in harmony or out of harmony with these larger systems and their evolution.

As we begin to re-relate to the planetary places in which we live, we begin to understand, for instance, why certain architectural styles have originated from particular regions. Local materials, regional climate, and human ingenuity are of course important factors, but each architectural style that authentically originated from a regional place, also has a mood and tenor that helps to capture and express the unique quality of experience that a cultural body of people have in relationship to the natural place in which they live. The building styles as such grow from the roots of a place, from its unique patterns and associations. The opportunity, therefore, is one of seeing self-actualization and human cultural expression as being a process that grows from and is in service to the larger planetary and cosmic systems of which we are a part (Fernandez 1998) as opposed to seeing these systems as merely feeding sources (which at best we seek to sustain) for our ongoing individual and collective human fancies (Deloria & Wildcat 2001).

Beyond the human good

Since the times of ancient Greece most of Western civilization's dialogue has revolved around these first two orders of psychological paradigms (Deloria & Wildcat 2001). Aristotle (1962), a thinker who profoundly influenced Western philosophy and thought, believed that *the good* existed within every person realizing his or her true nature. According to Aristotle, *summum bonum* (or the greatest good) was embodied in the collective social organization or state and was reflective of the degree to which each and every person within that society was able to realize his or her virtue to its fullest. According to Professor Daniel Wildcat (2001), a Yuchi member of the Muscogee Nation of Oklahoma, what has largely bounded this dialogue in Western thinking, is the defining of *greatest good* in largely human terms.

> By excluding the many other-than-human persons of the natural world from active full participation in determination of the greatest good, ecological catastrophe seems guaranteed. Whether intentional or not the result of this single idea has been to create a worldview where humans are thought to be above the rest of nature.... (It is) an idea that has brought us to the brink of global ecological crisis by reducing the question, the very idea, of the summum bonum to be about relationships among human beings. (Deloria & Wildcat 2001, p. 96)

In contrast to Aristotle and much of Western thought, many indigenous peoples define *persons* that contribute to the greatest good as including not just humans but also other planetary members such as plants, animals, and the physical elements of nature (Deloria & Wildcat 2001). Each has an inherent value or essence that contributes to the community or ecosystem as a whole. Therefore, society from many of these indigenous perspectives includes humans and other-than-humans in a mutual interrelated dialogue (Vasquez 1996; Fernandez 1998). Together, this dialogue cocreates the greatest good by continuously feeding life within the ecosystem and for the planet as a whole.

Interestingly enough, many scientists who are now working at the cutting edge of their fields are finding evidence and developing theories that corroborate such indigenous worldviews (Bohm 1985; Laszlo 2004; Peat 2002; Sahtouris 2000). Quantum physics has demonstrated the irreducible link between the observer and the observed and the implicit and undivided wholeness of all matter (Bohm 1985; Peat 2002). New biology and ecological sciences have stressed the interdependence of all life forms and the evolutionary trajectory of living systems toward higher orders of complexity and richness of interrelationship, thereby creating potential for new life forms and greater biodiversity on the planet (Sahtouris 2000). Transpersonal psychology and consciousness research studies have yielded impressive evidence regarding the effectiveness of telepathic and telesomatic information and energy transmission through the body-mind (Braud 1992; Byrd 1988; Dossey 1993; Playfair 2002). What all of these findings indicate is the seeding emergence of a new order worldview in science, one that is much more aligned and in concert with many ancient and indigenous wisdoms (Capra 1975; Laszlo 2004; Peat 2002). In this, a new order of psychology and psychological aims is also emerging in the Western society (Roszak 1993; Capra 1982). Abraham Maslow (1968) foresaw this emergence of a new psychology, one that went beyond humanistic purposes: "I consider Humanistic, Third Force Psychology to be transitional, a preparation for a still 'higher' Fourth Psychology, transpersonal, transhuman, centered in the cosmos rather than in human needs and interest, going beyond humanness, identity, self-actualization, and the like" (p. iii-iv).

Unfortunately, however, the field of Transpersonal Psychology (a field which Maslow helped to create) has to date remained largely couched in the paradigm of individual self-realization and transcendence to the detriment of a larger, ecological perspective (Fox 1995; Kremer 1998; Roszak 1993). As such, it has generally failed to move beyond a psychology of human potential to what in this chapter I call a psychology of living systems.

Living systems psychologies—System actualization and the development of value-adding roles

A third order of psychology moves beyond a focus on human actualization to that of seeing the greater systems of which we are a part, and for which we have a value-adding role to serve (Brown 1993; Hillman 1995; Suzuki & Knudtson 1992; Vasquez 1996). It relates to the capacity to coevolve in harmony with, and care for, the greater living systems of which a person is a part (Deloria & Wildcat 2001; Fernandez 1998; Roszak 1993). This order of psychology continuously seeks to address the question: What, as human beings, is our evolving role in the larger planetary and cosmic systems of which we are a part? It works to address this question not through detached logic and philosophical intellectualism (Abrams 1996) but rather through experience and understanding of natural living systems (Orr 1994) and their (and *our* as a part of them) capacity to grow life (Vasquez 1996). This involves the inner capacity to experience and develop understanding of the energic and systemic patterns of life that are continuously generating living systems of interrelationship and reciprocal nurturance (Capra 1996; Leopold 1949). Its primary focus, therefore, is not on individual development, but rather on the evolution of life systems and the development of our capacity to be in service of these systems (Fernandez 1998). This does not preclude self-development by any means, but rather helps to extend it by lifting up the greater wholes we are seeking to be in service of, and thereby further elucidating who we are called to be and become within these wholes (Metzner 1999; Roszak 1995). By better understanding the greater systems of which we are a part, we can develop and evolve our value-adding roles within these systems. Value adding in this sense is defined as that which increases systemic capacity for vitality, viability, and evolution of life (Krone 2005).

I have entitled this order of psychology, a *Psychology of Living Systems*, because it aims to develop coevolutionary partnership between humans and the greater living systems of which they are a part (Figure 4.4).

A psychology of living systems helps to enable the reciprocal maintenance and evolution of life on this planet (Anderson 2005; Brown 1993). As such, it is a critical modality of psychology that needs to be reintegrated into the field of modern psychology (Roszak 1993). Just like with the first two psychologies, a hazard lies in encapsulating this form of psychology in a closed-system paradigm. This hazard can be summed up as one of working on actualization without spiritualization (Krone 2005). Just as self-actualization cannot occur in a vacuum, but rather must occur in tandem with the actualization of the systems of which the individual self

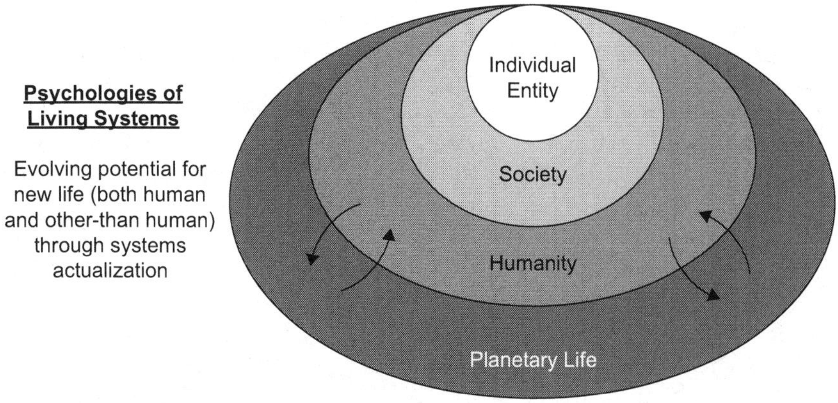

Figure 4.4 Psychologies of living systems (Mang 2009)

is a part (Sahtouris 2000), systems actualization also cannot occur in a vacuum, but rather takes place within a regenerating or degenerating field or plenum of life (Bennett 1964; Laszlo 2004; May 1982). In other words, we live within a bath of vitalizing energies that enable us to live and grow as individuals, as human societies, and as a planet as a whole (May 1982).

How living systems psychologies influence our relationship to a place

Beyond the development of functional spaces and human nurturing places, there is a third psychological modality for engaging with a place. This third order of work relates to our human value-adding role in natural and planetary place (Anderson 2005). A place, in this sense, can be defined as an evolving socioecological whole involving both people and the natural ecosystems in which they live (McMurray 2006). Rather than seeking to minimize our impact on the planet, at this level, humans actively work to help enable and even improve the workings of natural living systems. This is in fact the role that many indigenous peoples throughout the world have served in the past and, to the degree in which they can, do still today (Anderson 2005; Vasquez 1996). In studies on Native American knowledge and management of California's natural resources, Wildland Resource scientist Kat Anderson (2005) discovered that much of the ecological landscape of California today is the result of centuries of intentional management by Native Americans. Her findings dispel the oft-cited myth that California is as lush as it is because so few people lived on the land, thus largely leaving nature alone. Rather, her findings showed that Native American tribes were actively involved in the coevolution and development of California's natural fecundity and beauty. What Anderson's studies indicate is that humans have a value-adding role that they can serve in relationship to the land in which they live.

According to naturalist Aldo Leopold (1949), land is more than just the ground we live upon. Rather, it is a complex living energy system that includes soils, plants, animals, water, and air in a circulating flow of nutrients. The evolutionary trend of these energy systems is to continue to "elaborate and diversify the biota" (Leopold 1949, p. 216) on the earth. In these evolving energy circuits, humans are inextricably a part of them. As such, we can either help enable and feed this elaboration and diversification of biotic life in the places in which we live, or we can disenable and violently disrupt the energy flows and cycles of the land we live upon. The mounting ecological crisis that we now face (Brown 2008) is evidence that modern man has chosen to pursue the latter path. What is centrally missing in our Western culture today is an ethic regarding what Leopold (1949) describes as living rightly with the land. Living rightly, in this sense, would mean preserving and adding to "the integrity, stability, and beauty of the biotic community. It is wrong when it tends otherwise" (p. 224–225). The implication here is that we as humans can and should be living with the land in ways that work with and elevate its systemic generative capacity for life. According to wilderness guide Tom Brown's spiritual teacher, Grandfather, humanity's destined role on earth is to become good caretakers of natural creation: "Nature can exist without us, but it would struggle far more. Remember, we are here for a grand purpose, beyond the self. We are the caretakers" (Brown 1993, p. 74). This is the legacy that we as humans can grow on earth.

The implications for developing such a land ethic in our modernizing world are immense. First, it would require a complete transformation in the way in which we currently develop and settle upon land and in place. Currently our human developments block, disrupt, and harm the natural flows and energy cycles of a place. By developing understanding and mapping of these ecological flows and cycles, we can begin to discern how our developments can work with and re-enable these patterns (Thompson et al. 2000). In particular, we will need to tackle the daunting challenge of reintegrating our vast urban centers into these natural flows and cycles. This goes beyond the vision of creating green, energy self-sufficient cities to that of creating urban socioecological landscapes that are not isolated islands but rather integral members and contributors to the ecological systems in which they exist (McHarg 1971). This cannot be tackled by piecemeal solutions, such as creating wildlife trail bypasses over roads. What such a challenge requires is first a holistic understanding of the workings of the ecological systems in which we are situated (Mang and Haggard 2016). Only then can we see the uniquely situated place and the role that our settlements can serve within these larger systems. Second, we would need to transform the way that agriculture is currently practiced in our society (Jackson & Jackson 2002). Vast areas of the United States are decimated in terms of their biotic diversity and carrying capacity for life due to the ways in which we currently run our agro-businesses (Berry 1996; Brown 2008). We must therefore relearn how to cultivate food in ways that diversify and enrich biotic life as opposed to contributing to species extinction and the erosion of land (Jackson & Jackson 2002). Permaculture (Mollison 1997) and Biodynamics (Joly 2007; Steiner 2005) are two such agricultural systems that work toward these higher order ends. Third, we would need to transform the way we do industry (Hawken, Lovins & Lovins 1999). Currently, most of our resource-based industries are built on a model of extracting natural resources, transforming them into what we call "value-added" products that can be sold and used by humans, and then disposing of them after their life-term usage (Dale & Robinson 1996). The dangers of such a take-make-waste model are multifold. Currently we are stripping regions of their natural resources, transforming them into nonbiodegradable substances to be used, and then depositing them into landfills (Brown 2008). This process is destroying the natural cycles and flows of energy in our ecological systems (Brown 2008; Leopold 1949). What a land ethic calls for is the redevelopment of industry into an earth-to-earth process (Hawken et al. 1999), whereby natural resources are harvested in ways that do not strip a region of its resources but rather add to the health of its systems, they are then developed in ways that add not only to their human but also their biotic value, and then are returned to the earth in ways that reciprocally replenish the land that they came from.

To bring about such a new world requires, first and foremost, a transformation in our cultural values and our ways of interrelating with the socioecological places that we live within (Leopold 1949; Orr 1994). If we continue to view and relate to land through our analytical and objective mind, it will continue to be an object separate from us (Metzner 1999). To learn to love the land in which we live, we must learn to relate to it inwardly and bodily (Abrams 1996). Only when we experience the being of a place and what it energizes in our mind and speech, do we begin to see and relate to it as a truly living phenomenon that is to be respected, loved, and revered (Naess 1989). This is what is sometimes referred to as the spirit of a place (Swan 1990). It is what transforms our "caretaker" role of a place (Brown 1993) from one of onerous and daunting burden to one of inspired love and devotional caring. This concept of the spirit of a place will be addressed further in the next section.

Psychologies of spiritualization—Field regeneration and the spiritualizing of existence

Embedded within each of these three orders of psychology is a pervasive fourth order of psychology that aims toward the spiritualization and regeneration of life. Perennial philosophy teaches us that there is an oneness that unifies all of life in the universe (Huxley 1945; Wilber 2000). This has been corroborated with recent findings in physics, which indicate that all systems and fields are part of a greater plenum of life that maintains and regenerates the coherent patterns of life systems (Laszlo 2004). According to physicist Harold Puthoff (1991), electrons orbiting atomic nuclei and our planets orbiting around the sun would both collapse into the center if it is not for the continuous influx of energy from the zero-point field (i.e., the quantum vacuum energy field that fills all of the cosmic space). What this suggests is that the universe itself is in a process of continuous regeneration (Bohm 1985), sourced by a unifying field of energy (May 1982; Puthoff 1991). Many indigenous traditions, similarly, believe that the systems they help to actualize need to remain open and coherently attuned to this universal source (Deloria & Wildcat 2001). When the systems we help to actualize lose touch or become out of tune with this universal source, misinformation and entropy occurs (Bohm 1985). This then leads to the hazard of actualization without spiritualization, of working to develop and grow supportive living systems while losing sight of the pleromic field that sources and regenerates the life and spirit within those systems. Without this continuing renewal of spiritual will, what physicist David Bohm (2003) refers to as the supra-implicate, all living systems face entropy and eventual degeneration. Ultimately, therefore, regeneration is not just a matter of improving the working of a living system but also one of regenerating its spirit.

Throughout the ages and throughout different cultures, humans have continued to be drawn toward the creation of artifacts and ways of living that renew and breathe new life into the spiritual dimensions of existence. The act of spiritualizing life comes not from us but through us (May 1982). As such, we are not the source of spirit and godliness, but rather can become an instrument for this source and in so doing give back to it what it has given freely to us. I have entitled this order of psychology, a *Psychology of Spiritualization* (Figure 4.5).

How psychologies of spiritualization influence our relationship to a place

Every place or region on the planet has a unique spirit that energizes the mind and peoples who live there in a particular way (Swan 1990). According to Andean scholar, Eduardo Grillo Fernandez (1998), "every great people, each culture, each form of life has its own world. In this way the Andean world has its own peculiar mode of being and therefore experiences, in its own way, the events of its life" (p. 127). For instance, when we go to New Orleans and its surrounding bayou

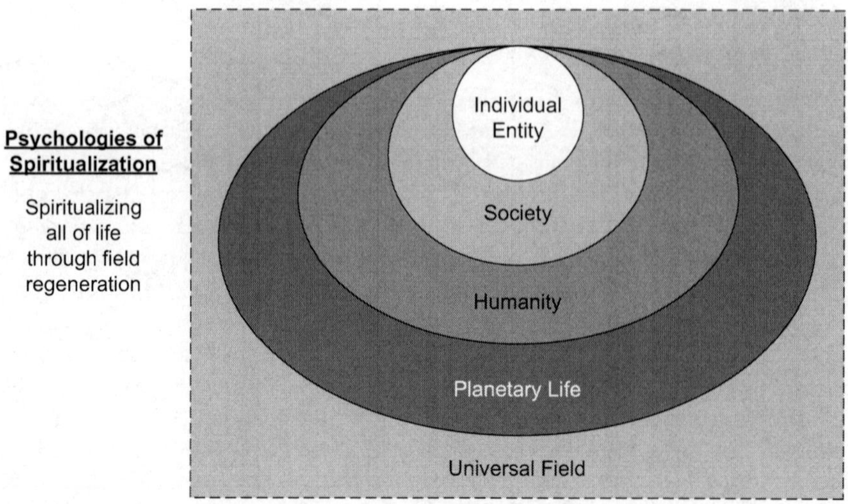

Figure 4.5 Psychologies of spiritualization (Mang 2009)

landscape, we experience a particular rhythm and quality of life that is very different from life on the San Francisco Bay or life in Tibet and the Himalayan mountain range. In our everyday hustle and bustle of modern life, it is sometimes difficult to slow down enough to consciously experience these rhythms of the place we live within, but when we do they are still pervasively present (Beatley 2004).

Many ancestral traditions the world over, designated sites for particular cultural and religious activities based on the unique energic qualities of the site (Lake 1991). The idea here, which is the basis of geomancy, is that with each place there is a right action that best coalesces with the spirit of that place. According to Greek wisdom, for instance, the famed temple at Delphi was sited not haphazardly in its location, but rather quite precisely based on the unique power and spirit of that place (Swan 1991a). In ancient Egypt, a number of distinct sites were designated as the spiritual origins of their world (Bonnemaison 2005; Swan 1990). They saw each of these sites as particularly strong focal centers for the different deities, or *Neters*, that they worshipped. According to Egyptologist Schwaller de Lubicz (1957), these *Neters* were not worshipped as Gods, in the way we use the term today, but rather as divine patterns or forces that create, generate, and regenerate the processes of our natural world. In certain locations, therefore, the Egyptians found a particular divine force to be more readily present and experienceable. It was there that they sited their temples in dedication to that particular *Neter*.

Beyond the intentional citing of their temples, ancient Egyptians also sought to structurally build their temples in alignment with the unique causative force they were intended to serve (Schwaller de Lubicz 1957). As such, they saw each temple as a resonant microcosm of that particular macrocosmic or universal force. They therefore used different scales of measurement and different coinciding proportions of forms to best match the patterns and rhythms of that particular *Neter*. According to environmental psychologist James Swan (1991a), this practice of building in resonance with the spirit of a place is in fact a common theme among spiritual architecture around the world; its purpose being to structurally embody and amplify the spirit of that place. Through these sacred temples and the unique rituals that developed within them, the ancient Egyptians sought to feed and regenerate the presence of the different spiritual forces that they believed created and sustained their world. If one of these forces became unduly weakened, the world would fall out of balance and into a state of degeneration, thus impairing the health of its members. Also, the Egyptians believed that at different times of the year and through different eras, different *Neters* (i.e., causative universal principle) ascended or receded in their influence over world tidings. This understanding was based in part on their astrological studies and the changing alignments of the planets. Therefore, at different times, a particular temple and spiritual *Neter* was held to be the center of the universe, based on their ascendancy in the cycle (Schwaller de Lubicz 1957).

This practice of balancing the causative forces of our world through right relationship to sacred places is evidenced not only in Egyptian wisdom but also in many other ancestral cultures of the world (Swan 1991b). What all of these ancestral traditions indicate (; Fernandez 1998; Lake 1991; Schwaller de Lubicz 1957; Swan 1990) is that (a) each place has a unique spirit that affects and influences our emotional/mental being in particular ways, (b) right action involves living in harmony with and in service of the spirit of the place in which one is situated, (c) the spiritual health of human and planetary community depends upon the continued contact and right balancing of these different sacred energy sources, and (d) humans have a role in maintaining and regenerating these spiritual forces.

Different people may find themselves connected and drawn to different places. As such, we may find that the place spiritually energizes us and gives us a sense of connection and purpose within a large whole (Deloria 1991; Lake 1991). As a Native American scholar, Vine Deloria, Jr (1991) stated it, "lands somehow call forth from us these questions [who we are, what our society is, where we came from, where we are going, and what it all means] and give us a feeling of being within something larger and more powerful than ourselves" (p. 30).

Places, from this regenerative perspective, can be seen as a nested system of ever more encompassing wholes. The urban and regional places on this planet that we feel called to connect and serve are themselves a part of and help to serve a larger living place called Gaia (Lovelock 1988). As an ending note to this section, I leave you with the eloquent and inspiring words of historian Donald Hughes (1991):

> That is really what we are challenged to do today: to find the places where we connect with the larger cosmos, to keep them free of the impedimenta that would block access to the spirit, and to open ourselves to the values that come from those places. When place is respected and treated properly, spirit is never used up; on the contrary, it becomes stronger. And the more one studies the past experience of sacred place in human history, the more one is impressed by the variety of values that can emerge from it. It is as if this vast organism in which we live, Gaia, the biosphere, and indeed the entirety of the planet earth, has a multitude of organs, of connections and nodes, no two exactly the same, and as we move among them, we give and receive, and subtract from her life or enhance it according to our attitude and our sensitivity. The place where natural ecosystems are intact and functioning in the full spectrum of their beauty is the place where spirit is most manifest. (p. 25)

Implications for regenerative development

This chapter has sought to develop a map of the territory of psychology by laying out four distinct meta-aims toward which psychological development has—and can be—directed. It posits that each level of aim encompasses and informs the prior levels, thus forming a holarchy. Furthermore, this chapter posits that the level of mind (i.e., our psychological paradigm) that we bring to development frames and influences the way we relate to and understand place and that a regenerative relationship to place must include all four of these levels. The implications

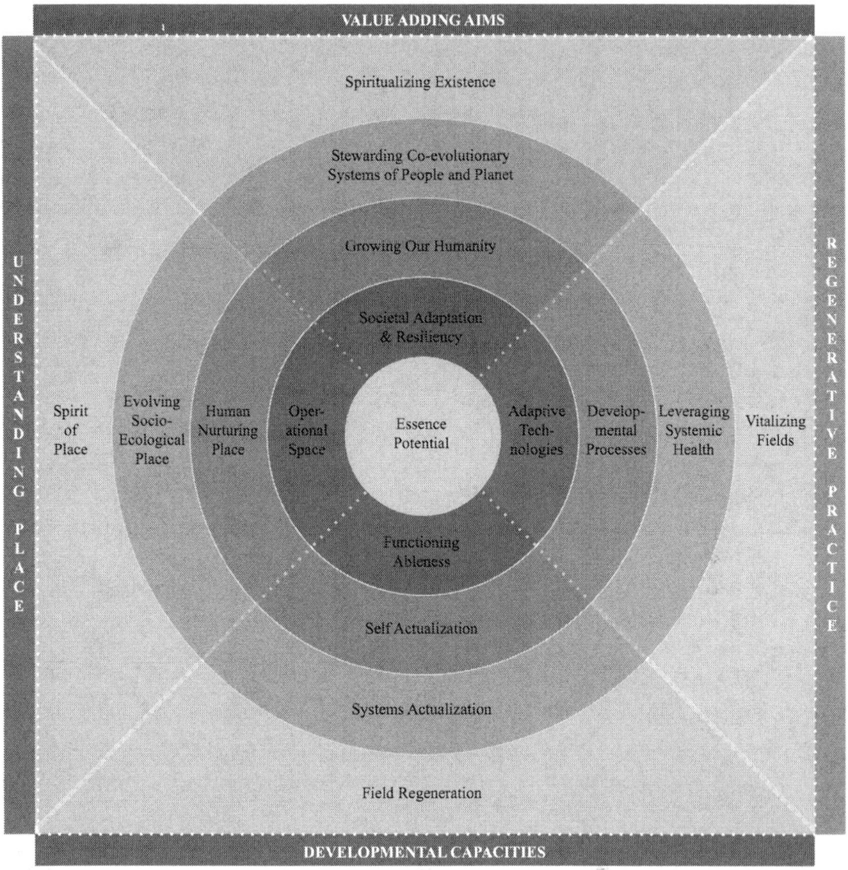

Figure 4.6 Psychology for regenerative development framework

of this for creating a regenerative development practice are depicted in Figure 4.6 and explored briefly in the next section.

Adaptive technologies

At one level of development, we as individual entities—be it as a person, an organization, or a society—seek to grow our adaptivity and resilience in the midst of a dynamic and continuously changing environment. To do so, we develop particular techniques, skills, and technologies (our collective know how) to better support and grow this resiliency. It is no accident that the *Green* movement is often associated with technologies that it has helped to spur the development of (e.g., energy-efficient lighting, recycling and waste-processing technologies, resource-efficient appliances, clean energy and water harvesting technologies, LEED building technologies, etc.).

The key question behind the technologies we develop is, "what nature and order of aim we are pursuing that informs the direction of our further innovations?" As natural resources become scarcer, for instance, and we develop technologies to adapt to this changing global dynamic, is our aim just to be more efficient in our resource use so that we can sustain the growth of our current lifestyles and development patterns into the future? Is it to adapt our lifestyles and patterns of development so that we utilize resources in ways that lessen and remediate our harmful impact to the environment? Or is it to improve the vitality, viability, and potential for evolution of life on this planet for all living beings? This chapter proposes that it is the latter aim that the field of regenerative development should be ultimately pursuing. This setting and defining of aims is not a matter of blue-skying, but rather it has real pragmatic implications for the direction we focus on as a field in developing knowledge, wisdom, and technologies. Such an aim helps to elevate the design challenges we set for ourselves and likewise exponentially elevate the value-adding potential of a project. For example, the design of water infrastructure through an urban development would look very different if we are working toward a design aim of just functional water provision versus that of conserving water through increased efficiency and reuse technologies versus one of seeing the urban water infrastructure as not separate from but rather a part of the overall working of the regional watershed, and therefore seeking to design in ways that maximize the quality and expression of life (both human and other-than human) that water flowing through the system can help support and nourish. In the latter case, for instance, we would design for greater exchange between water and life throughout the system—thus seeking to stack functions and increase the number and quality of microcycles (i.e., reuse, filtration, and recharge) throughout the course of the larger water cycle (Mollison 1997).

Developmental processes

Technology is only as good as the users that steward it. As anecdote, a friend and colleague of mine designs natural wastewater systems for a living. She told me of a particular project she worked on in which her team designed a very innovative wastewater system only to find, when she returned several years later, that it was in disrepair and disuse because the site manager never cared for the system and therefore had no motivation to learn how to correctly upkeep it. Ultimately, if people cannot see how a given technology is supporting them in furthering the actualization of their potential, they will lose value for it and move away from it. Critical for regenerative development practitioners, therefore, is the developmental engagement of stakeholders in a project in ways that foster and grow coevolving partnership.

A good example of this is Curitiba, Brazil's food-for-trash program (Hawken et. al. 1999; McKibben 1995; Ribeiro & Tavares 1992). In the 1970s, Curitiba was faced with a massive influx of migrants who were setting up favelas (i.e., "shanty" towns) on the outskirts of the city. Because these favelas were poorly built, densely populated, and without proper infrastructure, the city leaders realized they had mounting conditions for a possible disease outbreak and identified the most critical near-term hazard for this being the mounting waste and refuge that was building up in the favelas. Rather than approaching this issue as a technical and/or technocratic (e.g., policy) design problem to be merely solved, they saw it as an opportunity for beginning to build an ongoing codevelopmental relationship with these new stakeholders in their city, one that would allow for deepening cycles of actualization of people and place. By investing up front in developing relationship and more deeply and systemically understanding the context and constraints that these people were facing in actualizing their lives and livelihoods, the city leaders in partnership with these stakeholders came up with an extremely simple yet holistic solution. They traded food and transportation passes on their new rapid transit bus system for bags of trash collected by the residents. Furthermore, they purchased this food from small farmers in the surrounding region that were economically struggling to hold onto their land—a trend that they discovered was the primary cause for the recent influx in migration to the city. The result of this simple program was that it helped to keep farmers on the land, thus slowing down future migration rates, offered the people who had already migrated to the favelas an opportunity to provide food for their families and facilitate access to other parts of the city to find longer-term employment, and solved the immediate public health hazard. More importantly than all of this, however, was that the city leaders did this in a way that empowered these stakeholders to actualize their lives and livelihoods in developmental partnership with the city. This led to an evolving relationship through time that spurred the development of a rich ecosystem of civic partnership programs that helped create what the city leaders referred to as "an equation of co-responsibility" (McKibben 1995). This included innovative programs for elder-youth mentorship, youth-led tree nurseries and green space stewardship, mobile classrooms, owner-built housing, and craft making and upcycling entrepreneurism.

Leveraging systemic health

Beyond good, appropriate technologies and developmentally engaged stakeholders, any successful regenerative development project requires a living systems understanding of the place and its dynamic workings in order to best see how to leverage the improvement of its inherent health-maintaining and health-regenerating systems. Regenerative development, from this perspective, is not just about creating net positive benefit to individual stakeholders or capital stocks in a system but rather about leveraging resources to maximize the systemic workings of the socioecological system as a whole. This requires a mental reorientation from project-centric thinking (which sees the project as the central actualizing entity) to systems-actualizing thinking (which involves starting with seeking to better understand the nested living systems that the project is embedded within in order to more deeply inform how the project in its conception, design, and development can best serve—i.e., its regenerative role—in leveraging the health-maintaining and health-regenerating capacities of these larger systems).

A project example from the consulting firm Regenesis Group (that I work with) may help to illustrate this point. Brattleboro Food Co-Op originally sought Regenesis Group's help in replacing its existing facility with a new energy-saving LEED building that would enable it to expand. By first working with the co-op to better understand the nested systems they were a part of (i.e., their *place* in the world), it quickly became clear that merely focusing on building an energy-saving green building held little regenerative potential for Brattleboro and its region. Rather, re-envisioning the project's aim and role as one of "creating a regenerative marketplace" to help revitalize their region's foodshed and people's relationship to it would be helpful,

> (t)hey began to look for ways to grow the value-generating capability of local farmers, landowners, and food processors and also for ways to grow their own capability to truly serve all of their stakeholders. This led to the creation of a hundred-year plan as a source of direction. The co-op went to work on creating the leveraged programs and systems that they could see would make a difference right away, and they became intensely involved in community building with more than a dozen other cooperatives in their region. From this they began to envision a local credit union, an agricultural education program, ways to involve youth, a cooking school emphasizing local foods, and much more. As it turned out, producing food locally would save far more energy than the building alone. (Benne 2018, p. 1)

In the words of Mark Goehring (2005), Brattleboro Co-Op's Board president at the time, they came to realize that "systemic change requires more than a band-aid of simple green building concepts. It requires that we become engaged in how our entire nutrient or energy delivery systems work … it was apparent that greening the store's products and purpose … would have much more significant long-range impact than looking at the store by itself" (p. 1).

Almost two decades later, the new store is fully integrated into downtown Brattleboro. Its design and construction were a cooperative project of the food co-op, the Windham and Windsor Housing Trust, and Housing Vermont. It occupies an entire city block, with a natural foods market and deli on the ground floor, and on the floors above; the co-op offices; a commissary kitchen; a cooking classroom; and 24 residential apartments. There are solar panels on the roof, and the heating system for the entire structure is based on recycling the heat produced in the store by refrigeration. "But more than green and co-operative, the most important thing about the building is that it fully supports and helps to continuously regenerate this 6000-member co-op's commitment to regenerative community building, and a vital local food system" (Benne 2018, p. 1).

Vitalizing fields

We can all recognize when we step into a vitalizing field (e.g., a particularly vibrant restaurant or a lively neighborhood). We can also recognize its absence.

Ultimately, regenerative development requires the fostering of a vitalizing field of hope, faith, and caring between inhabitants and their place. I challenge the reader to try and think of a successful regenerative development example in which such a vitalizing field (in some form or other) was completely absent. What I can offer is an example of a promising yet ultimately unsuccessful development project that illustrates this point in reverse.

> In 2015 [...] a team of local developers with a reputation for successful green and community-minded projects [...] proposed a residential complex geared toward young people seeking affordable living spaces to rent. It offered a mix of market-rate and subsidized units. In addition, the project embraced high sustainability aspirations. It was centrally located with the city, directly adjacent to the Santa Fe River along an established bus route and a planned extension to the city's growing network of bike paths. It included rainwater catchment, graywater reuse, community gardens, edible landscaping, river trail access, super-low energy use, and a pair of electric vehicles on site for a resident car-share. (Mang & Haggard 2016, p. 179)

On paper, this project seemed an ideal project for Santa Fe—a thoughtfully designed housing development that helped the city address its pressing affordable housing needs while aiming to do so in a way that embodied the city's progressive values and sustainability aspirations. However, what the project failed to take into consideration was the spirit of the place (the historical Agua Fria Village neighborhood) that it was situated within. Agua Fria village is a semirural community with agricultural roots and a historical Hispanic community with families that have inhabited the area since the mid-1600s. It's an area that in addition to semirural single-family residences, hosts a native-plant nursery and a community farm. The site itself of the proposed new development was a previous center for teaching sustainable agriculture. What the developers failed to take into consideration in designing their project was the neighborhood's strong sense of identity and relationship to the place that they cared deeply about maintaining and growing (Mang & Haggard 2016). Rather than seeking to align with this spirit of the place and seeking to orient the design toward supporting and fostering this vitalizing identity, they created a design that was perceived as a direct threat to the faith, hope, and caring that residents in that neighborhood had worked hard in generating for their place. Ultimately, the city council voted to reject the project's request for a zoning-use variance, thus killing the project. The city council members all agreed that the project was critically needed for the city, but that to allow it to be developed in that particular place in the city would be a violation of the community's trust (Last 2016).

Critical to the practice of regenerative development, therefore, is the ability to deeply listen and be receptive to the spirit of a place and its inhabitants who care for and steward it and to use this as a regenerative source for enrolling others in a cocreative design process that grows a vitalizing field of shared hope, faith, and caring between people and their place. According to the renowned landscape architect, Lawrence Halprin (interviewed in Swan 1991c):

> The place isn't really mystical, but it has a quality. When we could get people to come together and work as a group to report on how it felt to them, we came up with a design which has worked and has inspired other people all around the world to try this kind of approach. This is the kind of design which generates a spirit all its own, which encourages the human spirit to soar. That's when real magic starts to happen. (p. 314)

What Halprin is speaking to is a process of direct perception followed by reflective dialog to help the design team experience and articulate the vitalizing energies of a place and use these energies as the source point for creating an inspiriting project. If the developers from Santa Fe had taken the time up front to engage in such a process, their process of engagement with the community would have ended up looking very different and would have informed a project that was much more in harmony with the character and spirit of that place.

Conclusion

In conclusion, it is the proposition of this chapter that internal and external development practices are inseparable (i.e., we can not work on one without the other) and that the meta-aims we pursue, the roles we define for ourselves, and the capacities we seek to build in order to serve those roles are all influenced by the developmental frames (of self, psyche, and world) that we adopt (either consciously or unconsciously). Towards this end, this chapter explores four distinct psychological paradigms that, subtly and not so subtly, influence the way we relate to and work on the development of place and their implications for defining a holistic regenerative development practice.

References

Abrams, D 1996, *The spell of the sensuous: Perception and language in the more-than-human world*, Vintage Books, New York.

American Psychiatric Association 1994, *Diagnostic and statistical manual of mental disorders* (4th ed.), Author, Washington DC.

Anderson, MK 2005, *Tending the wild: Native American knowledge and the management of California's natural resources*, University of California Press, Berkeley, CA.

Anthony, C 1995, "Ecopsychology and the deconstruction of whiteness", In T Roszak, ME Gomes & AD Kanner (ed.), *Ecopsychology: Restoring the earth, healing the mind*, Sierra Club Books, San Francisco, pp. 263–278.

Aristotle 1962, *The politics*. Penguin Books, London.

Baldursson, S 2009, *The nature of at-homeness*, viewed 13 April 2018 <http://www.phenomenologyonline.com/articles/baldursson.html>

Baldwin, JM 1913, *History of psychology: A sketch and an interpretation*. Watts, London.

Bayes, K 1970, *The therapeutic effect of environment on emotionally disturbed and mentally subnormal children*. Gresham Press, London.

Beatley, T & Manning, K 1997, *The ecology of place: Planning for environment, economy, and community*, Island Press, Covelo, CA.

Beatley, T 2004, *Native to nowhere: Sustaining home and community in a global age*, Island Press, London.

Benne, B 2018, *Building a place-sourced regenerative economy: Co-evolving relationships between a regenerative organization and its socio-ecological environment*, viewed 1 September 2018 <https://regenesisgroup.com/place-sourced-regenerative-economy>

Bennett, JG 1964, *A spiritual psychology*, CSA Press, Lakemont, GA.

Bennett, JG 1966, *The dramatic universe volume III: Man and his nature*, Claymont Communications, Charles Town, WV.

Berry, W 1996, *Unsettling of America: Culture and agriculture*, Sierra Club Books, San Francisco.

Bohm, D 1985, *Unfolding meaning: A weekend of dialogue*, Routledge, London.

Bohm, D 2003, *The essential David Bohm*, Routledge, New York.

Bonnemaison, J 2005, *Culture and space: Conceiving a new cultural geography*, I. B. Tauris, London.

Braud, WG 1992, "Human interconnectedness: Research indications", *Revision: A Journal of Consciousness and Transformation*, vol. 14, no. 3, pp. 140–148.

Brown, LR 2008, *Plan b 3.0: Mobilizing to save civilization*, W.W. Norton and Company, New York.

Brown, T 1993, *Grandfather: A native American's lifelong search for truth and harmony with nature*, Berkley Books, New York.

Byrd, RC 1988, "Positive therapeutic effects of intercessory prayer in a coronary care population", *Southern Medical Journal*, vol. 81, no. 7, pp. 826–829.

Capra, F 1975, *The tao of physics*, Shambhala, Boston.

Capra, F 1982, *The turning point: Scientific society and the rising culture*, Simon and Schuster, New York.

Capra, F 1996, *The web of life: A new scientific understanding of living systems*, Anchor Books, New York.

Cresswell, T 2004, *Place: A short introduction*, Wiley-Blackwell, Oxford, UK.

Dale, A & Robinson, JB 1996, *Achieving sustainable development: A project of the Sustainable development research institute*, University of British Columbia Vancouver, Canada.

Daniels, V 1997, *Psychology in Greek philosophy*, viewed 9 October 2018 http://www.sonoma.edu/users/d/daniels/Greeks.html

Day, C 1990, *Places of the soul: Architecture and environmental design as a healing art*, Aquarian Press, Wellingborough, Northamptonshire, England.

Deloria, V 1991, "Reflection and revelation: Knowing land, places, and ourselves", In JA Swan (ed.), *The power of place: Sacred ground in natural and human environments*, Quest Books, Wheaton, IL, pp. 28–40.

Deloria, V & Wildcat, DR 2001, *Power and place: Indian education in America*, Fulcrum Resources, Golden, CO.

Dernbach, JC 2002, *Stumbling toward sustainability*, Environmental Law Institute, Washington DC.

Dossey, L 1993, *Healing words of prayer and the practice of medicine*, Harper, San Francisco.

Easwaran, E 1978, *Gandhi the man*, Nilgiri Press, Petaluma, CA.

Fadiman, J & Frager, R 2002, *Personality & personal growth* (5th ed.), Prentice Hall, Upper Saddle River, NJ.

Fernandez, EG 1998, "Development or cultural affirmation in the Andes?", In F Apffel-Marglin (ed.), *The spirit of regeneration: Andean culture confronting western notions of development*, Zed Books, London, pp. 124–145.

Fox, W 1995, *Toward a transpersonal ecology*, State University of New York, New York.
Frankfort, H, Frankfort, MH, Wilson, JA & Jacobsen, T 1949, *Before philosophy*, Penguin Books, Baltimore, MD.
Freud, S 1961, *Civilization and its discontents*, Norton & Company, New York.
Fromm, E 1955, *The sane society*, Fawcett Publications, Greenwich, CT.
Fuerst, JS 2004, *When public housing was paradise: Building community in Chicago*, University of Illinois Press, Urbana, IL.
Goehring, M 2005, "Co-op as store becomes co-op as community", *Cooperative Grocer Network*, viewed 1 September 2018 <https://www.grocer.coop/articles/co-op-store-becomes-co-op-community>
Goleman, D 1993, "Psychology, reality, and consciousness", In R Walsh & F Vaughan (eds.), *Paths beyond ego: The transpersonal vision*, Perigee Books, Los Angeles, pp. 18–20.
Hawken, P, Lovins, A & Lovins, LH 1999, *Natural capitalism: Creating the next industrial revolution*, Little, Brown and Company, Boston.
Hillman, J & Ventura, M 1992, *We've had a hundred years of psychotherapy and the world's getting worse*, HarperSanFrancisco, San Francisco.
Hughes, D 1991, "Spirit of place in the western world", In JA Swan (ed.), *The power of place: Sacred ground in natural and human environments*, Quest Books, Wheaton, IL, pp. 15–27.
Huxley, A 1945, *The perennial philosophy*, Harper and Brothers, New York.
Jackson, DL & Jackson, LL 2002, *The farm as natural habitat: Reconnecting food systems with ecosystems*, Island Press, London.
Joly, N 2007, *What is biodynamic wine: The quality, the taste, the terroir*, Clairview Books, London.
Kaiser, EJ, Godschalk, DR & Chapin, FS 1995, *Urban land use planning*, University of Illinois Press, Urbana, IL.
Kim, U, Yang, KS & Hwang, KK 2006, *Indigenous and cultural psychology: Understanding people in context*, Springer, New York.
Kremer, JW 1998, "The shadow of evolutionary thinking", In D Rothberg & S Kelly (eds.), *Ken Wilber in dialogue: Conversations with leading transpersonal thinkers*, Quest Books, Wheaton, IL, pp. 237–258.
Krone, CG 1997, *West Coast Resource Development: Session Notes #124*, Unpublished transcription of dialogue by members of the Institute for Developmental Processes.
Krone, CG 2005, *West Coast Resource Development: Session Notes #186*, Unpublished transcription of dialogue by members of the Institute for Developmental Processes.
Kunstler, JH 1993, *The geography of nowhere: The rise and decline of America's man-made landscape*, Simon & Schuster, New York.
Lake, MG 1991, "Power centers", In JA Swan (ed.), *The power of place: Sacred ground in natural and human environments*, Quest Books, Wheaton, IL, pp. 48–60.
Last, TS 2016, "A new city plan may address housing concerns", *Albuquerque Journal*, viewed 1 September 2018, <https://www.abqjournal.com/801542/a-new-city-plan-may-address-housing-concerns.html>
Laszlo, E 2004, *Science and the akashic field: An integral theory of everything*, Inner Traditions, Rochester, VT.
Leopold, A 1949, *A Sand county almanac: And sketches here and there*, Oxford University Press, New York.
Lovelock, JE 1979, *Gaia: A new look at life on earth*, Oxford University Press, New York.
Lovelock, JE 1988, *The ages of Gaia: A biography of our living earth*, W.W. Norton & Co, New York.

Mack, JE 1993, "Forward", In R Walsh & F Vaughan (eds.), *Paths beyond ego: The transpersonal vision*, Perigee Books, Los Angeles, pp. xi–xiv.

Mang, NS 2009, *Toward a regenerative psychology of urban planning*, Proquest Dissertations Publishing, San Francisco, CA.

Mang, P & Haggard, B 2016, *Regenerative development and design: A framework for evolving sustainability*, John Wiley & Sons, Hoboken, NJ.

Maslow, AH 1968, *Toward a psychology of being* (2nd ed.), Von Nostrand Reinhold, New York.

Maslow, AH 1993, *The farther reaches of human nature*, Arkana, New York.

Maslow, AH 1998, *Maslow on management*, John Wiley & Sons, New York.

May, GG 1982, *Will and spirit: A contemplative psychology*, Harper & Row, San Francisco.

McHarg, IL 1971, *Design with nature*, Doubleday & Company, Garden City, NY.

McKibben, B 1995, *Hope, human and wild: True stories of living lightly on the earth*, Hungry Mind Press, Saint Paul, MN.

McMurray, A 2006, *Community health and wellness: A socio-ecological approach*, Elsevier, Chatswood, Australia.

Metzner, R 1999, *Green psychology: Transforming our relationship to the earth*, Inner Traditions Press, Rochester, VT.

Mollison, B 1997, *Permaculture: A designers' manual*, Tagari Publications, Tyalgum, Australia.

Mumford, L 1961, *The city in history: Its origins, its transformations, and its prospects*, Harcourt Brace Jovanovich, San Diego, CA.

Nabhan, GP 1997, *Cultures of habitat: On nature, culture, and story*, Counterpoint, Washington, DC.

Naess, A 1989, *Ecology, community, and lifestyle*, Cambridge University Press, Cambridge, UK.

NewUrbanism.org 2006, *New urbanism*, viewed 16 December 2018, <http://www.newurbanism.org/>

Orr, DW 1994, *Earth in mind: On education, environment, and the human prospect*, Island Press, Covelo, CA.

Ouspensky, PD 1981, *The psychology of man's possible evolution*, Vintage Books, New York.

Peat, FD 2002, *Blackfoot physics*, Weiser, Boston.

Playfair, G 2002, *Twin telepathy: The psychic connection*, Vega Books, London.

Puthoff, HE 1991, "Zero-point energy: An introduction", *Fusion Facts* 3, vol. 3, no. 1, p. 29.

Ribeiro, J.A. & Tavares, N 1992, *Curitiba: A revolução ecológica* (N. Torres, Trans.), Prefeitura Municipal de Curitiba, Curitiba, Brazil.

Relph, E. (1976). *Place and placelessness*. Pion Limited, London.

Rogers, CR 1980, *A way of being*, Houghton Mifflin, Boston.

Roszak, T 1993 *The voice of the earth: An exploration of ecopsychology*, Touchstone, New York.

Roszak, T 1995, "Where psyche meets Gaia", In T Roszak, ME Gomes & AD Kanner (eds.), *Ecopsychology: Restoring the earth, healing the mind*, Sierra Club Books, San Francisco, pp. 1–20.

Sahtouris, E 2000, *Living systems, the internet, and the human future*, viewed 9 October 2018, <http://www.ratical.org/LifeWeb/Articles/LSinetHF.html>

Schwaller de Lubicz, RA 1957, *The temple of man*, Inner Traditions, Rochester, VT.

Seamon, D 2000, "Phenomenology, place, environment, and architecture: A review", *Environmental & Architectural Phenomenology Newsletter*, viewed 16 December 2018, <http://www.arch.ksu.edu/seamon/articles/2000_phenomenology_review.htm>

Steiner, R 2005, *What is biodynamics?: A way to heal and revitalize the earth*, Steiner books, Herndon, VA.

Suzuki, D & Knudtson, P 1992, *Wisdom of the elders: Honoring sacred native visions of nature*, Bantam, New York.

Swan, JA 1990, *Sacred places*, Bear, Santa Fe, NM.

Swan, JA 1991a, "Introduction", In JA Swan (ed.), *The power of place: Sacred ground in natural and human environments*, Quest Books, Wheaton, IL, pp. 1–14.

Swan, JA 1991b, "The spots of the fawn: Native American sacred sites, cultural values and management issues", In JA Swan (ed.), *The power of place: Sacred ground in natural and human environments*, Quest Books, Wheaton, IL, pp. 63–81.

Swan, JA 1991c, "Design with spirit: Interview with Lawrence Halprin", In JA Swan (ed.), *The power of place: Sacred ground in natural and human environments*, Quest Books, Wheaton, IL, pp. 305–314.

Thompson, JW, Sorvig, K & Farnsworth, CD 2000, *Sustainable landscape construction: A guide to green building outdoors*, Island Press, Washington DC.

Van Putten, M 1996, "Living up to our legacy", *National Wildlife*, vol. 34, no. 5, p. 6.

Vasquez, GR 1996, "Culture and biodiversity in the Andes", *Development*, vol. 4, pp. 34–38.

Wilber, K 1993, "Psychologia perennis: The spectrum of consciousness", In R Walsh & F Vaughan (eds.), *Paths beyond ego: The transpersonal vision*, Perigee Books, Los Angeles, pp. 21–34.

Wilber, K 2000, *Integral psychology: Consciousness, spirit, psychology, therapy*, Shambhala, Boston.

5 Regenerative development and environmental justice

David N. Pellow

Literature and thesis

Much of the scholarship on green design, sustainability, and sustainable development has focused on minimizing harm to ecosystems and human well-being (McDonough & Braungart 2002). These frameworks have emphasized the efficient usage of ecological materials, essentially arguing for a slowing down of the rate of degradation of the earth's life support systems (see Cole 2012b; World Commission on Environment and Development [WCED] 1987). Scholars and practitioners promoting regenerative development and design contend that this is not nearly enough. Instead, they argue that, in addition to slowing down and reversing activities that are harmful to ecosystems and human health, we must place much more focus on designing human settlements in ways that co-evolve with ecosystems, generating mutual benefits for human and more-than-human communities (Girardet 2014; Hemenway 2015; Hes & du Plessis 2015; Mang, Haggard & Regenisis 2016; Palazzo & Steiner 2011). As some design scholars like John Tillman Lyle put it, sustainable design is "merely breaking even while regenerative design renews earth's resources" (Mang & Reed 2012b). The earlier ideas and practices of sustainability grew out of a framework rooted in "technological sustainability" (e.g., engineering-based, eco-efficiency-focused, green building focused work, etc.), while regenerative development and design emerged out of a framework rooted in "ecological sustainability" (e.g., living systems principles). Some of the regenerative development literature includes a focus on social equity, primarily through the encouragement of broad stakeholder involvement in the process of design and development (Cole 2012a, 2012b; Gabel 2015; Hoxie, Berkebile & Todd 2012; Tainter 2012), a form of community engagement aimed at being inclusive toward building solutions and capacities with broad participation. However, the regenerative design literature generally pays far less attention to the fact that social inequalities and power imbalances are the root drivers of our ecologically unsustainable societies (Lyle 1994; for an exception see Fullerton 2015). And even when inequality is considered, it is generally framed in broad economic terms, suggesting that social class is the most important category of inequality to be addressed, to the neglect of racism, patriarchy, nativism, colonialism, and other systems of power. That is, we need a form of "intersectional" environmentalism that links and confronts various forms of inequality to build deeper analyses of the problems and stronger coalitions for charting solutions. So, while including stakeholders in building a regenerative future is certainly critical, the underlying vision of the problem has been largely one of design rather than of unjust imbalances of power. Thus, while including diverse stakeholders is a good idea, one must first consider that the exclusion and dispossession of various populations from society on a daily basis must be addressed, and that encouraging multiple stakeholder voices in any given project may not be sufficient. In this chapter, I suggest that this particular component of regenerative development and design should receive more attention, by drawing from the scholarship on environmental justice (EJ).

The scholarship on EJ (and climate justice) reveals that, with few exceptions, human populations that already exist at the social, political, economic, and/or cultural margins of most societies tend to also experience additional burdens associated with disproportionate exposure to environmental threats (Bullard & Wright 2012; Caniglia, Vallee & Frank 2017). This includes people of color, indigenous people, immigrants, ethnic/religious minorities, women, and low-income populations being forced to live in or near flood plains, residing in neighborhoods in close proximity to landfills, power plants, waste incinerators, mining and other extractive operations, industrial agricultural facilities, and/or working in hazardous occupations (Bullard & Wright 2012; Harrison 2011; Taylor 2014). Moreover, while those trends persist in the United States and globally, they are reinforced by the scourge of environmental privilege—those landscapes marked by wealth and near-exclusive access to spaces where air, land, and water are relatively cleaner, safer, and unsullied by the excesses of capitalism (Park & Pellow 2011; Taylor 2009). The impacts of climate change follow the same general patterns, as people of the global South generally contribute far less to climate disruption than the global North, yet they pay the greatest price in terms of risks of extreme weather, flooding, excessive heat, threats to food systems, disease, and state violence that often accompany these phenomena (Ciplet, Roberts & Khan 2015). Thus, the literatures on EJ studies reveal that where we find social inequalities we tend to also observe environmental inequalities. They also suggest quite clearly that those communities where marginalized populations are protected and defended via progressive public policies and supportive cultural practices tend to also have strong environmental and climate protection policies and practices (Boyce 1994; Downey 2015; Ergas & York 2012).

EJ scholar Julian Agyeman has done important work on these topics, beginning with his bridge building concept of "just sustainability"—which fused EJ and sustainability by defining this idea as "the need to ensure a better quality of life for all, now and into the future, in a just and equitable manner, while living within the limits of supporting ecosystems" (Agyeman, Bullard & Evans 2003, p. 5). In other words, sustainability without justice is limited, so limited in fact that it is actually not sustainable. Justice is a key ingredient to ensure that our life support systems are healthy.

There are numerous EJ studies that grapple with defining different forms of justice with respect to integrating EJ into community planning and public policy making (Banerjee 2017), which would seem to be a key overlap with the concern for stakeholder inclusion in the regenerative development literature. Recent EJ scholarship draws on philosophy and political theory to specify that justice can be defined as distribution, inclusion, participation, recognition, fairness/procedure, or capabilities (Harrison 2011; Schlosberg 2004, 2007; Walker 2012). Years earlier, sociologist Stella Čapek (1993) laid the groundwork for these ideas when she introduced the "environmental justice frame," which views EJ as a lens that offers a way of constructing meaning and social significance for activists. According to Čapek, the EJ frame consists of six key claims, including the right to accurate information from authorities concerning environmental risks, public hearings, democratic participation in decision-making regarding the future of any threatened community, compensation for injured parties from those who inflict harm on them, expressions of solidarity with survivors of environmental injustices, and a call to abolish environmental racism/injustice. According to Čapek, environmental injustice was not just about the threats associated with disproportionate hazards facing marginalized communities; it was about ordinary people demanding recognition that their analysis of environmental injustice in their communities was legitimate, while gaining access to democratic processes and exercising power. That is, EJ was ultimately focused on reordering power relations among stakeholders involved. These arguments were later expanded by David Schlosberg (2007) and others, who concluded that much of the EJ literature's focus on justice was limited in that scholars and activists should also emphasize the power structures and social systems that give rise to environmental inequalities to begin with.

Where there is strongest overlap between EJ studies and regenerative development is the EJ scholarship focused on *procedural* justice (see Schlosberg 2004, 2007). Arising from the idea of participatory democracy, procedural justice shifts the lens from distributive outcomes to decision-making processes and the importance of "recognition." That is, proponents of procedural justice maintain that a focus on mere distribution of environmental risks and benefits is incomplete, so they argue for a closer examination of group recognition. Group recognition involves the practice of acknowledging a community's existence, its right to speak, and the validity of its claims for justice. Recognition includes those political and cultural practices involving the acknowledgment and inclusion of marginalized groups in decision-making (Young 2001). These issues have particular salience in communities of color and Indigenous communities globally where dominant state forces and elites have denied residents the opportunity to participate in decisions regarding environmental impacts that shape their lives (Bullard & Wright 2012; Whyte 2017).

While the field of EJ studies pioneered research demonstrating that working class neighborhoods and communities of color face disproportionately high levels of pollution, waste, and other toxic industrial detritus by comparison with more affluent communities, there is still much work needed that could explore the planning and positive political possibilities that might emerge from such difficult social and ecological terrain. Isabelle Anguelovski's book, *Neighborhood as Refuge: Community Reconstruction, Place-Remaking, and Environmental Justice in the City* (2014), is an exemplary study that offers a number of bridges between the literatures on regenerative development and EJ. Anguelovski concentrates her work on proactive efforts to produce sustainable urban spaces and new identities. The activists featured in her research are involved in placemaking as much as defending neighborhoods against potential toxic threats (Anguelovski 2014). Through this emphasis on placemaking and identity work, what Anguelovski offers is a critical intervention in that she addresses the importance of community revitalization and restoration as methods and practices that allow residents to combat the stigma of living in neighborhoods characterized by socio-environmental disamenities. Anguelovski's emphasis that place and placemaking has a strong overlap with regenerative development stresses the importance of context and refuses universal approaches to solutions across diverse geographic and cultural landscapes (see Frampton 1983; Mang & Reed 2012a). Anguelovski's work also overlaps nicely with regenerative design scholarship in that both emphasize the importance of *stories* about the places we seek to defend, protect, and/or regenerate. The story Anguelovski tells is one that includes the traditional EJ studies narrative that marginalized populations often face environmental threats, but she also tells stories that suggest that some of the most unhealthy and precarious places can be transformed into sites of rejuvenation and health-sustaining practices and material relations both within human societies and between humans and the more-than-human world.

Another terrain on which to consider the overlaps and distinctions between regenerative development and EJ is the Principles of Environmental Justice, a sort of founding document of the EJ movement, penned at the First National People of Color Environmental Leadership Summit in 1991. Let us consider some of these EJ Principles that primarily emphasize social justice and equity, because those are values that are at the core of EJ but not always central to regenerative development. These principles include the following:

- Principle #2: EJ demands that public policy be based on mutual respect and justice for all peoples, free from any form of discrimination or bias.
- Principle #4: EJ calls for universal protection from nuclear testing, extraction, production, and disposal of toxic/hazardous wastes and poisons and nuclear testing that threaten the fundamental right to clean air, land, water, and food.
- Principle #5: EJ affirms the fundamental right to political, economic, cultural, and environmental self-determination of all peoples.
- Principle #9: EJ protects the right of victims of environmental injustice to receive full compensation and reparations for damages as well as quality health care.

- Principle #10: EJ considers governmental acts of environmental injustice a violation of international law, the Universal Declaration on Human Rights, and the United Nations Convention on Genocide.
- Principle #11: EJ must recognize a special legal and natural relationship of Native Peoples to the US government through treaties, agreements, compacts, and covenants affirming sovereignty and self-determination.
- Principle #13: EJ calls for the strict enforcement of principles of informed consent, and a halt to the testing of experimental reproductive and medical procedures and vaccinations on people of color.

The above principles make it clear that core concerns among EJ scholars and activists center around confronting racism, white supremacy, colonization, genocide, and systems of discrimination, as they impact human populations and ecosystems. It is fair to say that these issues are not given prominence in the regenerative development literature. The EJ Principles that reflect key concerns within the scholarship on regenerative development include the following:

- Principle #1: EJ affirms the sacredness of Mother Earth, ecological unity, and the interdependence of all species, and the right to be free from ecological destruction.
- Principle #3: EJ mandates the right to ethical, balanced, and responsible uses of land and renewable resources in the interest of a sustainable planet for humans and other living things.
- Principle #7: EJ demands the right to participate as equal partners at every level of decision-making, including needs assessment, planning, implementation, enforcement, and evaluation.
- Principle #12: EJ affirms the need for urban and rural ecological policies to clean up and rebuild our cities and rural areas in balance with nature, honoring the cultural integrity of all our communities, and provided fair access for all to the full range of resources.
- Principle #17: EJ requires that we, as individuals, make personal and consumer choices to consume as little of Mother Earth's resources and to produce as little waste as possible; and make the conscious decision to challenge and reprioritize our lifestyles to ensure the health of the natural world for present and future generations.

The above EJ Principles suggest clear and fertile ground for building common cause with scholars and practitioners of regenerative development.

Another area where common ground between EJ studies and regenerative development could be built is on the terrain of culture. Agyeman and others have also made the argument that ecological sustainability is supported in robust ways by strengthening our commitment to cultural diversity, because the "diversity advantage" in cities, communities, and workplaces that are marked by demographic richness includes mutual learning and joint growth, acquisition of skillsets that enable problem-solving across wide range of contexts, and a pluralist transformation of public space (Agyeman 2016; Wood & Landry 2008). This suggests a line of thinking and action that is quite important for regenerative development: activities that seek to mutually promote human and environmental health will need to include a strong commitment to cultural, racial, and social diversity. In the rest of the chapter, I seek to bring key ideas from regenerative development and EJ together and I apply these insights to the case of the fires that ravaged much of the California coastline in the fall and winter of 2017.

Case study: The California wildfires of 2017

2017 was the deadliest and most destructive year for wildfires in the state of California, resulting in dozens of deaths and more than 8,000 structures destroyed (Repogle 2017). And while fire season traditionally extends from June into October, experts are now declaring that California's fire season is "year round." In April of 2017, Governor Jerry Brown declared the state's historic six-year drought over, issuing an executive order that lifted the majority of drought emergency efforts in most of California's counties. But scholars who study this issue closely contend that things are not expected to improve anytime soon. For example, University of California Los Angeles (UCLA) geographer Glen MacDonald stated, "In California, we are always going to have drought … This is the same thing with fire … Fires are always going to be present with us, and there are always going to be ones that we simply cannot be able to control" (Calfas 2017: 16).

Beginning in early October 2017, wild fires devastated numerous communities in Northern California's wine country, causing numerous deaths and widespread property destruction, and impacting urban areas directly in ways that few if any residents or scholars have seen in living memory. In December 2017, another series of wildfires broke out in Southern California (the Creek Fire, Rye Fire, Skirball Fire, Lilac Fire, and Liberty Fire), including the Thomas Fire, which quickly became the state's largest wildfire on record (and has already been surpassed as of this writing, by the Mendocino Complex Fire in summer 2018). The Southern California fires also produced damage in urban areas unlike anything observers had seen, particularly that late in the calendar year (Livingston & Panzar 2017). Hundreds of thousands of people were evacuated and official air quality measurements were at "very hazardous" levels for days. Scientists conclude that these fires are linked to climate change through a pattern of extremely dry and wet winters, followed by the growth of grasses and other plants that become very dry during what are becoming longer and hotter summers and autumns. Those grasses and plants then provide fuel for increasingly more intense wildfires. Researchers also predict that average temperatures in Northern California will rise to match those in Southern California within a decade (Romero 2017). UCLA geologist Aradhna Tripati stated, "This will create a whole cascade of problems, from loss of life and homes, to health problems associated with the poor air quality, to rising insurance premiums, to housing shortages, to movement of people due to these climate change–linked natural disasters" (UCLA Institute on the Environment and Sustainability [IOES] 2017).

In the aftermath of this latest and more intense round of wildfires, community leaders are expected to address the question of how to prepare for, minimize, and hopefully prevent future conflagrations. In the next sections, I offer different ways of thinking about and responding to the California fires from an EJ perspective, a regenerative development orientation, and an integrated approach that draws on ideas from both frameworks.

An environmental justice perspective on the California fires

The California fires of 2017 produced incalculable harm to human and environmental health and had a disparate impact on already marginalized populations, including low-income persons, immigrants, and people of color (KCSB News 2017; PBS News Hour 2017). This is a pattern we see in other "natural" disasters around the globe. As one of the landmark studies of "natural" disasters concluded, "In the United States the key characteristics that seem to influence disaster vulnerability most are socioeconomic status, gender, and race or ethnicity" (Mileti 1999, p. 122). This is largely because existing inequalities in society differentially position groups in ways that privilege and protect some populations and place others at greater risk, a phenomenon that Caniglia, Vallee and Frank (2017) call "injustices in waiting" (see also Caniglia and Frank 2017).

The story took an unexpected turn early on, though. On October 17, 2017, the conservative and openly racist website known as Breitbart, published a story claiming that an undocumented person was arrested in Sonoma County on suspicion of starting the Northern California fires, and that story went viral among right-wing news services (Breitbart is run by Steve Bannon, Donald J. Trump's former chief strategist). Sonoma County Sheriff's office spokesperson, Sergeant Spencer Crum, described this story as "completely false, bad, wrong information." The Sheriff's office did confirm that an undocumented homeless man was indeed arrested for starting a small fire that was unconnected to the larger fires, which was quickly extinguished (Nichols 2017). Despite the fact that PolitiFact rated Breitbart's story false, the right wing media had manufactured yet another reason to target and discriminate against immigrants. There is a long history of blaming and targeting marginalized communities for fires, including the stories claiming that undocumented immigrants contributed to the massive 2007 fires in San Diego County, California (Pellow 2009) and the long-standing view that Rome's Emperor Nero blamed Christians (a minority religious sect at the time) for the great urban fire that consumed most of Rome in July AD 64 (Gray-Fow 1998). These particular stories around fires and blame reflect the larger problem that James Hoggan (2016) calls the pollution of the public square, via the toxic narratives and propaganda that emerge to polarize and paralyze public debate around a spectrum of otherwise legitimate and pressing social concerns.

In many ways, while the narrative about climate change's link to the fires did in fact gain occasional traction in the media, it was often overshadowed by the scapegoating of immigrants and homeless persons, so that the deeper, underlying causes of climate change and the encroachment of urban human settlement into wild lands were given insufficient consideration. Ironically, while immigrants and marginalized communities were being blamed for the fires in some media outlets, it was precisely these communities that faced the greatest harm during the conflagration. For example, there were several confirmed reports that farmworkers—the vast majority of whom are immigrants—were still obligated to work during the wildfires, and did so for several days (during the times of the worst air quality) without protective masks. In response, community groups like the Central Coast Alliance United for a Sustainable Economy (CAUSE) passed out more than 15,000 free N95 protective facemasks to workers in the fields and to people in low-income neighborhoods throughout Santa Barbara and Ventura counties (CAUSE 2017). Another problem was that government agencies were issuing emergency notifications (including evacuation orders, safety notices, emergency shelter location announcements, etc.) only in English, despite a large Spanish-speaking population that resides in Southern California. This is a continuing problem in Latinx and Asian immigrant communities whose residents often are excluded from access to public participation and critical safety information around environmental threats when key documents and announcements are made available in English only (Cole & Foster 2000). Fortunately, as CAUSE (2017, p. 2) later reported, "After a difficult start with the vast majority of disaster information inaccessible in Spanish [...] now both Ventura and Santa Barbara County have Spanish emergency websites, Spanish social media posts, and Spanish safety notices."

Finally, we saw that, while professional firefighters made up the bulk of persons combatting the blazes (and did so with great courage and success), many of the people putting their bodies on the line were prisoners who were given the opportunity to join in the effort. This has become controversial because many people question if this arrangement is exploitative and unsafe since prisoners have less training than professional firefighters (two weeks on average), may feel coerced to participate, may abscond from the scene, and have little chance of being hired as firefighters after their release from prison. Prisoners make up about one third of the California firefighting force and earn $2/day and $1/hour for their labor (Frankel 2017). Thus, the storytelling around the fires includes a narrative that lays the blame for the disasters on vulnerable and marginalized communities, but also includes a narrative that suggests that the driving forces of social and environmental injustice can be confronted through a greater commitment to equity and democracy.

If these were the primary EJ concerns that arose during the fires, what lessons can be learned for the future? The work of groups like CAUSE and other concerned community leaders reveals that, from an EJ perspective, the California wildfires reflect the need for greater attention to the ways that social and racial injustices shape the contours of our societies and make certain populations more vulnerable to "natural" disasters than others. This case also suggests quite clearly that a deeper commitment to democracy, fairness, and inclusion is critical to confronting historical and ongoing injustices as we rebuild in the wake of the fires. An EJ approach to rebuilding those California communities harmed by the fires would then center on the "diversity advantage" associated with being the most demographically diverse state in the nation, to build on that reality as a resource for strengthening the health of the community and encouraging skills that will enhance the state's ecosystems (particularly the flora and fauna in the coastal mountains and forests). The diversity advantage and democracy would work hand in hand with ecological sustainability because, as EJ Principle #7 states, we would expect everyone to "participate as equal partners at every level of decision-making, including needs assessment, planning, implementation, enforcement and evaluation" which would support the goal of EJ Principle #12: "urban and rural ecological policies to clean up and rebuild our cities and rural areas in balance with nature, honoring the cultural integrity of all our communities, and provided fair access for all to the full range of resources." Thus, an EJ perspective on responding to the fires would consider a number of issues not typically included in regenerative development practices (racism, social injustice, democracy, and equity), but would also include concerns that are most certainly at the center of that emergent tradition (inclusion, participation, storytelling). As inspiring and effective as this EJ approach to the fires might be, it would undoubtedly be strengthened if key themes of regenerative development were included, such as a deeper focus on the goals of coevolution, bioregionalism, and ecological whole/living systems thinking.

A regenerative development orientation to the California fires

A regenerative development approach to the California wildfires would put less emphasis on the importance of cultural politics and historical and contemporary oppression, and put more weight on community engagement, coevolution, bioregionalism, and storytelling that reflects systems thinking and the importance of and respect for place. What might that look like in the case of the aftermath of the California fires?

Community engagement would likely involve a series of public conversations aimed at achieving broad inclusion of various voices and perspectives in charting a course forward for the bioregion and place in which this tragedy unfolded. This would place an emphasis on allowing stakeholders to articulate their stories, narrating what they believe a more sustainable, resilient, and regenerative future could look like, with a particular attention to the unique context and character of the bioregions of Southern and Northern California, and the living systems of which they are a part.

Bioregionalism is a commitment "to developing communities integrated with their surrounding ecosystems. Rather than legally defined regions, bioregionalism considered geographical province with a marked ecological and often cultural unity, often demarked by the watersheds of major river systems" (Cole 2012a, p. 3). Using this definition, bioregionalism is, in my view, of critical importance to regenerative development because it refuses state-imposed borders—the kinds of limitations and restrictions that governments produce that are disruptive and violent with respect to ecosystem and human health. California's Mediterranean climate, extensive coastline, and vast desert and mountain ranges constitute a unique set of bioregions that require a specific response to ensure human and ecological health and a coevolution (Mang & Reed 2012a).

Coevolution is a concept that pushes beyond protecting ecosystems and human populations as if those were separate agendas to ensure that both are linked in mutually beneficial ways well into the future. Mang and Reed (2012a) describe coevolution as follows:

> It means shifting human communities and economic activities back into alignment with life processes. It implies every human settlement organizing itself around evolving its watershed's capacity to support life. The creative and economic activities of human communities can be directed toward the development of human potential through harmonization of and with the dynamic energies of nature. This is not preservation of an ecosystem, nor is it restoration. Instead, it is the continual evolution of culture in relationship to the evolution of life. This defines the work of regeneration. (p. 26)

This approach to regenerative development calls for an unprecedented determination to coexist: to shift away from excessive consumption and economic models reliant on systemic injustice, away from fossil fuels, and away from market-based strategies that cannot address the challenge of a year-round fire season—toward circular economic models, reduced consumption, healthy ecosystems, and participatory democracy to promote a future marked by preparedness for the next fires and practices that reduce their frequency and intensity (Donath 2018).

Whole systems and living systems are concepts that reflect the view that we live in networks of multiscalar relationships in which the whole is much more than the sum of its parts. This is an idea developed by Charles Krone (1992) to enable practitioners to see that individuals, organizations, and communities are made up of, nested within, and always connected to multiple systems that are constituted through reciprocal relationships. In other words, this is a basic yet profound recognition of one of the main principles of ecology—that everything and everyone is connected, so that what affects one individual, organization, or community has implications for the rest. That means that any successful regenerative development effort with respect to a fire-affected community in California must proceed with an understanding of that area's relationship to its many micro and macro socio-ecological networks.

Place is a concept that is inseparable from bioregionalism and whole/living systems. As Mang and Reed (2012a) explain,

> Place is defined here as the unique, multilayered network of living systems within a geographic region that results from the complex interactions, through time, of the natural ecology (climate, mineral and other deposits, soil, vegetation, water and wildlife, etc.) and culture (distinctive customs, expressions of values, economic activities, forms of association, ideas for education, traditions, etc.). The regenerative paradigm asserts that development can and should contribute to the capacity of all of the natural, cultural and economic systems that it affects in a place (to grow and evolve their health and ongoing viability) (p. 28).

A regenerative development approach to fire-affected communities in California would be different than an approach that might work elsewhere, because each place is unique, and that state has a particular set of histories and sociocultural terrains.

Storytelling is central to regenerative development because stories allow people to articulate how they—as individuals and as members of larger groups—actually fit into the complex relationships that constitute whole and living systems. Stories facilitate people's ability to deepen those relationships and strengthen their understanding of and commitment to place. Finally, storytelling enables people to share those narratives and to imagine different futures, empowering people through continuous learning that makes possible the productive coevolution of humans with their environment (Denning 2005; Korten 2007; Mang & Reed 2012a, p. 29). In order to heal and build a regenerative future, Californians must articulate and share their stories of the fires, including their causes and consequences, as well as how they will address those risks in the future.

Tying each of these concepts together is quite a challenge, but a regenerative development approach to the California fires would draw on and integrate each of them to provide spaces where people could convene and work collectively to engage in storytelling that reflects their connections to the places and bioregions of California, to the whole and living systems that are the subcomponents of those communities and in which those communities are nested, and that enable the articulation of a transformative vision and practice that facilitates sustainable coevolution of human and more-than-human populations, bodies, and places. More specifically, this process would have to uncover the root causes of the fires and offer a way to build capacity to change the cultural, spatial, and material identities and relationships that produced the fires so that in the future such possibilities will be minimized and that community resilience would be enhanced. As powerful and generative as such an approach would be, from an EJ perspective, it would still overlook the specific ways in which white supremacy, nativism, patriarchy, and racial capitalism produce vastly unequal exposures to environmental harms associated with fires, and that these realities would need to be fully and openly integrated into any efforts at capacity building for the future. I consider these possibilities in the conclusion.

Toward an integrated regenerative environmental justice framework

Low, Taplin and Scheld's (2003) work on the intersections of power and public space speaks to the importance of emphasizing equity in any consideration of regenerative development. Specifically, they demonstrate that the design of space across a range of contexts often produces highly consequential exclusions of certain populations who are therefore unable to access the benefits that others enjoy. The implication of this work is that, like sustainable development, regenerative development without an equity and justice focus can actually be deployed as a tool of oppression. I want to stress the point that justice and equity are of paramount importance because it is possible to imagine regenerative design efforts that ignore how living systems intersect with broader social, cultural, and political relations, even projects that promote, for example, what appear to be ecologically healthy but are also fundamentally oppressive and even militaristic practices. Justice and equity are not peripheral to regenerative development. In fact, there can be no regenerative development without justice and equity.

EJ studies has a primary concern with questions of power, justice, equity, inequality, democracy, and diversity, and considers addressing those topics as central to the production of our socio-ecological crises, as well as necessary to moving forward on future planning and policy making. Regenerative development has a primary focus on ensuring that human activities enhance ecosystem and human health. There are strong overlaps between EJ and regenerative development ideas and practices with respect to (1) an overall concern with the unsustainability of our current economic and political systems, (2) the recognition that diverse stakeholders must be included in decision-making at all scales, and (3) a prime value on the importance of place and storytelling as a way of narrating and framing problems and solutions. What regenerative development and design scholarship offers EJ studies is a greater emphasis on defending and protecting human populations and habitats to ensuring that both are linked in mutually beneficial ways well into the future, through attention to the particular contexts and places in which people and the ecosystems they rely upon coexist and coevolve. What EJ studies offers regenerative development is a much stronger emphasis on the role of social inequality, power, justice, and democracy—with a specific focus on racism and other forms of hierarchy—in understanding the driving forces behind ecological harm and for thinking and acting through solutions. In other words, while regenerative development offers EJ an emphasis on whole systems thinking from an ecological perspective, EJ offers regenerative development an emphasis on whole systems thinking through an intersectional justice perspective that centers on racism, colonization, empire, patriarchy, nativism, and their links to ecological violence. And rather than "rebuilding" and "restoring" what we had prior to the fires, an integrated EJ and regenerative approach to the wildfires would focus on reimagining a California marked by improved urban planning and human settlements that looked and functioned quite differently, or as EJ Principle #3 states, we would embrace "ethical, balanced and responsible uses of land and renewable resources in the interest of a sustainable planet for humans and other living things."

If scholars and practitioners could be attentive to these key points of overlap and common ground between EJ and regenerative development—while staying true to the core of both of those traditions—I believe the possibilities for transforming and enhancing human and more-than-human communities would be greatly expanded. The gap in EJ studies most relevant here is that there is insufficient focus on the deep linkages between marginalized human populations facing threats of environmental harm and the ways in which justice for those groups must be achieved through close attention to the health of specific ecosystems. The gap in regenerative development is what some scholars have called the "equity deficit"—the need to emphasize how those actions and practices intended to encourage the mutual improvement in ecological and human health may be made more challenging by the role of social inequalities that not only shape people's participation in such initiatives, but that actually undergird unsustainable economies in the first place. The idea of blending EJ and regenerative development practices is exciting because: (1) it offers the promise that we can push far beyond *reducing* racial/social/environmental injustices and ecologically harmful policies and practices by actually *strengthening* democracy and ecosystems; and (2) it facilitates the realization that EJ and regenerative development are inseparable and that their integration is actually required for achieving either goal. My hope and intention is to pursue such a goal in California in the years ahead and I look forward to learning and receiving feedback and guidance from leaders from both of these important schools of thought and action.

References

Agyeman, J 2016, "Cultivating food justice", Paper presented at "Whose Globalization? Whose Justice?" Workshop, sponsored by the Orfalea Center, UC Santa Barbara, February 26.

Agyeman, J, Bullard, RD & Evans B 2003, *Just sustainabilities: Development in an unequal world*, The MIT Press, Cambridge, MA.

Anguelovski, I 2014, *Neighborhood as refuge: Community reconstruction, place-remaking, and environmental justice in the city*, The MIT Press, Cambridge, MA.

Banerjee, D 2017, *Conceptualizing environmental justice: Plural frames and global claims in land between the rivers, Kentucky*, Lexington Books, Lanham, MD.

Boyce, JK 1994, "Inequality as a cause of environmental degradation", *Ecological Economics*, vol. 11, pp. 169–78.

Bullard, RD & Wright, B 2012, *The Wrong complexion for protection: How the government response to disaster endangers African American communities*, New York University Press, New York.

Calfas, J 2017, "California's drought is over. But this wildfire season will still be severe." *Time Magazine*, viewed 29 June 29, 2018, <http://time.com/4838989/california-drought-wildfire-fire-season/>.

Caniglia, BS & Frank, B 2017. "Revealing the resilience infrastructure of cities: Preventing environmental injustices-in-waiting", In BS Caniglia, M Vallee & B Frank (eds.), *Resilience, environmental justice & the city*, Routledge, Abingdon, UK, pp. 57–76.

Caniglia, BS, Vallee, M & Frank B 2017, *Resilience, environmental justice & the city*, Routledge, Abingdon, UK.

Čapek, S 1993, "The "environmental justice" frame: A conceptual discussion and an application", *Social Problems*, vol. 40, no. 1, pp. 5–24.

CAUSE 2017, Email message to supporters, December 28. Oxnard, California.

Ciplet, DJ, Roberts, T & Khan, MR 2015, *Power in a warming world: The new global politics of climate change and the remaking of environmental inequality*, The MIT Press, Cambridge, MA.

Cole, L & Foster, S 2000, *From the ground up: Environmental racism and the rise of the environmental justice movement*, New York University Press, New York.

Cole, R 2012a, "Regenerative design and development: Current theory and practice", *Building Research & Information*, vol. 40, no. 1, pp. 1–6.

Cole, R 2012b, "Transitioning from green to regenerative design", *Building Research & Information*, vol. 40, no. 1, pp. 39–53.

Denning, S 2005, *The leader's guide to storytelling: Mastering the art & discipline of business narrative*, Jossey-Bass, San Francisco, CA.

Donath, J 2018, "After the flood", viewed 29 June 2018, <https://groundswell.blog/2018/02/24/after-the-flood/>.

Downey, L 2015, *Inequality, democracy and the environment*, New York University Press, New York.

Ergas, C & York, R 2012, "Women's status and carbon dioxide emissions: A quantitative cross-national analysis", *Social Science Research*, vol. 41, no. 4, pp. 965–976.

Frampton, K 1983, "Towards a critical regionalism: Six points for an architecture of resistance", In Foster H. (ed.), *The anti-aesthetic: Essays on postmodern culture*, Bay Press, Port Townsen, WA.

Frankel, J 2017, "California fires: Meet the prisoner firefighters who are battling the flames in southern California", *Newsweek*, viewed 29 June 2018, <https://www.newsweek.com/california-fires-meet-prisoner-firefighters-who-are-battling-flames-southern-748618>.

Fullerton, J 2015, *Regenerative capitalism: How universal principles and patterns will shape our economy*, Capital Institute, Greenwich, CT.

Gabel, M 2015, "Regenerative development: Going beyond sustainability", *Kosmos Journal*, viewed 29 June 2018, <https://www.kosmosjournal.org/article/regenerative-development-going-beyond-sustainability/>.

Girardet, H 2014, *Creating regenerative cities*, Routledge, Abingdon, UK.

Gray-Fow, MG 1998, "Why the Christians? Nero and the great fire", *Latomus*, vol. 57, pp. 595–616.

Harrison, JL 2011, *Pesticide drift and the pursuit of environmental justice*, The MIT Press, Cambridge, MA.

Hemenway, T 2015, *The permaculture city: Regenerative design for urban, suburban, and town resilience*, Chelsea Green Publishing, Hartford, VT.

Hes, D & du Plessis, C 2014, *Designing for hope: Pathways to regenerative sustainability*, Routledge, Abingdon, UK.

Hoggan, J 2016, *I'm right and you're an idiot: The toxic state of public discourse and how to clean it up*, New Society Publishers, Gabriola Island, BC.

Hoxie, C, Berkebile R & Todd, JA 2012, "Stimulating regenerative development through community dialogue", *Building Research & Information*, vol. 40, no. 1, 65–80.

KCSB News 2017. "The social implications of California wildfires", December, viewed 29 June 2018, <https://soundcloud.com/kcsbfm/the-social-implications-of-california-wildfires>.

Korten, D 2007, *The great turning: From empire to earth community*, Kumarian, Bloomfield, CT.

Krone, C 1992, Notes from Singular Integrated Developmental Business Series, Session 2. Unpublished notes, Institute for Developmental Processes, Carmel, CA.

Livingston, M & Panzar J 2017, "Thomas fire becomes largest wildfire on record in California", *Los Angeles Times*, December 23, viewed June 29 2018, <http://www.latimes.com/local/lanow/la-me-thomas-fire-size-20171222-20171222-htmlstory.html>.

Low, S, Taplin, D & Scheld S 2003, *Rethinking urban parks: Public space and cultural diversity*, University of Texas, Austin, TX.

Lyle, JT 1994, *Regenerative design for sustainable development*, John Wiley & Sons, Hoboken, NJ.

Mang, P, Haggard B & Regenisis, 2016, *Regenerative development & design: A framework for evolving sustainably*, Wiley Books, Hoboken, NJ.

Mang, P & Reed, B 2012a, "Designing from place: A regenerative framework and methodology", *Building Research & Information*, vol. 40, no. 1, pp. 23–38.

Mang, P & Reed, B 2012b, "Regenerative development and design", In RA Meyers (ed.) *Encyclopedia of sustainability science and technology*, Springer, New York, NY, chapter 303.

McDonough, W & Braungart, M 2002, *Cradle to cradle: Remaking the way we make*, North Point, New York.

Mileti, D 1999, *Disasters by design: A reassessment of natural hazards in the United States*, Joseph Henry Press, Washington, DC, United States.

Nichols, C 2017, "Breitbart's False and inflammatory claim about deadly Wine Country fires", PolitiFact, October 20, PolitiFact.com.

PBS News Hour 2017, "Wind-driven fires force California evacuations, turn homes to ash", December 7, viewed 29 June 2018, <https://www.pbs.org/newshour/show/wind-driven-fires-force-california-evacuations-turn-homes-to-ash>.

Palazzo, D & Steiner FR 2011, *Urban ecological design: A process for regenerative places* (2nd ed.), Island Press, Washington, DC.

Park, LS & Pellow DN 2011, *The slums of aspen: The war on immigrants in America's Eden*, New York University Press, New York.

Pellow, DN 2009, "'We didn't get the first 500 years right, so let's work on the next 500 years': A call for transformative analysis and action." *Environmental Justice*, vol. 2, no. 1, pp. 3–8.

Repogle, J 2017, "2017 is California's worst year for wildfires on record", *KPCC Radio*, Southern California, December 6.

Romero, D 2017, "Global warming creates 'the worst of all possible worlds' for California fires", *LA Weekly*, October 19.

Schlosberg, D 2004, "Reconceiving environmental justice: Global movements and political theories", *Environmental Politics*, vol. 13, pp. 517–540.

Schlosberg, D 2007, *Defining environmental justice: Theories, movements, and nature*, Oxford University Press, Oxford, UK.

Tainter, J 2012, "Regenerative design in science and society", *Building Research & Information*, vol. 40, no. 3, pp. 369–372.

Taylor, D 2009, *The environment and the people in American cities, 1600s–1900s: Disorder, inequality, and social change*, Duke University Press, Durham, NC.
Taylor, D 2014, *Toxic communities: Environmental racism, industrial pollution, and residential mobility*, New York University Press, New York.
UCLA Institute on the Environment and Sustainability (IOES) 2017, "UCLA experts explain why California is burning in December", December 12, viewed 26 September 2018, <https://www.ioes.ucla.edu/article/ucla-experts-explain-california-burning-december/>.
Walker, G 2012, *Environmental justice: Concepts, evidence, and politics*, Routledge, Abingdon, UK.
Whyte, KP 2017, "The Dakota access pipeline, environmental injustice, and U.S. colonialism", *Red Ink*, vol. 19, no. 1, pp. 154–169.
Wood, P & Landry C 2008, *The intercultural city: Planning for diversity advantage*, Earthscan, New York.
World Commission on Environment and Development (WCED) 1987, *Our common future*, Oxford University Press, New York.
Young, IM 2001, "Activist challenges to deliberative democracy", *Political Theory*, vol. 29, no. 5, pp. 670–690.

6 Governing regenerative development

Thomas Dietz

Sustainable development and the newer idea of regenerative development call for corrections to the kinds of social and environmental transformation that have been occurring at least since World War II or even since the rise of colonialism, industrialization, and global capitalism in the 16th, 17th, and 18th centuries, respectively (Dietz 2015; Portney 2015; Worster 2016). But while there is a general sense that the current trajectory of coupled human and natural systems (CHANS) is highly problematic, the appropriate goals of regenerative development are less clear. The term regenerative suggests that we will "create again" (Brown 1993). But because human transformations of the biosphere are so long standing and pervasive and will continue for centuries, a program of regenerative development cannot be guided by a vision of restoration to a past state or even by the intent to conserve present conditions. Past and near-term human actions ensure decades to centuries of further anthropogenic global environmental change, including climate change, alteration of biogeochemical cycles, sharp changes in biodiversity, and widespread dispersal of persistent toxics. Further, the emerging complex of Nano, Information, Bio, and Cognitive technologies (NIBC) technologies is likely to be as transformative as the industrial revolution (Garden & Winickoff 2018; Righetti Madhavan & Chatila 2018; Schwab 2017; US National Research Council 2016; US National Research Council 2017).

It follows that a central problem, perhaps the central problem of regenerative development, is governance—finding effective and ethical ways of making decisions. There is a substantial literature on governance in environmental decision-making that can provide some ideas for establishing processes to guide regenerative development. Here, I will consider the nature of the governance problem in regenerative development, invoking the literature on analytic deliberative processes for adaptive risk governance. Analytic deliberative processes link scientific analysis with public deliberation; I will describe them below. Moving beyond that literature, I will suggest eight questions that should be considered in designing and implementing processes to govern regenerative development efforts.

Sustainability discourse as background

The discourse on sustainability emerged in the late 20th century in response to concerns about the effects of global transformations on human well-being and the environment, including the well-being of other species (Grober 2012; Portney 2015; Robinson 2004). The idea that resources should be preserved to ensure the well-being of future generations stretches back to the earliest treatises on natural resource management. The first legal statement of the ideas of sustainability in the United States may be in the National Park Service Enabling Act of 1916: "which purpose is to conserve the scenery and the natural and historic objects and the wild life therein and to provide for the enjoyment of the same in such manner and by such means as will leave them unimpaired for the enjoyment of future generations" (National Park Service Enabling Act of 1916 §1, 16 U.S.C. 1). The idea of using ecology to help plan the use of natural resources, taking into account future generations, appears in a UN Economic and Social Council resolution in 1959 (Macekura 2015, p. 76). The International Union for the Conservation of Nature (IUCN) (International Union for the Conservation of Nature 1980), under the leadership of Lee Talbot, used the term sustainable development and brought the concept into policy circles. But the popularity of sustainable development as a concept can be traced to the famed Bruntland report, *Our Common Future* (World Commission on Environment and Development 1987).

The modern discourse of sustainability can reasonably be traced to the confluence of two concerns that emerged in large part from settler colonialism. One was with the transformation of ecosystems and, in particular, threats to ecosystems that were the home to charismatic megafauna. The other was a concern with economic growth in poor areas of the formerly colonized parts of the world, motivated in part by the political unrest that poverty and inequality entrained. In its prescient report, the IUCN suggested that concern with development and with conservation must be combined into an integrated approach and that scientific analysis was a necessary, if not sufficient, underpinning for what would come to be called sustainable development (International Union for the Conservation of Nature 1980).

While the needs of future generations have been the most common ways of defining sustainability, I believe it is more useful to consider the two streams of concern that have formed the normative basis of sustainability from the IUCN report. One is that we want to improve human well-being. The other is that we want to reduce human impact on the biosphere and especially on other species. This is the logic that underpins the Millennium Ecosystem Assessment and subsequent work (Alcalmo et al. 2003; Reid et al. 2005; Yang et al. 2013a; Yang et al. 2013b). These two objectives permeate the UN Sustainable Development Goals (http://www.un.org/sustainabledevelopment/sustainable-development-goals/). It is from this context that the ideas of regenerative development have emerged.

Regenerative development discourse

The core ideas behind regenerative development seem to be (see Chapter 1 for references):

- Catalyzing increased prosperity and health of human and natural environments through holistic design and meaningful community participation
- Fostering positive feedback loops where excess human and natural resources are reabsorbed by the system to create mutually beneficial relationships that self-replicate to build inclusive resilience
- Having respect and deep consideration to local contexts, whether economic, cultural, or ecological, so that development is properly adapted to local ecosystem and cultural and economic circumstances

This framing is largely consistent with the concerns that drove the emergence of sustainability arguments: the need to improve human well-being while reducing impact on the environment and other species. Key elements of these core ideas were already present in environmental discourse in the 1970s. The idea of holistic design and system analysis that take account of feedback loops was a key theme in the structural functional ecosystem theory of the 1960s, as well as in the sociological functionalism of that period (McLaughlin 2012; Odum 1969; Parsons 1966; Richerson 1977). For example, the idea of a "wicked problem" seems to emerge from frustration in trying to apply holistic thinking and systems analysis to complex adaptive systems such as human influenced ecosystems (Churchman 1967; Rittel & Webber 1973). The focus on the local was strongly present in the appropriate technology movement and in the idea that "small is beautiful" (Dickson 1975; Pursell 1993; Schnaiberg 1982; Schnaiberg 1983; Schumacher 1973). All this is not surprising since the ideas of regenerative development seem to have emerged from exactly this context in the late 1970s.

How then does regenerative development differ from the idea of sustainability? Clearly the difference is a matter of emphasis since the narrative of regenerative development is largely consistent with most narratives of sustainability. But I see two areas where the regenerative development literature seems to differ somewhat from most sustainability discussions: one in terms of spatial scope and the other in terms of temporal scope. Of course not all discussions of regenerative development differ in spatial and temporal scope from sustainability discussions, so my arguments are most salient where there are such distinctions.

A common, albeit not universal, regenerative development trope is the strong emphasis on the local, with the implication that larger goals can be achieved by local processes. A local focus is important for several reasons. First, the substantial literature on commons has noted that it seems easier to develop sustainable approaches to resource management at the local to regional level (Ostrom et al. 2002). Second, while scientific expertise is essential for effective decision-making, the science has to be applied to specific contexts. That usually increases uncertainty and requires the engagement of a diverse set of experts, including experts in indigenous and local knowledge (ILK) (Dietz 2013a; Nelson & Shilling 2018; Whyte 2013; Whyte, Brewer II & Johnson 2015). Third, research on deliberative processes and consensus formation has provided an array of understandings and tools that allow for effective deliberative processes, but most of that knowledge is from experience at the local to regional level (US National Research Council 2005). Fourth, local efforts in regenerative development provide a sort of Jeffersonian laboratory for experiments on how best to reduce environmental stress and increase human well-being, providing variation in approaches that can facilitate social learning (Henry 2009; Vandenbergh & Gilligan 2017: Chapter 3).

However, the local focus in most regenerative development efforts has to be tempered with explicit concerns with the regional and global scale. First, all CHANS are teleconnected via ecological and human processes—climate change and global trade are the obvious examples (Liu et al. 2013). (The term "teleconnected" is used to span the use of "telecoupled" in the physical sciences and "interconnected" and "world system" in the social sciences.) In many parts of the world, the rationale for limiting greenhouse gas emissions or creating net greenhouse gas sinks rests not in local impacts of climate change but in the argument for making a contribution to the solution of a global problem. Second, many technologies, forms of expertise, and materials cannot be sourced locally; rather, locally designed solutions must often draw on external resources. For example, solar photovoltaics, high-efficiency windmills, and advanced battery storage will often prove a useful approach to providing energy at the local level, but the technology, expertise, and raw materials needed will usually not be available locally. And even where expertise and technology is available locally, some attention should be given to comparative advantage that may, in some circumstances, make it desirable to import what is needed. Third, local solutions must always be attentive to cumulative effects. Indeed, one of the major conceptual breakthroughs in environmental policy in the 1970s was the growing acknowledgment of cumulative effects—the tragedy of the commons (Dietz, Ostrom & Stern 2003). Actions in one location may have minimal adverse impacts in that they don't overtax the resilience of an ecosystem. But replication of the same seemingly innocuous action in many places can lead to catastrophic effects. Finally, there is the question of whether local efforts can "scale" to transform a global industrial capitalist economy that seems strongly committed to conventional economic growth. Whether or not local efforts can aggregate in both their reductions in stress placed on the environment and as examples that allow political and cultural mobilization is a key theoretical challenge for locally focused efforts such as regenerative development. Thoughtful arguments have been deployed on the issues of scaling up reforms versus (or in addition to) systems transformation (Anderson 1976; Mol, Spaargaren & Sonnenfeld 2014; Schnaiberg 1980; Vandenbergh & Gilligan 2017; York, Rosa & Dietz 2010; York 2012). The regenerative development discourse would benefit from a thoughtful reading of them.

In addition to the spatial challenge of balancing the local and the global, governance of regenerative development involves a temporal challenge. The history of human transformation of CHANS makes clear that the goals for regenerative development cannot meaningfully reference past conditions. Global environmental change moots any simple notion of preservation even in protected areas (Dietz 2017). At least since the rise of agriculture, humans have been active agents of ecosystem change and have coevolved with many animal and plant species. Further, large-scale human migrations have altered ecosystems and landscapes in ways that will play out for centuries, and climate change induced now will play out for millennia (Solomon et al. 2009). Massive technological change will be occurring at the same time. Regenerative development requires decisions not only about how to design systems but also about the goals for those systems. And design and goal setting must acknowledge uncertainty about the future and the need to shift approaches and goals over time.

In an insightful essay comparing "regenerative sustainability" and "regenerative development and design," Robinson and Cole note that the goal of regenerative approaches is to formulate "approaches for successful co-evolution of human and natural systems" (Robinson & Cole 2015, p. 133). They emphasize in particular that the concept of "regenerative sustainability" is grounded in procedural sustainability, and that: "regenerative development ... precludes predetermined outcomes while regenerative sustainability suggests that it precludes predetermined goals as well" (Robinson & Cole 2015, p. 135). They note that "sustainability cannot be defined scientifically or in absolute terms but finds different expression in different times and places. In turn this leads to the view that sustainability can usefully be thought of in procedural terms as the emergent property of a conversation" (Robinson & Cole 2015, p. 135). (For simplicity, I will continue to use the term "regenerative development" although I use the conceptualization that Robinson and Cole refer to as "regenerative sustainability.")

I want to follow their lead in arguing that some of the key challenges for regenerative development are in establishing an adequate ethical framework for setting and modifying goals. Like them, I argue that a deliberative approach is the most appropriate ethical frame for governance of regenerative development. The idea of "meaningful community participation" in regenerative development seems to call for a deliberative approach. But I will also suggest that, especially given the tendency of the regenerative development literature to emphasize the local, some key questions must be addressed in deliberative processes intended to guide regenerative development.

The normative bases for governing regenerative development

Projects to advance regenerative development or even simply "traditional" sustainability rest on both positive (scientific) analysis and on normative (ethical) analysis (Dietz 2013b). That is, decisions require an assessment of the facts and a consideration of what is ethically appropriate, values. Ethical analysis is essential because there are nearly always tradeoffs between desirable goals. For example, arguing that governments and private sector organizations ought to adhere to a "triple bottom line" of economic, community, and environmental goals is admirable but tells us nothing about how to make tradeoffs across the three bottom lines when conflicts arise, as they often do (Norman & MacDonald 2004; Pava 2007). More technical approaches such as strong and weak sustainability involve both scientific and ethical arguments (Neumayer 2010). Unfortunately, aside from scholars who write specifically about the ethics of sustainability (e.g., Norton 2005; Thompson 2010), many scholars and practitioners engaged in the sustainability discourse are mute about the ethical theories that underpin their views. If the regenerative development program is to make progress, it would be useful to acknowledge that there are a variety of defensible ethical theories that can be brought to bear and to explore and perhaps even advocate for appropriate ethical frames.

A number of ethical traditions, including utilitarianism and libertarianism, are often deployed in policy debates (Dietz 2003; Raffaelle, R, Wade, R & Evan, S (eds.) 2010; Thompson 2010). But, like Robinson and Cole, many who have worked on environmental and sustainability problems have argued for an approach that links scientific analysis to public deliberation. The tradition stretches back to Dewey (Dewey 1923) and was advanced by Habermas (Habermas 1970; Habermas 1990; Habermas 1993). It has been proposed as a basis for environmental and sustainability decisions for more than 30 years (Brulle 1993; Dietz 1984; Dietz 1987; Dryzek 1987; Renn, Webler & Wiedemann 1995; Tuler & Webler 1995; Webler 1993). The core argument is that decisions should be based on fair and competent deliberation by the public, where the public is defined in Dewey's (1923) terms as interested or affected parties. But starting with Dewey's work, there is wide acknowledgment that while public engagement is essential, there is also a need to engage a variety of forms of expertise in the process of designing and evaluating sustainability, and by implication, regenerative development, efforts (Dietz 2013b; Dietz 2017; Renn 2017; Rosa, Renn & McCright 2013).

Regenerative development projects can build on a substantial body of experience and literature in designing processes that link diverse forms of expertise with public deliberation. In the United States, versions of this approach, linking scientific analysis to public deliberation in an iterative process, have long been advocated by the US National Academies of Science as a way of approach complex public policy problems (U.S. National Research Council 1996; U.S. National Research Council 2008). Figure 6.1 provides a schematic of a linked process of analysis and deliberation. The core idea is that interested and affected parties interact with each other and with those conducting scientific analysis in an ongoing discourse. The process informs the analysis being done while the analysis informs the deliberation. The overall process is co-designed by the participants. When applied to governance, depending on legal mandates, the process either advises the decision makers or is empowered to make governance decisions. The International Risk Governance Council has called for a similar approach (International Risk Governance Council 2005). It will be useful to note a few features of the approach now, and I will return to them next in suggesting key questions for governing regenerative development.

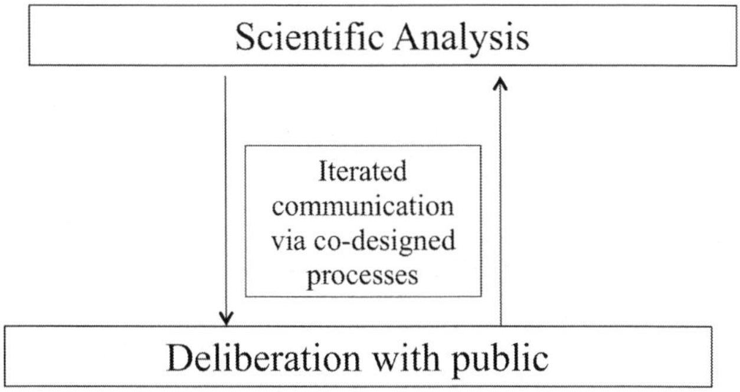

Figure 6.1 Schematic of a linked process of analysis and deliberation

First, the process of linking scientific analysis and public deliberation is a method of adaptive risk governance (Renn 2017; U.S. National Research Council 2010). Adaptive risk governance embraces uncertainty and social learning as part of governance. Since decisions are inherently about the future, uncertainty is inevitable and so decision-making is best conceptualized as dealing with risk. Further, if projects and programs are thoughtfully designed, they will allow for learning from experience; thus, risk governance can adapt to new circumstances and new understandings over time. But such adaptive learning will occur only if the projects are designed to ensure learning from experience and incorporate ways of modifying actions to build on that learning. Governance is an ongoing and active process.

Second, multiple forms of expertise are essential for effective governance. As Rosa has articulated, when science needs to be applied to local contexts, uncertainty multiplies, and unless that uncertainty is acknowledged and engaged, decision can be ill-calibrated and mistrust can emerge (Rosa 1998). Analytic deliberative processes emerged from concern with how best to reach decisions about environmental health and safety risks and ecosystem management in circumstances where more technocratic approaches were problematic (U.S. National Research Council 1996; U.S. National Research Council 1999). In particular, bringing expertise in ILK into deliberation is a key way of both building trust and ensuring that analysis will be useful in the local context (Nelson & Shilling 2018).

Third, there is substantial literature examining obstacles to deliberative processes and evaluating what does and does not work in addressing those challenges. A comprehensive review by the US National Academies suggests diagnostic questions and design principles as guidance in structuring analytic deliberative processes that can be adapted to processes for designing and governing regenerative development (U.S. National Research Council 2008). In Tables 6.1 and 6.2, I have revised the diagnostic questions and design principles for thinking about regenerative development efforts. Note that the tables are structured around three roles in the process: those who conduct analysis and design in support of

Table 6.1 Diagnostic questions to assess challenges to deliberative governance

Questions about scientific context

1 What information is currently available on the issues? How adequate is available information for giving a clear understanding of the problem? Do the various parties agree about the adequacy of the information?
2 Is the uncertainty associated with the information well characterized, interpretable, and capable of being incorporated into the assessment or decision?
3 Is the information accessible to and interpretable by interested and affected parties?
4 Is the information trustworthy?

Questions about convening and implementing agencies

1 Where is the decision-making authority? Who would implement any agreements reached? Are there multiple forums in which the issues are being or could be debated and decided?
2 Are there legal or regulatory mandates or constraints on the convening agency? What laws or policies need to be considered?

Questions about the abilities of and constraints on the participants

1 Are there interested and affected parties who may have difficulty being adequately represented?
   a What does the scale of the problem, especially its geographic scale, imply for the range of affected parties?
   b Are there disparities in the attributes of individual potential participants that may affect the likelihood of participation?
   c Are there interests that are diffused, unorganized, or difficult to reach?
   d Are there disparities across groups of participants in terms of their financial, technical, or other resources that may influence participation?
2 What are the differences in values, interests, cultural views, and perspectives among the parties? Are the participants polarized on the issues?
3 Are there substantial disparities across participant groups in their power to influence the process?
4 To what degree can the individuals at the table act for the parties they are assumed to represent?
5 Are there significant problems of trust among the agency, the scientists, and the interested and affected parties?
   a Are there indications that some participants are likely to proceed insincerely or to breach the rules of the process?
   b Are some participants concerned that the convening agency will proceed in bad faith?
   c Do some participants view the scientists as partisan advocates and so mistrust them?

the regenerative development project, those with the authority to make decisions about the project, and the public of interested and affected parties. Who takes on these roles will differ from project to project, so the diagnostic questions and design principles have to be altered to match the context. Designing a governance process should start by diagnosing the problems being addressed and the

Table 6.2 Design principles for governing regenerative development

1 Agencies should proceed with:

- i Clarity of purpose
- ii Commitment to use the process to inform actions
- iii Adequate funding and staff
- iv Appropriate timing in relation to decisions
- v Focus on implementation
- vi Commitment to self-assessment and learning from experience

2 The process must be:

- i Inclusive
- ii Collaborative in problem formulation and process design
- iii Transparent
- iv Based on good-faith communication
- v Find ways to compensate for power differentials across participants

3 The process must attend to uncertainty by:

- i Ensuring transparency of decision-relevant information and analysis
- ii Paying explicit attention to both facts and values
- iii Promoting explicitness about assumptions and uncertainties
- iv Including independent review of official analysis and/or engaging in a process of collaborative inquiry with interested and affected parties
- v Allowing for iteration to reconsider past conclusions on the basis of new information

context in which decisions will be made and actions taken. An adequate diagnosis can then be used to deploy the design principles in a process that is responsive to the context of a particular regenerative development effort.

Finally, our best understandings of analytic deliberative processes as a method of adaptive risk governance are calibrated at the local to regional level. This of course matches the emphasis of regenerative development efforts on the local scale. But as I have noted, consideration of the local is necessary but not sufficient for enhancing the well-being of humans and other species and protecting the environment. And some issues, such as climate change and existential risks to human civilization clearly require some form of global deliberation for decision-making. Incorporating the global scale remains a major challenge for the deliberative approach although recent literature is tackling this difficult issue (Gunderson 2016; Gunderson and Dietz 2018; Payne, Shwom & Heaton 2015; Shwom & Kopp 2016).

Key Questions for the Governance of Regenerative Development

A reasonable criticism of deliberative ethics is that it stops with a call for a process and is silent on what issues should be given attention in the process. It depends wholly on those participating to identify what issues should be discussed. While I endorse the idea that norms and goals have to emerge from the process, there is always the danger that a deliberative process becomes focused on only a subset of perspectives. Thus it may be useful to offer some substantive criteria for adaptive risk governance in addition to the process principles (Table 6.2). I offer a list of eight criteria, explored through questions below, derived from earlier work on what constitutes a good decision (Dietz 2003). These are issues that should be raised periodically in any ongoing regenerative development effort. Some reenforce concerns about process; some are perspectives that should be considered even if in the end the deliberation and analysis decide to give them little weight. My argument is that it is important to "seed" the discussion with some considerations that probably deserve attention in an effort to develop goals and strategies for regenerative development.

Criterion 1: How will we define the well-being of humans, other species, and the environment?

Regenerative development is about "increased prosperity and health of human and natural environments" (*This volume, Chapter 2, 8 and 14*). A first step is to achieve some understanding of what the interested and affected parties feel should be included in the ideas of prosperity and health—what I would term well-being—for both humans and the environment. For projects that are primarily local in their impacts, the interested and affected parties need to discuss and define what they consider to be human well-being within their communities. It may be useful to draw on a growing literature on how to assess the well-being of humans and incorporate those assessments into policy debates (Fox & Erickson 2018; Kahneman & Krueger 2006; OECD 2013; Stiglitz, Sen & Fitoussi 2009). Many nations and some US states are experimenting with measures of human well-being as supplements to the economic accounts, such as gross domestic product per capita, that are routinely invoked in policy discussions (Durand 2015; Krueger & Stone 2014). The problem of how to conceptualize the well-being of the environment and other species is also key. I will return to it below. And as Question 4 emphasizes, well-being will have to be defined in the face of differences in values across interested and affected parties and even uncertainty about what values to deploy in making decisions.

Criterion 2: Are we allocating resources efficiently?

Concerns with resource constraints drove much of the discourse that underpinned the rise of the 1970s environmental movement and the sustainability discourse (Worster 2016). Thus there is a common ground between sustainability and standard economic policy analysis—both are concerned with finite resources and how to best deploy them. Economic policy analyses are about efficiency—finding the most output per unit input. While efficiency analyses will never suffice as a basis for governing regenerative development projects, neither should efficient resource allocation be ignored. For example, while regenerative development efforts emphasize doing things locally, as noted above, there are times when comparative advantage suggests that it may be desirable to import resources, technology, or expertise. The notion of "reabsorbing" "excess human and natural resources" touches the issue of efficiency. Environmental economics has developed a variety of tools, especially risk-benefit-cost analysis, to aid in such analyses (Atkinson & Mourato 2008; Brent 2017). Of course, there are many concerns with these formal approaches (Fischhoff 2015): How uncertainty is handled, the problem of discounting future costs and benefits, and the lack of attention to equity issues, among others. There is also a tendency for "bad numbers to drive out good paragraphs"—leaving out of the formal analysis what is hard to quantify, including many ecosystem services and the value of existing cultural forms. But however skeptical a deliberative process might be of a formal quantitative efficiency analysis, it seems prudent to always ask how resources might be better allocated to produce more well-being and less harm to the environment.

Criterion 3: Is the analysis taking adequate account of uncertainty about facts?

Science is always uncertain. When we have to take scientific understanding developed as generalizable abstractions and apply it to a particular geographic and substantive context, uncertainty almost always increases substantially (Dietz 2013a; Rosa 1998). The need to model or otherwise project future consequences also increases uncertainty (York 2013). So, regenerative sustainability projects will have to acknowledge that the information available for decisions is uncertain.

Unfortunately, humans have a great deal of difficulty processing uncertainty (Kahneman, Slovic & Tversky 1982; Marx & Weber 2011). Care must be taken in dealing with uncertainty in both formal analysis and in discussion. There are processes for analyzing uncertainty and for working with interested and affected parties to help them cope with uncertainty (Arvai et al. 2012; Bessette, Arvai & Campbell-Arvai 2014; Renn & Schweitzer 2009; Renn & Klinke 2015). These can help compensate for the challenges humans have in processing uncertainty and can be built into the governance process for regenerative development. Viewing regenerative development projects as adaptive risk governance emphasizes that decisions are being made in the face of uncertainty and may have to be revisited. (See Criterion 8.)

Criterion 4: Are we identifying and acknowledging value uncertainty and value conflict?

Many sustainability and development processes raise new challenges that may be difficult to map onto existing values. For example, climate change impacts mean that many protected ecosystems, such as national parks, will undergo radical transformations unless managers make substantial interventions, but such interventions contradict the long-standing ethos of preserving these areas with minimal human interference (Dietz 2017).

Even when the value implications are clear, people will differ in both their interests and in the values they hold. In some cases, consensus statements, such as the UN Sustainable Development Goals, might form the starting point for identifying values, but even then it will often be the case that the goals will be in conflict (http://www.un.org/sustainabledevelopment/sustainable-development-goals/). What might enhance progress toward one goal will neglect or even retard progress toward another goal. Further, the implications of global environmental change, changing culture, and changing technology present challenges to existing value stances (Garden & Winickoff 2018; Righetti, Madhavan & Chatila 2018; Schwab 2017; U.S. National Research Council 2016; U.S. National Research Council 2017). With rapidly improving capabilities in NIBC technologies, regenerative development projects will have an array of strategies and approaches unprecedented in human history. It will not be obvious how to map them onto existing values.

Governing regenerative development projects will require dealing with the traditional problem of conflicting values (Dietz 2001). For example, a shift to renewable energy can reduce greenhouse gas emissions but may raise energy prices especially for low income households that can least afford them. But governance will also have to acknowledge that many interested and affected parties will have to struggle with the implications of future courses of action for their values. Of course, one strong argument for deliberative approaches is that such communicative engagement can allow participants to better assess how their values apply to the choices being faced. Deliberation facilitates the evolution of values and norms. So, as Robinson and Cole emphasize, goals themselves must evolve in regenerative development.

Criterion 5: Are we engaging the full range of useful expertise?

The complexity of increasing well-being while reducing harm to the environment in the context of an evolving CHANS inevitably requires diverse forms of expertise (Dietz 2013a; Dietz 2013b). These include:

- Scientific expertise on the substantive problems being considered
- Scientific expertise on values and beliefs of the interested and effected parties and on processes for deliberating in the face of conflict and uncertainty
- Local expertise that engages ILK, including those who can articulate the values of the local communities
- Political expertise on the institutional and policy context in which actions will have to be taken

As Harding (2015) has noted, such diversity in perspective helps enhance our factual understanding. Of course, effective communication across these different perspectives and cultures requires design of a process that encourages fair and competent communication and encourages synthesis and consensus. Diversity in expertise is also important when addressing uncertainty and diversity in values. Everyone interested in or impacted by a decision has a legitimate right to have their values considered in decision-making. As noted, there will often be conflicts in values, and for many projects the implications for values are uncertain. An effective discourse about values cannot make conflicts disappear, but can help build mutual understanding, can help clarify value positions, and can indicate modifications in plans that do a better job of aligning with the diverse values of the public. Designing processes to do this effectively may be substantially aided by bringing another form of expertise to the process—expertise about decision-making in the face of uncertainty and conflict about facts and values.

Criterion 6: Are both the process of governance and the outcome fair to all?

Power differentials have been a part of human society at least since the rise of agriculture (Fukuyama 2011; Mann 1986). There will be substantial power differential among those who are interested in and/or affected by a regenerative development project. These power differentials can lead to inequities in both the process of governing regenerative development projects and in their outcomes. A common concern for deliberative processes is that they must be attentive to power differentials in the deliberative processes itself, a point noted in the Design Principles. Centuries of settler colonialism, discrimination based on gender, sexual identity nd class,, and other manifestation of power cannot be swept away in a deliberative process. But acknowledging these problems can allow for process design that works to ensure that all voices can be heard in a way that respects differences in perspectives and allows this diversity to influence decisions. We are still learning how to do this effectively. But emerging literature is engaging the problem, including the crucial issue of how to effectively incorporate both "Western" science and ILK into governance discussions (Nelson & Shilling 2018; Reo 2011; Whyte 2012; Whyte 2013; Whyte, Brewer II & Johnson 2015).

Fairness in process can be a major guarantor of fair outcomes and is one of the rationales for arguing that many forms of expertise are needed in the deliberation. But questions about the differential impacts of a regenerative development project must always be asked, giving special attention to the multiple ways in which outcomes can be unfair. There can be unfair distribution of costs, benefits, and risks across traditional lines of social stratification such as race, ethnicity, class, sexual identity, and gender. There also can be unfair distributions across space and time: A key argument in the Bruntland definition of sustainability is that we must guard against being unfair to future generations.

The emphasis on local solutions in some regenerative development narratives requires special attention to extra-local impacts and processes, as I have noted earlier. In some cases, it may be sufficient to raise the issue of extra-local consequences and the potential for the tragedy of the commons as local actions accumulate. But to ensure that such perspectives are engaged, it will often be useful to have some members of the governance discourse who are charged with representing a regional or global perspective. In support of these concerns, analyses of global and other large scale actions must consider not only global but also local and regional consequences of decisions.

Another voice almost certain to be absent for the deliberative process is that of non-human animals impacted by the decisions made. One might interpret the concern with the environment in sustainability from a purely anthropocentric perspective—we care about the environment only to the extent that the state of the environment influences human well-being. But most accounts of sustainability, including for example the Millennium Ecosystem Assessment, acknowledge the importance of the non-human even if we can assess that importance only from a human perspective (Reid et al. 2005). This is a challenge for the deliberative approach—Who will speak for wolf? As with extra-local interests, the best available solution seems to be processes that explicitly ask some participants to represent the interests of other species and biosphere itself (Eckersley 1999; Eckersley 2000; Gardner 2016a,b).

Criterion 7: Does the process build on human cognitive strengths and compensate for weaknesses?

Human cultural evolution has made us very skilled at pattern recognition, language processing, and communicating in small groups. It has left us less skilled at handling the kinds of uncertainties and nonlinear dynamics that characterize contemporary complex adaptive systems. In addition, we are subject to homophily—associating with those who resemble us and avoiding those that do not (Gerber, Henry & Lubell 2013; Henry, Prałat & Zhang 2011; Henry & Vollan 2014). Homophily in turn exacerbates motivated reasoning and biased assimilation—we take up information consistent with what we already believe and are skeptical of or ignore contradictory information (Bolsen, Druckman & Cook 2014; Druckman, Leeper & Slothuus 2018; Hoggan & Litwin 2016; Lord, Ross & Lepper 1979).

The scientific method is a process by which a community of researchers, via public deliberation over time, tries to overcome these biases to develop better understandings of facts. The process is never infallible, but it is self-correcting. Often, strong consensus emerges from this process, although some uncertainty always remains. Such a consensus, when treated as "facts" but with an acknowledgment of uncertainty, is useful guidance in decision-making. One rationale for linking scientific analysis with public deliberation is that it helps temper the scientific discourse with other perspectives that help "get the science right" for local contexts and also helps "get the right science" to ensure that the concerns of interested and affected parties are being supported with analysis. In a sense, analytic deliberative processes are expanding the discourse that underpins the scientific method to ensure analysis that is broadly informed and while providing a way to incorporate the analysis into the broader deliberation.

This in turn requires a process that takes account of our cognitive strengths and weaknesses. Deliberative processes can benefit from careful guidance and structured tools to deal with uncertainty in both current factual understandings and our projections of what will happen under different projects for the future (Arvai, Gregory & McDaniels 2001; Bessette, Arvai & Campbell-Arvai 2014;

Bessette, Campbell-Arvai & Arvai 2015; Florig et al. 2001; Morgan et al. 2001; Moss 2016). The processes also have to be designed so that the deliberation overcomes both preexisting cliques and biases and avoids the formation of such cliques and biases in the process itself.

Criterion 8: Is the project designed to learn from experience?

The idea of *adaptive* risk governance emphasizes that uncertainty about the future will require that we alter regenerative development projects as they move forward. To do that in ways that enhance human well-being and protect the environment means that we have to learn from experience (Henry 2009). To learn from experience requires more than simply collecting data on the project over time. Such information is essential, of course. But data collection has to be designed in a way that allows for attribution of effects to the regenerative development efforts. That is, we have to assess what has caused observed changes in CHANS, including assessing the impact of the regenerative development efforts. To know if a regenerative development effort brought about change means we have to have information on both the system that was the target of the effort and on one or more systems that were not part of the effort. That requires either a formal experiment with randomization or utilizing "natural" experiments, including pre- and post-project measurements and data on systems not influenced by the regenerative development effort. Happily, we know a great deal about how to assess causality in both experimental and non-experimental situations (Campbell 1969; Campbell 1998; Frank et al. 2013; Morgan & Winship 2014; Pearl & Mackenzie 2018). But without attention to proper design of data collection, the data need for social learning from experience will simply not be available.

Governing regenerative development

As the essays in this volume make clear, the idea of regenerative development is producing a great deal of creative thought about how to address the problems of protecting the environment while enhancing human well-being. My argument here is that a central problem for effective regenerative development projects will be developing an ethical basis for governing them. The core arguments around regenerative development invoke a deliberative governance approach. The theoretical bases for deliberative governance are well developed, and there is strong body of evidence about the use of deliberative processes in environmental assessment and decision-making. In particular, processes that link scientific analysis with deliberation among interested and affected parties seems well suited to governing regenerative development projects. Thus key findings from the literature on deliberation, summarized in Tables 6.1 and 6.2, may be of help in designing governance for regenerative development.

However, deliberative processes can be a "black box" with the contents left to those deliberating. At some level, ceding substantial power to deliberators is the essence of deliberative ethics. But I suggest that there are some issues that should always be part of the discussion when deliberative processes are used to govern regenerative development projects. I offer eight such questions.

The utility of both the existing literature on analytic deliberative processes and the specific questions that I suggest will of course have to be tested in practice. The diagnostic questions of Table 6.1, the design principles of Table 6.2, and the use of the eight questions must be treated as hypotheses about how to effectively govern regenerative development. It is clear that the success of specific regenerative development projects, and of the idea of regenerative development, will depend crucially on careful thinking about how to effectively govern such efforts. The diagnostic questions and design principles can be thought of as hypotheses that can be tested in application, and thus help guide that thinking.

References

Alcalmo, J, Ash, NJ, Butler, CD, Callicot, JB, Capistrano, D, Carpenter, SR, Castilla, JC, Chambers, R, Chopra, K, Cropper, A, Daily, GC, Dasgupta, P, de Groot, R, Dietz, T, Duraiappah, AK, Gadgil, M, Hamiltion, K, Hassan, R, Lambin, EF, Lebel, L, Leemans, R, Jiyuan, L, Malingreau, JP, May, RM, McCalla, AF, McMichael, AJ, Moldan, B, Mooney, H, Naseem, S, Nelson, GC, Wen-Yuan, N, Noble, I, Zhiyun, O, Pagiola, S, Pauly, D, Percy, S, Pingali, P, Prescott-Allen, R, Reid, WV, Rickets, TH, Samper, C, Scholes, R, Simons, H, Toth, FL Turpie, JK, Watson, RT, Wilbanks, TJ, Williams, M, Wood, S, Shidong, Z & Zurek, MB 2003, *Ecosystems and human well-being: A framework for analysis*, Island Press, Washington, DC.

Anderson, CH 1976, *The sociology of survival: Social problems of growth*, Dorsey Press, Homewood, Illinois.

Arvai, J, Gregory, R, Bessette, D & Campbell-Arvai, V 2012, "Decision support for developing energy strategies", *Issues in Science and Technology*, vol. 28, no. 4, pp. 43–52.

Arvai, J, Gregory, R & McDaniels, T 2001, "Testing a structured decision approach: Value-focused thinking for deliberative risk communication", *Risk Analysis*, vol. 21, no. 6, pp. 1065–1076.

Atkinson, G & Mourato, S 2008, "Environmental cost-benefit analysis", *Annual Review of Environment and Resources*, vol. 33, pp. 317–344.

Bessette, DL, Arvai, J & Campbell-Arvai, V 2014, "Decision support framework for developing regional energy strategies", *Environmental Science & Technology*, vol. 48, no. 3, pp. 1401–1408.

Bessette, DL, Campbell-Arvai, V & Arvai, J 2015, "Expanding the reach of participatory risk management: Testing an online decision-aiding framework for informing internally consistent choices", *Risk Analysis*, vol. 36, no. 5, pp. 992–1005.

Bolsen, T, Druckman, JN & Cook, FL 2014, "The influence of partisan motivated reasoning on public opinion", *Political Behavior*, vol. 36, no. 2, pp. 235–262.

Brent, RJ 2017, *Advanced introduction to cost–benefit analysis*, Edward Elgar Publishing, Cheltenham UK.

Brown, L 1993, *The new shorter Oxford English Dictionary on historical principles*, Clarendon Press, Oxford.

Brulle, RJ 1993, "Environmentalism and human emancipation", In SD Wright, T Dietz, R Borden, G Young & G Guagnano (eds.), *Human ecology: Crossing boundaries*, Society for Human Ecology, Ft. Collins, CO, pp. 2–12.

Campbell, DT 1969, "Reforms as experiments", *American Psychologist*, vol. 24, pp. 409–429.

Campbell, DT 1998, "The experimenting society", *The Experimenting Society: Essays in Honour of Donald T. Campbell*, vol. 11, pp. 35–68.

Churchman, CW 1967, "Guest editorial: Wicked problems", *Management Science*, vol. 14, no. 4, pp. 141–142.

Dewey, J 1923, *The public and its problems*, Henry Holt, New York.

Dickson, D 1975, *The politics of alternative technology*, Universe Books, New York.

Dietz, T 1984, "Social impact assessment as a tool for rangelands management", In National Research Council (ed.), *Developing strategies for rangelands management*, Westview, CO, pp. 1613–1634.

Dietz, T 1987, "Theory and method in social impact assessment", *Sociological Inquiry*, vol. 57, pp. 54–69.

Dietz, T 2001, "Thinking about environmental conflict", In L Kadous (ed.), *Celebrating scholarship*, George Mason University, Fairfax, VA, pp. 31–54.

Dietz, T 2003, "What is a good decision? Criteria for environmental decision making", *Human Ecology Review*, vol. 10, no. 1, pp. 60–67.

Dietz, T 2013a, "Epistemology, ontology, and the practice of structural human ecology", In T Dietz & AK Jorgenson (eds.), *Structural human ecology: Essays in risk, energy, and sustainability*, WSU Press, Pullman, WA, pp. 31–52.

Dietz, T 2013b, "Bringing values and deliberation to science communication", *Proceedings of the National Academy of Sciences*, vol. 110, no. 10, pp. 14081–14087.

Dietz, T 2015, "Prolegomenon a structural human ecology of human well-being", *Sociology of Development*, vol. 1, no. 1, pp. 123–148.

Dietz, T 2017, "Science, values, and conflict in the national parks", In SR Beissinger, DB Ackerly, H Doremus & GE Machlis (eds.), *Science, conservation, and national parks*, University of Chicago Press, Chicago, IL, pp. 247–274.

Dietz, T, Ostrom, E & Stern, PC 2003, "The struggle to govern the commons", *Science*, vol. 301, no. 5652, pp. 1907–1912.

Druckman, JN, Leeper, TJ & Slothuus, R 2018, "Motivated responses to political communications", In H Lavine & CS Taber (eds.), *The feeling, thinking citizen: Essays in honor of Milton Lodge*, Routledge, New York, pp. 125–150.

Dryzek, JS 1987, *Rational ecology: Environment and political economy*, Oxford University Press, Oxford, England.

Durand, M 2015, "The OECD better life initiative: How's Life? and the measurement of well-being", *Review of Income and Wealth*, vol. 61, no. 1, pp. 4–17.

Eckersley, R 1999, "The discourse ethic and the problem of representing nature", *Environmental Politics*, vol. 8, no. 2, pp. 24–49.

Eckersley, R 2000, "Deliberative democracy, ecological representation and risk: Towards a democracy of the affected", In M. Saward (Ed.), *Democratic innovation: Deliberation, representation and association* (pp. 117–132). London: Routledge.

Fischhoff, B 2015, "The realities of risk-cost-benefit analysis", *Science*, vol. 350, no. 6260, aaa6516.

Florig, HK, Granger Morgan, M, Jenni, KE, Fischoff, B, Fischbeck, PS & DeKay, ML 2001, "A deliberative method for ranking risks (I): Overview and test-bed development", *Risk Analysis*, vol. 21, no. 5, pp. 913–921.

Fox, M-J V & Erickson JD 2018, "Genuine economic progress in the United States: A fifty state study and comparative assessment", *Ecological Economics*, vol. 147, pp. 29–35.

Frank, KA, Maroulis, S, Duong, MQ & Kelcey, B 2013, "What would it take to change an inference? Using Rubin's causal model to interpret the robustness of causal inferences", *Educational Evaluation and Policy Analysis*, vol. 35, pp. 437–460.

Fukuyama, F 2011, *The origins of political order: From prehuman times to the French Revolution*, Farrar, Straus & Giroux, New York.

Garden, H & Winickoff, D 2018, *Issues in neurotechnology governance*, OECD Publishing, Paris.

Garner, R. (2016a). Animal rights and the deliberative turn in democratic theory. *European Journal of Political Theory*, 0(0), 1–21.

Garner, R. (2016b). Animals and democratic theory: Beyond an anthropocentric account. *Contemporary Political Theory*, *16*(4), 459–477.

Gerber, ER, Henry, AD & Lubell, M. 2013, "Political homophily and collaboration in regional planning networks", *American Journal of Political Science*, vol. 57, no. 3, pp. 598–610.

Grober, U 2012, *Sustainability; a cultural history*, Translated by R Cunningham, Green Books, Devon, UK.

Gunderson, R. (2016). "Why Global Environmental Governance Should Be Participatory: Empirical and Theoretical Justifications and the Problem of Scale". *International Sociology*, 33(6), 715–737.

Gunderson, R., & Dietz, T. (2018). "Deliberation and Catastrophic Risks". In A. Bächtiger, J. Mansbridge, M. E. Warren, & J. Dryzek (Eds.), *Oxford Handbook of Deliberative Democracy* (pp. 768–789). Oxford: Oxford University Press. (pp.) Under review.

Habermas, J 1970, *Towards a rational society*, Beacon Press, Boston, Massachusetts.

Habermas, J 1990, *Moral consciousness and communicative action*, Beacon Press, Boston.

Habermas, J 1993, *Justification and application: Remarks on discourse ethics*, The MIT Press, Cambridge, Massachusetts.

Harding, S (2015) *Objectivity and Diversity: Another Logic of Scientific Research*, University of Chicago Press, Chicago, Illinois.

Henry, AD 2009, "The challenge of learning for sustainability: A prolegomenon to theory", *Human Ecology Review*, vol. 16, no. 2, pp. 131–140.

Henry, AD, Prałat, P & Zhang, CQ 2011, "Emergence of segregation in evolving social networks", *Proceedings of the National Academy of Science, USA*, vol. 108, no. 21, pp. 8605–8610.

Henry, AD & Vollan, B 2014, "Networks and the challenge of sustainable development", *Annual Review of Environment and Resources*, vol. 39, pp. 583–610.

Hoggan, J & Litwin, G 2016, *I'm right and you're an idiot: The toxic state of public discourse and how to clean it up*, New Society Publishers, Grabiola Island, British Colombia.

International Risk Governance Council 2005, *Risk governance: Towards an integrative approach*, International Risk Governance Council, Geneva, Switzerland.

International Union for the Conservation of Nature 1980, *World conservation strategy*, International Union for the Conservation of Nature, Gland, Switzerland.

Kahneman, D &Krueger, AB 2006, "Developments in the measurement of subjective well-being", *Journal of Economic Perspectives*, vol. 20, no. 1, pp. 3–24.

Kahneman, D, Slovic, P & Tversky, A (eds.) 1982, *Judgement under uncertainty: Heuristics and biases*, Cambridge University Press, Cambridge, England.

Krueger, AB &Stone, AA 2014, "Progress in measuring subjective well-being", *Science*, vol. 346, no. 6205, pp. 42–43.

Liu, J, Hull, V, Batistella, M, DeFries, R, Dietz, T, Fu, F, Hertel, TW, Izaurralde, RC, Lambin, EF, Li, S, Martinelli, LA, McConnell, W, Moran, EF, Naylor, R, Ouyang,

Z, Polenske, KR, Reenberg, A, Rocha, G de Miranda, Simmons, CA, Verburg, PH, Vitousek, P, Zhang, F & Zhu, C 2013, "Framing sustainability in a telecoupled world", *Ecology and Society*, vol. 18, no. 2, p. 26.

Lord, CG, Ross, L & Lepper, MR 1979, "Biased assimilation and attitude polarization: The effects of prior theories on subsequently considered evidence", *Journal of Personality and Social Psychology*, vol. 37, no. 11, pp. 2098–2109.

Macekura, SJ 2015, *Of limits and growth: The rise of global sustainable development in the twentieth century*, Cambridge University Press, New York.

Mann, M 1986, *The sources of social power, Vol. I: A history of power from the beginning to A.D. 1760*, Cambridge University Press, Cambridge.

Marx, SA & Weber, E 2011, "Decision making under climate uncertainty: The power of understanding judgment and decision processes", In T Dietz & D Bidwell (eds.), *Climate change in the great lakes region: Navigating an uncertain future*, Michigan State University Press, East Lansing, Michigan.

McLaughlin, P 2012, "The second Darwinian Revolution: Steps toward a new evolutionary environmental sociology ", *Nature and Culture*, vol. 7, no. 3, pp. 231–258.

Mol, APJ, Spaargaren, G & Sonnenfeld, DA 2014, "Ecological modernization theory: Taking stock, moving forward", In *Routledge international handbook of social and environmental change*, Routledge, New York.

Morgan, KM, DeKay ML, Fischbeck PS, Morgan, MG, Fischoff B & Florig HK 2001, "A deliberative method for ranking risks (II): Evaluation of validity and agreement among risk managers", *Risk Analysis*, vol. 21, no. 5, pp. 923–937.

Morgan, SL & Winship C 2014, *Counterfactuals and causal inference*, Cambridge University Press, Cambridge, UK.

Moss, RH 2016, "Assessing decision support systems and levels of confidence to narrow the climate information 'Usability Gap'", *Climatic Change*, vol. 135, no. 1, pp. 143–155.

Nelson, MK & Shilling, D 2018, *Traditional ecological knowledge: Learning from indigenous practices for environmental sustainability*, Cambridge University Press, Cambridge, UK.

Neumayer, E 2010, *Weak versus strong sustainability: Exploring the limits of two opposing paradigms*, Edward Elgar, Cheltenham, Gloucester, UK.

Norman, W & MacDonald, C 2004, "Getting to the bottom of 'Triple Bottom Line'", *Business Ethics Quarterly*, vol. 14, no. 2, pp. 243–262.

Norton, BG 2005, *Sustainability: A philosophy of adaptive ecosystem management*, Oxford University Press, New York.

Odum, EP 1969, "The strategy of ecosystem development", *Science*, vol. 164, pp. 262–270.

OECD, March 2013, *OECD guidelines on measuring subjective well-being*, OECD Publishing. <http://www.oecd-ilibrary.org/economics/oecd-guidelines-on-measuring-subjective-well-being_9789264191655-en> [Last accessed September 4, 2013].

Ostrom, E, Dietz, T, Nives, D, Stern, PC, Stonich, S & Weber, E (eds.) 2002, *The Drama of the Commons*, National Academy Press, Washington, DC.

Parsons, T 1966, *Societies: Evolutionary and comparative perspectives*, Prentice Hall, Englewood Cliffs, New Jersey, NJ.

Pava, MA 2007, "A response to "Getting to the Bottom of 'Triple Bottom Line'"", *Business Ethics Quarterly*, vol. 17, no. 1, pp. 105–110.

Payne, CR, Shwom R &Heaton S 2015, "Public participation and norm formation for risky technology: Adaptive governance of solar-radiation management", *Climate Law*, vol. 5, no. 2–4, pp. 210–251.

Pearl, J & Mackenzie D 2018, *The book of why: The new science of cause and effect*, Basic Books, New York.

Portney, KE 2015, *Sustainability*, MIT Press, Cambridge, Massachusetts.

Pursell, C 1993, "The rise and fall of the appropriate technology movement in the United States, 1965–1985", *Technology and Culture*, vol. 34, no. 3, pp. 629–637.

Raffaelle, R, Wade, R & Evan, S (eds.) 2010, *Sustainability ethics: 5 Questions*, Automatic Press, Copenhagen, Denmark.

Reid, WV, Mooney HA, Cropper A, Capistrano D, Carpenter SR, Chopra K, Dasgupta P, Dietz T, Kumar Duraiappah A, Hassan R, Kasperson R, Leemans R, May RM, McMichael T(AJ), Pingali P, Samper C, Sholes R, Watson RT, Zakri AH, Shidong Z, Ash NJ, Bennett E, Kumar P, Lee MJ, Raudsepp-Hearne C, Simons H, Thonell J & Zurek MB 2005, *Ecosystems and human well-being: Synthesis*, Island Press, Washington, DC.

Renn, O 2017, *Risk governance: Coping with uncertainty in a complex world*, Routledge, London.

Renn, O & Klinke, A 2015, "Risk governance and resilience: New approaches to cope with uncertainty and ambiguity", In *Risk Governance*, Springer, pp. 19–41 Berlin, Germany.

Renn, O &Schweitzer, PJ 2009, "Inclusive risk governance: Concepts and application to environmental policy making", *Environmental Policy and Governance*, vol. 19, no. 3, pp. 174–185.

Renn, O, Webler, T & Wiedemann, P (eds.) 1995, *Fairness and competence in citizen participation: Evaluating models for environmental discourse*, Kluwer Academic Publishers, Dordrecht.

Reo, NJ 2011, "The importance of belief systems in traditional ecological knowledge initiatives", *International Indigenous Policy Journal*, vol. 2, no. 4.

Richerson, PJ 1977, "Ecology and human ecology: A comparison of theories in the biological and social sciences", *American Ethnologist*, vol. 4, pp. 1–26.

Righetti, LQ-C, Madhavan, PR & Chatila, R 2018, "Lethal autonomous weapon systems [Ethical, Legal, and Societal Issues]", *IEEE Robotics & Automation Magazine*, vol. 25, no. 1, pp. 123–126.

Rittel, HWJ &Webber, MM 1973, "Dilemmas in a general theory of planning", *Policy Sciences*, vol. 4, no. 2, pp. 155–169.

Robinson, J 2004, "Squaring the circle? Some thoughts on the idea of sustainable development", *Ecological Economics*, vol. 48, pp. 369–384.

Robinson, J & Cole, RJ 2015, "Theoretical underpinnings of regenerative sustainability", *Building Research & Information*, vol. 43, no. 2, pp. 133–143, doi: 10.1080/09613218.2014.979082.

Rosa, EA 1998, "Metatheoretical foundations for post-normal risk", *Journal of Risk Research*, vol. 1, pp. 15–44.

Rosa, EA, Renn, O & McCright, AM 2013, *The Risk Society revisited: Social theory and governance*, Temple University Press, Philadelphia.

Schnaiberg, A 1980, *The environment: From surplus to scarcity*, Oxford University Press, New York.

Schnaiberg, A 1982, "Did you ever meet a payroll? Contradictions in the structure of the appropriate technology movement", *Humboldt Journal of Social Relations*, vol. 9, no. 2, pp. 38–62, doi: 10.2307/23261947.

Schnaiberg, A 1983, "Redistributive goals versus distributive politics: Social equity limits in environmental and appropriate technology movements", *Sociological Inquiry*, vol. 53, no. 2–3, pp. 200–315, doi: 10.1111/j.1475-682X.1983.tb00034.x.

Schumacher, EF 1973, *Small is beautiful: A study of economics as if people mattered*, Perennial Library, New York.

Schwab, K 2017, *The fourth industrial revolution: What it means, how to respond.* Retrieved from Cologny, Switzerland: World Economic Forum, Crown Business.

Shwom, RL & Kopp, RE 2016, *Long-term risk governance: When do societies act before crisis?*, Department of Human Ecology, Rutgers University, New Brunswick, New Jersey, NJ.

Solomon, S, Plattner, GK, Knutti R & Friedlingstein, P 2009, "Irreversible climate change due to carbon dioxide emissions", *Proceedings of the National Academy of Sciences*, vol. 106, no. 6, pp. 1704–1708.

Stiglitz, J, Sen, A & Fitoussi, J-P 2009, *The measurement of economic performance and social progress revisited*, Commission on the Measurement of Economic Performance and Social Progress, Paris.

Thompson, P 2010, *The agrarian vision: Sustainability and environmental ethics*, University of Kentucky Press, Lexington, KY.

Tuler, S & Webler, T 1995, "Process evaluation for discursive decision making in environmental and risk policy", *Human Ecology Review*, vol. 2, pp. 62–71.

U.S. National Research Council 2005, "Panel on public participation in environmental assessment and decision making", In *Public participation in environmental assessment and decision making*, National Academy Press, Washington, DC.

U.S. National Research Council 1996, *Understanding risk: Informing decisions in a democratic society*, Edited by PC Stern & H Fineberg, National Academy Press, Washington, DC.

U.S. National Research Council 1999, *Perspectives on biodiversity: Valuing its role in an everchanging world*, National Academy Press, Washington, DC.

U.S. National Research Council 2008, *Public participation in environmental assessment and decision making*, Edited by T Dietz and PC Stern, National Academy Press, Washington, DC.

U.S. National Research Council 2010, *Advancing the science of climate change*, National Academies Press, Washington, DC.

U.S. National Research Council 2016, *Gene drives on the horizon: Advancing Science, navigating uncetainty, and aligning research with public values*, National Academies Press, Washington, DC.

U.S. National Research Council 2017, *Human genome editing: Science, ethics, and governance*, National Academies Press, Washington, DC.

Vandenbergh, MP & Gilligan, JM 2017, *Beyond politics: The private governance response to climate change*, Cambridge University Press, Cambridge, England.

Webler, T 1993, "Habermas put into practice: A democratic discourse for environmental problem solving", In SD Wright, T Dietz, R Borden, G Young & G Guagnano (eds.), *Human ecology: Crossing boundaries*, Society for Human Ecology, Ft. Collins, CO, pp. 60–72.

Whyte, KP 2012, "Indigenous peoples, solar radiation management, and consent", In CJ Preston (ed.), *Engineering the climate*, Lexington Books, Lanham, MD, pp. 65–76.

Whyte, KP 2013, "On the role of traditional ecological knowledge as a collaborative concept: A philosophical study", *Ecological Processes*, vol. 2, no. 1, pp. 1–12.

Whyte, KP, Brewer JP II & Johnson, JT 2015, "Weaving indigenous science, protocols and sustainability science", *Sustainability Science 11*(1), 25–32.

World Commission on Environment and Development 1987, *Our common future*, Oxford University Press, Oxford, England.

Worster, D 2016, *Shrinking the earth: The rise and decline of American abudance*, Oxford University Press, New York.

Yang, W, Dietz, T, Kramer, DB, Chen, X & Liu, J 2013a, "Going beyond the millennium ecosystem assessment: An index system of human well-being", *PLoS One*, vol. 8, no. 5, e64582.

Yang, W, Dietz, T, Liu, W, Luo, J & Liu, J 2013b, "Going beyond the millennium ecosystem assessment: An index system of human dependence on ecosystem services", *PLoS One*, vol. 8, no. 5, e64581.

York, R 2012, "Do alternative energy sources displace fossil fuels?", *Nature Climate Change*, vol. 2, pp. 441–443.

York, R 2013, "Metatheoretical foundations of post-normal prediction", In T Dietz & AK Jorgenson (eds.), *Structural human ecology: New essays in risk, energy and sustainability*, State University Press, Pullman, WA: Washington, DC, pp. 19–29.

York, R, Rosa, EA & Dietz, T 2010, "Ecological modernization theory: Theoretical and empirical challenges", In MR Redclift & G Woodgate (eds.), *The international handbook of environmental sociology*, (2nd ed.), Edward Elgar Publishing Limited, Cheltenham, UK, pp. 77–90.

7 Regenerative development and environmental ethics: Healing the mismatch between culture and the environment in the third millennium

Thomas J. Burns, Tom W. Boyd, and Carrie M. Leslie

We exist at a precarious place in human history. The need for regenerative development and environmental ethics is eminently pressing because of how quickly ecological and social problems have escalated, from being barely recognizable to potentially catastrophic regarding human survival within the last century (Burns & Caniglia 2017; McNeill 2000). This is particularly true in regards to environmental threats including, but not limited to, global climate change; deforestation; desertification; declining fish stocks in marine systems; extinct and endangered animal species; air, water, and noise pollution; increases in "diseases of civilization" (Garrett 1994) or rapid onset of new modern diseases; the dramatic increase in toxic "sacrifice zones" (Lerner 2010) or inhabited areas destroyed by pollution; and the melting of polar ice caps as a result of global warming (Burns & Caniglia 2017; Burns et al. 1994, 2003; Jorgenson and Burns 2004, 2007; Kick et al. 1996; Klein 2014; LaDuke 1999; Ponting 2008; Sawyer 2004; Smil 2003). Is it possible for an ethical system and consciousness to catch up with these runaway problems, such that humankind can work to unwind them sufficiently, and in time, to live sustainably (Burns 2009a) and regeneratively?

We explore the complexities of the interrelation between humans and the natural environment—or lack thereof, and then propose ways to facilitate harmony on a number of levels, between the individual and community, as well as nationally and internationally to promote the prevalence of regenerative development and environmental ethics. The following chapter will examine the causes of this discord between humanity and the natural environment—or the mismatch between social institutions and the environment, as well as alienated individualism as a cultural force detrimental to the environment. We address cultural lag and ecological overshoot in terms of how they hinder the development of global environmental ethics, and we offer transitional steps to best prepare society to live in a regenerative partnership with our natural world, concentrating on education, both formal and social, with the aim of encouraging environmental ethics of ecological peace and regenerative development.

The problem

First, we will investigate what is perpetuating the disconnect between humans and the natural environment, and make the case that with environmental issues, there is a considerable unsettled ethical arena that is often directly influenced by an elite agenda. A wave of recent qualitative work indicates a tendency for people to base environmental decision-making not on objective information, but on peers' opinions (Vance 2016). These peers often are reinforcing a certain way of seeing the world using sources that they preselect based on expectations—and not necessarily conscious ones—such that their preexisting views will be reinforced rather than challenged (Bageant 2007; Berger & Luckmann 1967; Hochschild 2016). Moral positions also can be relative to the situation where one action is condemned while other equally harmful actions are supported simply because of the framework in which they exist (Burke 1969a, 1969b; Burns 1999, 2009a; Burns & LeMoyne 2003). For example, some individuals may condemn infidelity, yet see no problem in working for an industry that does not provide transparency for their clients regarding environmental hazards. Why then are these ethical parameters differentiated in this way? Why does our modern society *not* have a more robust set of environmental norms, let alone laws that create a broad base of legitimacy and support that safeguard the natural environment from harmful pollution? What can lead to a significant rise in environmental consciousness (Caniglia et al. 2014), sufficient to address this problem in more than superficial ways?

Many societies (the United States being a prime exemplar) are multicultural now, and by that we are *not* talking in a purely demographic sense, but more so in the ethical sense. There are different spheres of discourse, with different norms and different belief systems underpinning them (Burns 1999, 2009; Burns & LeMoyne 2003). There is a fractionation in the polity that is almost hard-wired for distrust, with one side not just disagreeing with the other but going a step further and *doubting the legitimacy* of the other side (Hunter 1992). Perhaps, regenerative development and environmental ethics' initiatives could be the bridge between these different belief systems since we all share our natural environment and depend on its viability.

More ominous, however, are the findings that voters come to support antienvironmental candidates, often without reflection. These and other related ideas (e.g., anti-feminism) come to be accepted as part of class nonconsciousness of the late industrial era. As Gramsci (1971), among others, have pointed out, the biggest sign of cultural hegemony is the widespread dispersion of elite ideas that come to be regarded as "the way it is." The masses come to use the tropes, summary symbols, and, eventually, even the prepackaged ideas of elites (Burns 1999; Burns & LeMoyne 2003; Gramsci 1971; Mills 1956; Wolff 2016) that have been "framed" in particular ways (Benford and Snow 2000; Goffman 1974; Snow et al. 1986; Snow and Benford 1988; Wheeldon and Faubert 2009). This begs the question though, as to what becomes the draw among non-elites for buying into antienvironmentalism and/or voting for virulently antienvironmental regimes in the first place. As has been pointed out by a number of observers, there is ample evidence of people who have lost health care, who live downstream from serious ongoing environmental contamination, and yet who continue to vote for candidates who favor the gutting or even abolition of the bales (Bageant 2007; Frank 2005; Hochschild 2016; Isenberg 2016; Vance 2016).

Part of the reason for this may be the categorization of environmental problems as "post-political." By "post-political" we are alluding to *which* issues are in the political arena at a given time. For example, there is not a hot debate right now about whether there should be slavery or feudalism, or whether a person should be able to force another's labor without paying for it. These and related concerns were contested in times past. The US Civil War in the mid-19th century was fought over these major issues. The majority of humanity has settled on an ethical position regarding slavery and moved on from its contestation. If someone would bring up proslavery legislation, one would expect, from all sides of the political spectrum, vehement opposition. This is the sort of issue that can be deemed "post-political."

In the United States, the environment has gone from what seemed like a post-political issue for many people to a politicized one again. In the early 1970s, when the EPA, the Clean Water Act, and Clean Air Act were created, all received widespread support from both Democrats and Republicans. Yet, many issues have become contested again precisely because they get re-politicized—wrested out of the "post-political," back into the political sphere by contentious parties. We can see how coercion and influence operate, particularly when following money trails; Dunlap and McCright (2015) investigate this, as do Antonio and Brulle (2011). The consensus and take-away findings from this research are that well-organized, moneyed interests can and do have profound effects, not only on political outcomes, but also in shaping what is seen as political in the first place. It is these perceptions influenced by powerful actors that hinder a general acceptance of the urgency for promoting regenerative development practices and environmental ethics.

The problem of moral mismatch

What are the existential problems embedded in the different cross-sections of modern people? How might we best convince people that the problem of environmental crisis is urgent and offer ways to resolve it? We are not implying that individuals are *immoral* as such. Rather, it is that the commonly shared morality of people is virtually a mismatch with reality, particularly with the impending threat of environmental collapse that we share—our predominant moral framework is simply too narrowly focused. A common conception is that morality is exclusively about human interactions and the integrity of choices that influence our behavior and our relationships with other people. There are at least two principal reasons for this. First, most people do not grasp that a deeper understanding of the human condition must be expanded to include our relationship to our natural environment (Beck 2010; Burns & Caniglia 2017). Why? Human survival depends on a regenerative and sustaining natural environment. This, we may think, is self-evident, but it is not. How do we address this lack of awareness and practice?

Second, at a more profound level most people unknowingly suffer from estrangement or detachment to their fellow humans and environment (Burns 2009a; Beck 2010). People in both rural and urban areas have lost an immediate sense of participation in the very environment that sustains them. They dwell in a wonderfully contrived world developed through human ingenuity and effort, and this world virtually floats above and remains in many ways disconnected from the natural world (Brown 2003). However, this does not thwart humanity's constant draw on that natural world as a boundless resource to sustain our seemingly contrived world. Even more concerning, in many ways, is the waste produced that is then dumped back into that natural world, often polluting the water, air, and land (Johnson 2013). In brief, another predominant conception is to see the natural world as exclusively for human use (Burns & Caniglia 2017; Mayer 2016). Answering these questions pose the first challenges of our work.

Another problem is that many differing perspectives on how to address the environmental crises that we face are locked in a negative dialectic of mistrust and lack of cooperation (Hochschild 2016; Hunter 1992). We have these disagreements in mind as we take on the task of encouraging regenerative development and environmental ethics that more adequately engage with the ethical predispositions of most people. We do this because without a *sustainable* habitat, humanity's future is threatened on a number of levels (Sikora and Barry 1978). It is vital for society to move toward environmental ethics that are workable and general enough to create a critical mass of legitimacy. This will help catalyze an important rise in environmental consciousness and the promotion of regenerative development practices.

Norms, laws, and ethics

Ethical systems take time to develop—typically over the course of generations. Put another way, ethics tend to be born of the collective experience of people, but that experience unfolds unevenly over long periods of time (Ogburn 1932/1961; Singer 1988). For example, regarding slavery, in some cases, laws preceded the ethical shift in moral consciousness. Regardless, the ethical shift was gradual to take hold as a predominant cultural norm.

What happens when change takes place more rapidly? History shows that there often has been social dislocation when this occurs. Different subcultures adapt in different ways, and early adopters tend to experience pushback from the more predominant value system (Burns, Boyd & Dinger 2003). In Aristotelian terms, an individual internalizes crucial aspects of the broader, macro-level ethic which in some sense can be seen as character formation (e.g., Hunter 2000). That ethic then is expressed—lived out really—on the individual level as *phronesis*, or practical wisdom, where micro-level decisions reflect and embody this ethic (Burns 1999; Burns & LeMoyne 2003).

We see ethics, as did Aristotle (2009), as having a relatively stable character. The Aristotelian distinction between *Hexis* and *Diathesis* is useful here. *Hexis* (ἕξις in Greek) is a relatively stable arrangement or disposition (Stamatellos 2015). An example might be a person's semantic knowledge about "the way things are," that person's stable self-knowledge or in more classical terms, of one's "character." As Gerth and Mills (1964) point out, this character comes into its own in dialectical response to the social structure it encounters.

Diathesis comes from the Greek word διάθεσις This is rendered into English in a few different ways, but the most typical is "disposition" or "conception" or even "preconception." As Aristotle pointed out, while *Hexes* tend to be stable, diatheses do not. Not unlike the waves of the sea resting on the sand, *Diatheses* rely on the more stable Hexis to give them basic direction and shape. Without a stable Hexis underpinning it, *Diathesis* cannot retain its shape.

The environmental ethics that we hope to see arise are of the *Hexis* type. While framing, or *Diathesis*, can perhaps help, in the short run, to enter a cognitive cycle to get ideas in the running in the first place, ultimately *Hexis* operate on a different, deeper, level than does framing (Burns 1999). The trade-off then is that an ethic, once in place, can become a bedrock of human society—even a society undergoing rapid and chaotic change. An ethic takes a long time to develop, and once in place it typically takes a long time to change. Approaches would necessarily need to take place in a variety of places, from the micro to the macro levels, in order to integrate fully environmental ethics into the consciousness of people more generally (Peterson 1999).

Scale, scope, speed, and cultural lag

Environmental problems manifest at all different scales, from the microscopic to the global. When thinking and discussing environmental issues, it is important to keep this in mind. It is often the case that when the quantity or *scale* of a phenomenon changes, its *qualitative properties* also change, or its spatial parameters shift.

Temporal considerations are crucial here as well regarding the evolutionary scope of our planet. Nature may be able to recycle virtually anything—but in what time frame? If that frame is in millennia, humans could not survive (Diamond 2005). Power and affluence disparities cause extreme exploitation of resources which is occurring rapidly, both locally and globally. This degradation of the planet causes environmental exploitation to take place across generations as well effecting the particular histories of people (Thompson 1971; Wolf 1982).

Throughout history, in every part of culture and its respective institutions, there has been the phenomenon of "cultural lag" (Ogburn 1932/1961). "Cultural lag" is when older cultural beliefs, norms, or values tend to hold on, long after their optimal usefulness has passed for people. It is also related to changes in the material conditions of a society at a given place and time, and the adaptive cultural response capable of handling those new conditions (Polanyi 2001). That is to say, when there are rapid changes in material conditions (e.g., the invention and growing use of industrial machinery), the culture tends to adapt by changing what it values over time. Nevertheless, those changes often do not fit the new material conditions exactly, because there is a lag time in that adaptation (Burns & Caniglia 2017; Grubler 1991; McNeill 2000). The lag time tends to be less significant in cosmopolitan places (e.g., world cities), and greater in areas where communication, transportation, and population density are less widespread (e.g., rural areas). Regarding environmental problems, the scale, scope, and speed of their development has produced in many cases a lag of cultural values or environmental ethics that could address solutions more efficiently (MacIntyre 2007; Taylor 2012). Regenerative development is one solution that can act as a totem or guide to transform more rapidly an emphasis on environmentally conscious values and ethics.

Institutional mismatch with the natural environment

The degree of environmental degradation since the Industrial Revolution, and more particularly post World War II, is unprecedented in human history (Burns & Caniglia 2017; McNeill 2000). The problem is how do people situate regenerative development and environmental ethics at the forefront of their realities? Or rather how do these developmental practices and ethics become adopted or "internalized" more generally? These are important questions—vital really. To address them, let us consider the problem for a moment, not so much in the abstract, but in the concrete regarding certain institutions created by human beings. On a macro-level, the legacy of late modernity is a series of *institutional mismatches*. In many cases, institutions that evolved to facilitate life and human flourishing have gone out of date, at best matching the material conditions of times long gone (Brown 2003; Inglehart 1990; Inglehart & Baker 2000; Swidler 2001). In other words, they do not necessarily match what is the greatest collective good that could be provided for the people who belong to them. The two social institutions that we will focus on are the geopolitical economy and religion, particularly Christianity. These mismatches are possibly getting more acute and chronic over time, as the scale, scope, and speed of commerce and extractive exploitation increases endangering our ecological balance (Taylor 2012). By way of analogy, consider the rise of "diseases of civilization," such as type II diabetes, cancers, cardiovascular problems, asthma, spectrum disorders, and obesity, all related to the mismatches between how our bodies evolved as hunter-gatherers and the lives we lead in our late industrial society (Colborn, Dumanoski & Myers 1997; Nesse & Williams 1994; Steingraber 2010). This also includes environmental toxins and stressors for which our bodies and temperaments are ill-equipped. If we wait the requisite time for cultural values and norms to catch up, given the breadth and depth of the problems we are creating, it may very likely be too late.

Macro-level environmental problems and the geopolitical economy

The global economic market system is the strongest, most robust social institution to date, establishing a worldwide reach, and affecting virtually every aspect of human and planetary life at multiple levels. As various authors have pointed out (e.g., Mandel 1998; Robinson 2004), the global market system that has evolved, has far outstripped the logic and governance of the nation-state. While transnational corporations still typically are headquartered in core states and enjoy the protection of the military and political apparatus of those states, their reach is no longer constrained by the nation-state. As Marx (1867/1967) foresaw, capital can move quickly between places and polities, finding its most advantageous place, exploiting the exchange, and then moving on to the next opportunity; this was predicted as one of the competitive advantages of capitalism over feudalism. However, the drawback to this, noted by a number of scholars in the Marxist tradition, is that it has led to a widely unequal ecological exchange as far as resource accessibility (Foster 1999; Foster, York & Clark 2011; Moore 2000, 2003).

Environmental problems do not match the culture and institutions of the geopolitical economy (Archer 1988; Brown 2003; Burns 2009a). They also compete for attention with other more profit-driven economic and political priorities. Even when there is a will to address them, the focus and resolve can be deflected with infighting that often makes the eventual action less effective than it would be optimally (Burns & LeMoyne 2001). At the global level, there is precious little environmental governance; rather, the truly global institutions of markets, sometimes accompanied by military intervention, tend to trump most other institutions (Hornborg 2001, 2016; Robinson 2004) and environmental initiatives. In sum, moving into the Third Millennium, the logic of global capitalism threatens to overwhelm natural ecological balances by overpowering and engulfing the institutions that would keep it in check (Burns 2009a; Foster, York & Clark 2011; Gould et al. 2008; Hornborg 2001; Mayer 2016; McNeill 2000; O'Connor 1994; Schnaiberg 1980; Schnaiberg & Gould 1994).

What is the best hope for preventing environmental catastrophe? Ratcheting back on huge scales of production may be a step toward saving grace (Burns & Rudel 2015; DeFries 2014; Holling 1973; Schumacher 1973), but possibly difficult with a growing global population. This is extremely important in areas around historically important resources that are being depleted faster than they can replenish themselves (Antonio 2009; Foster et al. 2011; O'Connor 1994, 1998). Learning to live with the abundance that nature provides, while at the same time being aware of and respecting natural limits, is crucial as society moves further into the 21st century. More broadly, the solution is implied in the statement of the problem. Global capitalism and the geopolitical economy are not going to be dismantled overnight, and it would be unrealistic to expect that. However, we could work toward an unwinding of their enormous scale and influence and promote a disengagement with the global market economy's connection to military force. Perhaps the rise of an international body, such as one of the Breton Woods Institutions (e.g., World Bank; IMF), but with a charter specifically to promote regenerative development and environmental ethics, would deescalate these problems (Anderies et al. 2004; Antonio 2009; Burns & Caniglia 2017, p. 148; Burns & Rudel 2015; Foster, York & Clark 2011; Moore 2015; O'Connor 1994, 1998).

Late modernity, rise of the environmental crisis, and the influence of religion

In the 1960s, the stirrings of concern for the environment emerged into social consciousness, albeit slowly. Rachel Carson's *Silent Spring* (1962/2002) became a definitive early warning sign that environmental pollution can be subtle—but still dangerous—and was soon followed by other voices. In 1967, scholar Lynn

White wrote a landmark essay, "The Historical Roots of Our Ecological Crisis," in which he maintains that the origins of current environmental crises date back to the origins of Judaism and Christianity, particularly to the creation myth and the designation of humans as having dominion over nature (Hitzhusen 2007). A spate of books and articles soon followed, some written by philosophers and especially ethicists, others by Christian theologians who believed that they needed to join those interpreting the environmental crisis in theological terms. Most of these writers understood that traditional Christian writings, especially Scripture, needed to be radically reinterpreted with the current situation in mind. In general, this form of theology came to be known as *constructive theology*—a deliberate reconstruction of traditional theology integrated into the concerns appropriate to the time in question. Some of the most noted voices in this movement included John Cobb's *Is It Too Late? A Theology of Ecology* (1971/1995). Cobb subsequently wrote other works relating theology to the environment, such as *The Earthist Challenge to Economism* (1998). This was done in the larger context of Cobb's constructive reinterpretation of God within Christianity known as *process theology* which is connected more to current happenings. In 1993, theologian Sallie McFague published *The Body of God: An Ecological Theology*. Again, her work was done in the context of a constructive approach to reinterpreting God for late modernity. Since the late 1960s, a significant movement of constructive theologians has been highly vocal in addressing the environmental crisis. This involved the coupling of theology with a direct and deep appreciation for nature. Langdon Gilkey (1970, 2001), for example, declares that nature is inherently valuable in its own right *because* humanity is a product of nature (evolution) and bears the image of God along with nature bearing that image (also see Bellah 2011; Bellah and Joas 2012).

In certain regards, conservative Evangelical, along with Fundamentalist, Christians have resisted these constructive theologies. They have done so on two principal grounds. First, a belief in a literal interpretation of Biblical scripture is a core value. One example of this is related to perceptions of Darwinian evolution. Many in this tradition see Darwin's findings as a perversion of truth due to the fact that Biblical creation contravenes evolution (Dennett 1995). This has been a long and continuing battle since the late 19th century and is one of the principal reasons that today there is such harsh anti-scientific thinking in our American culture and its body politic (Foster et al. 2011). Some more moderate evangelicals have insisted on changing this message in more accommodating ways, especially the work of Katherine Hayhoe, Climate Scientist and scientific advisor to the Evangelical Environmental Network. She and her husband, Andrew Farley, also an Evangelical pastor wrote *A Climate for Change: Global Warming Facts for Faith-Based Decisions* (Hayhoe & Farley 2011).

Second, and in close alignment with the first, is the intense belief among many Evangelicals and Fundamentalist Christians that the end of time is imminent and could take place at any moment. This is known popularly as *The Second Coming* or *The Apocalypse*. People holding this belief are inclined to maintain that care for the environment is simply irrelevant. The idea has continued to spread among some evangelical and fundamentalist Christians, particularly in "middle-America," or small-town America, in the central part of the United States. To understand this element, one helpful source is *Strangers in Their Own Land* by Arlie Russell Hochschild (2016), where she explores the cultural disconnect and antagonism between liberals and conservatives in the United States due to differences in value systems.

Let us conclude by going back to Lynn White's (1967) seminal piece on the religious sources of our ecological crisis. Near the end of the article, he discusses the life and ministry of St. Francis of Assisi as a key example of the bonding of Christianity to nature, as Pope Francis did in *Laudato Si'* (2015) in which he implores humanity to have greater concern and empathy for the environmental problems that we face and to engage in "a new dialogue about how we are shaping the future of our planet" (p. 14). In the midst of this discussion, White drops hints for our current dilemma over the environment by pointing out the following. First, he insists that we will not solve this environmental crisis "until we find a new religion to rethink our old one" (White 1967, p. 1206). We believe that while constructive theologians have sought to do this, it is also important to integrate diverse religious perspectives in their analysis (e.g., Burns 2012, 2014). Second, he very briefly references "the Irish saints" (this could include the Celtic Christians). They were more esoteric in their approach, approaching animism, which mainstream Christianity has rejected, but their historical connection to nature is still strong.

The third hint, near the end of Lynn White's essay, he declares that the growth and continuation of the Franciscan idea (1967):

> Cannot be understood historically apart from the distinctive attitudes toward nature which are deeply grounded in Christian dogma. The fact that most people do not think of these attitudes as Christian is irrelevant. No new set of basic values has been accepted in our society to displace those of Christianity. Hence we shall continue to have a worsening ecologic crisis until we reject the Christian axiom that nature has no reason for existence save to serve man. (p. 1207)

Finally, among other things White says that the core of the problem of Christianity in relation to nature and the environment relies on modern science being seen as an extrapolation of natural theology. White declares near the end of his article (1967):

> Since the roots of our trouble are so largely religious [that is, human exceptionalism and its domination of nature], the remedy must also be essentially religious, whether we call it that or not. We must rethink and refeel our nature and destiny. The profoundly religious, but heretical sense of the primitive Franciscans for spiritual autonomy of all parts of nature may point in that direction. We propose Francis as a patron saint of ecologists. (p. 1207)

The Franciscan read on Christian scripture, or the Sufi read on the *Qur'an*, can thrive in a rising environmental consciousness (Hekmatpour, Burns & Boyd 2017;

Hitzuhusen 2007). There is so much richness to recover from religious texts in terms of counsel for environmental stewardship. This particular purpose has been there for centuries (Burns 2014), but not necessarily accessed. It will, however, take a new generation of scholars to work to uncover this wisdom for the Anthropocene Age.

Since religion has historically borne much the values that continue to inform us, the resolution necessarily will include religious affirmation and promotion. People's history is not fixed in stone. Rather, its historiography, canons, and memories change in response to its slowly changing values. This is why encouraging regenerative development within a religious context can also create an acceptance of environmental ethics more predominantly—to unwind this particular institutional mismatch.

Regenerative development—Limitations and progressive growth

We want to usher in a new "axial age" globally, where human consciousness generates pivotal and transformative value systems that will sustain it moving forward (Bellah 2011; Bellah and Findlay 2012; Findlay 1970; Jaspers 1953/2011; Rescher 2005) and promote environmental regeneration. This will likely involve a new interdisciplinarity in the academy (Burns 2009b), a new combination of different beliefs on a spiritual level (Burns & Boyd 2015), and institutional regulations to reflect a sustainment of ecological welfare (Burns & Caniglia 2017). This transformation also will necessitate a new discourse focusing on the centrality of an egalitarian environmentalism at the interpersonal and collective levels (Brulle 2000). We also can look toward facilitating *The Drama of the Commons*, rather than the "tragedy of the commons," a key concept for the environment (Anderies et al. 2004; Dietz, Ostrom & Stern 2003; Hardin 1968; National Research Council et al. 2002) that will include a collective understanding on a humanistic level (Burns, Boyd & Burns 2014; Coleman 1986; Olson 1965) of the importance of environmental preservation. This also will cause the arise of a connectedness integrating experts and stakeholders, non-exclusively and transparently, which is vital to encouraging regenerative development practices and environmental ethics across diverse communities.

When money and power are combined with politics, there can be threatening environmental and social problems. In many ways, we now have an economic system where power begets more power, money begets more money, and greed begets more greed. These powerful individuals and entities often exploit not only people, but also the environment as well. Particularly, this power elite hinder people's ability to share equally in environmental resources, such as fresh air, clean water, thriving ecosystems, and local natural resources (MacLean 2017). More than anything, we need to adjust our normative system of ethics concerning the environment, and do it quickly, to stanch the current runaway wanton abuse of our environment's resources. It is crucial to establish that environmental abuses are directly connected to human rights abuses in many instances (Bales 2016;

Steingraber 2010). In this section, we will discuss the limitations ("metabolic" and *ecological rift* and ecological *overshoot*) to establishing regenerative development, and the progressive growth (praxis of reconnecting to the environment) of regenerative approaches that has already begun.

Metabolic and ecological rift

Marx and Engels's incorporation of Von Leibig's ideas about "metabolic rift" into *Capital* became widely recognized through the work of latter-day Marxist scholars, such as Jason Moore and John Bellamy Foster's work on *ecological rift*. A responsible reading of *Capital* now could not ignore the crucial role of "metabolic rift," or the ever-widening chasm between humans and nature as far as a lack of recognized interdependence. This rift also influences the *unequal* ecological exchange of critical resources between more and less "developed" countries and regions. Those exchanges further affect the ecological balance of the planet and the power imbalances among people. Thomas Berry's, *The Great Work: Our Way into the Future* (1999) reminds us that our natural environment is not a commodity to be exploited, and we must learn how to appreciate its vastness and seek ways to promote its continual regeneration.

Another key problem related to "metabolic rift," or the disconnect between humanity and the natural world by the dominance of capitalist processes is that the political has been made apolitical, and the apolitical has been made political. In other words, many people tend to view environmental problems as apolitical or not critically important. To complicate the matter further, antienvironmental lobbies are backed primarily by powerful individuals and corporations who prefer more lenient environmental protection regulations. For example, there is currently a political debate about whether or not there should even be an EPA and how much it should be allowed to regulate industrial environmental standards. These debates are dominated by profit-driven individuals who are able to control certain members of the US Congress, think tanks, and much of public opinion with their financial and political influence (MacLean 2017; Mayer 2016; Sandel 2013).

Overshoot and uneven ecological exchange

The question of ecological overshoot (Catton 1994) is a perplexing one to be sure. In biological systems where overshoot occurs, a species can hang on, as it exists, for varying lengths of time. This is true for human beings as well, but when, in addition to population, we consider other variables such as affluence and the accumulation of wealth, metabolic rift, "diseases of civilization," a culture of consumption, and the scale, scope, and speed of environmental degradation, the picture becomes staggering. What this implies is that certain members of the human species will hold on at the expense of others. An example of this is elites using their accumulated power, wealth, and technology to extract ever greater levels of energy, food, and other resources, further externalizing the costs to the rest of the world (Patel & Moore 2017).

Ecological Marxists (e.g., O'Connor 1994, 1998) have warned of the "second contradiction of capitalism" in which *ecological overshoot* is exacerbated greatly by these wealth and power dynamics. When combined with the technological capability and the culture of accumulation geared to make ever more profound incursions into the earth to externalize waste in the most expedient of ways, it is not difficult to envision the likelihood of the entire species coming to an irrevocable state of overshoot (Antonio 2009; Harvey 2015; Streeck 2016). From a sociological perspective, there is a fundamental flaw in social organizations in general. It goes by a number of names, but the most common is "the iron law of oligarchy" (Michels 1911/1949). The basic problem is that, above some tipping point, social systems in general tend to have a positive feedback loop for power and influence. The rich get richer; the powerful get more powerful. This keeps happening in human systems, to the point where a handful of people have a preponderance of wealth and power, and often tend to use it in ways that increase their own interests in the short run, redounding to the detriment of others and the overall planet in the long run. There is a built-in problem even with changing it, because as neo-Machiavellians (e.g., Mosca 1939; Pareto 1935/1961) have pointed out, even if there is a revolutionary sort of change, there often results in what amounts to a "circulation of elites" where the system itself goes on pretty much as before, albeit with a different group in charge.

Institutional fixes by themselves have always been, and will forever be, inadequate (Burns 2009a). While the accumulative nature of capitalism is particularly problematic (Antonio 2009; Foster, York & Clark 2011; Moore 2015), this iron law of oligarchy tends to hold regardless of whatever punitive system may have been imposed institutionally. A possible key to curbing these inequalities is that an emergent culture of environmental consciousness arises. In this, the environment becomes the central organizing principle, or the primary summary symbol (Burns 1999; Burns and LeMoyne 2001, 2003) for the culture itself (Berry 1999; Burns 2009a; Gore 1993). This, of course, is easier said than done. In the meantime, we can do what we can to contribute to the birthing of this idea. This includes acting in environmentally conscious ways, choosing vocations where we can make a difference bringing conscious ways of seeing, thinking, and acting into the work we do. These individual processes gradually will begin to transform institutional systems, such as the geopolitical economy and religious structures of belief. As the Italian Social Theorist, Antonio Gramsci (1971) noted cultural ideas must work their way through each of the respective institutions. In the European student movements of the 1970s, a variant of Gramsci's idea became the basis for a marching slogan. For the revolution to occur, the *war of position* (or the struggle to define ideas in a conscious way) must first move slowly through the institutions (Gramsci 1971). In the same way, transforming our institutions through a shift in consciousness toward regenerative development and environmentally based ethics can occur, even during the precarious time in which we exist, and it is essential for our survival.

There is a constant need for evolving solutions. For example, certain corporate business campaigns *green-wash*, or implicitly excuse environmentally unjust practices by promoting images of environmentalism or humanitarianism through a guise of environmentally friendly rhetoric. Suzana Sawyer points out a similar type of humanitarian and environmentally conscious advertising evident in the campaign introduced by Chevron Corporation in 2007 called *Human Energy* (Sawyer 2010, p. 68; also see Sawyer 2004). This does little to address actual, immeasurable environmental catastrophes such as in Alaska, the Exxon Valdez oil spill of 10 million gallons over fragile wetlands in 1989; in the New Orleans area, after the Hurricane Katrina disaster, a 7-million-gallon oil spill in 2005; and in the Gulf Coast, BP/Deep Water Horizon spilled 134 million gallons in 2010. These are but a few examples of the dozens of major oil spills in the US coastal waters since 1969 (NOAA 2018). Researchers have pointed out ways in which companies may be held more accountable. After all, companies are made of people, yet the reification of companies has been brought to new heights with the Supreme Court decision of Citizens United versus the US Federal Elections Commission. Since that decision in 2010, corporations have been deemed to have the rights of individuals without asking them to embody an accountability to the environment or affected people. In fact, a major role of the environmental sociologist is brought to light here; that is to act, not only as social analyst, but as social critic (Horkheimer & Adorno 1944; Marcuse 1964; Stearns and Burns 2011), particularly regarding the politicized communication, discourse, and rhetoric about the environment and its protection.

The praxis of reconnecting to the environment—Progressive growth

There is wide-scale disempowerment toward impacting environmental issues positively, possibly due to their scale and the powerful individuals protecting the entities committing the environmental atrocities. People need to be reengaged in these processes and to feel a sense of empowerment that they can make a difference. This can start on the local level through environmental education initiatives creating a more immediate feedback loop. This widespread feeling of alienation or disconnect to the environment is connected to a culture of individualism—especially in Western societies—in which people do not feel bound to a larger collective or society. Many see this also as a lack of connection to a spiritual tradition or code of ethics. We reiterate that despite the social problems caused by politicized capitalism, a resurgence toward regenerative development emphasizing environmental ethics can help promote environmentally regenerative practices (United Nations 2018).

It is crucial to reconnect with nature not just in the abstract, but in everyday life. This makes it incumbent upon society to conserve—where still possible in the late Anthropocene Age—natural resources in ways that make it possible to be in contact with nature on a daily basis. National parks, forests, and preserves are crucial for conservation. Local protected areas also can be accessed by citizens due to their closeness to urban settings. In Germany, for example, many of the World War II tank trails became the basis of hiking trails. Specifically, in Stuttgart, an industrial as well as military city, much of the surrounding area still has accessible, wooded trails for hiking and communing with nature. It must be noted that we do not condone "Green Gentrification" (Gould & Lewis 2016)—or participating in "green" development at the expense of people of lower socio-economic classes forcing them to relocate in the process; "green" regenerative development must enrich every member of society and the natural world.

Urban, community, and small yard gardens are important for people to be able to cultivate and reconnect with the possibility of raising their own food. This dovetails well with smaller scale agriculture and urban farming in order to downscale the use of agribusiness that overuses pesticides, herbicides, and artificial fertilizers. More generally the ability to grow things in small-scale locally while networking globally will give people a sense of empowerment, while helping to revitalize and regenerate the connection between community and the environment. Planting trees to help regenerate lost green spaces, but also eliminate the level of carbon in the atmosphere is another great example of regenerative development that the Greenbelt Movement started by Wangari Maathai in Kenya practiced. Since their beginning, they have planted over 35 million trees primarily by women (Maathai 2006).

Another example of regenerative development on a local level is the work of *Earth Rebirth* in Norman, Oklahoma, started by Andrew Sartain. It is a nonprofit environmental education organization whose goal this year is to help install gardens at all of the public schools in the community. This is a project that they have nearly completed already. It is an invaluable resource to the children who get to learn about gardening and how to grow organic vegetables from seeds. *Earth Rebirth* helps maintain and construct all of the gardens with the children and teachers on site as well. These tangible shifts in everyday life will encourage an environmental consciousness to arise in all people which will transform the influential institutional structures within which we live. Regenerative development can be a totem and guide to transformation of ethics more generally, from the local to the geopolitical.

Environmental ethics as pathways to regenerative development

Environmental problems are not only daunting, but in many cases are still escalating. The theoretically-based calls to action that follow can facilitate the re-centering of environmental ethics as deeply held, stable, and durable components for human survival. They also focus on the importance of regenerative development in this transformative process.

1. Environmental ethics often have been subordinate to other value systems, and this has become more pronounced over time. Regenerative development and environmental ethics are essential for human survival, as they help restore humanity's connection to the natural environment.
2. The promotion of regenerative development can bring people together in ways that may not otherwise happen, in helping to raise consciousness that humanity itself depends on the viability of the natural environment. This can act as a bridge for contentious political parties and can also expose environmentally hazardous practices by calling attention to policies that heretofore may have been widely accepted but which endanger ecosystems and human livelihood.

3 Environmental protection has become a politically charged issue, but one that is deeply connected to economically driven interests from the power elite, which represent the environment as already protected. This turns attention away from awareness of the severity of environmental degradation and the will to address it.
4 The importance of our natural environment must be re-centered, ethically and morally, so that it becomes a cardinal ethical value.
5 Regenerative development can offer invaluable guidelines for humankind to address the estrangement, alienation, and detachment experienced in the modern world.
6 Environmental ethics carry influence to the extent they are based in *Hexis*, as it forms the bedrock of human society. At a time of rapid and chaotic change, the scale, scope, and speed of environmental problems need an equally encompassing shift in cultural values and ethics that incorporate regenerative development models to revitalize natural environments.
7 Regenerative development can help restore the interconnectivity and balance found in the natural world to humankind.
8 Environmental ethics at the forefront of consciousness can create a unified resistance against politicized capitalism and the resulting communication, discourse, and rhetoric of fossil fuel and extractive industries that *greenwashes*, or implicitly excuses fraud, environmental pollution, and the endangerment of people and the ecosystems in which they function.
9 Ecological peace is facilitated by regenerative development practices and by placing environmental ethics at the forefront of consciousness engagement.
10 Regenerative development is vital to the transformation toward environmentally conscious ethics for humankind and its institutions, from the local to the geopolitical.

As Victor Hugo (1877/2001) observed, the one thing more powerful than all the armies in the world is an idea whose time has come. We are betting everything we have that the time indeed has come for a rising environmental consciousness, necessary and sufficient to bring about both robust and solid practices of regenerative development and environmental ethics for the Anthropocene Age and the Third Millennium. We place our bet, not because we are sure of winning—indeed we wish we could be more certain about the odds at this point—but because losing is not a viable option.

References

Anderies, JM, Jansen, MA & Ostrom E 2004, "A framework to analyze the robustness of social-ecological systems from an institutional perspective", *Ecology and Society*, vol. 9, pp. 1–18.

Antonio, RJ 2009, "Climate change, the resource crunch, and the global growth imperative", In HF Dahms (ed.), *Current perspectives in social theory*, Emerald, Bingley, UK, pp. 3–73.

Antonio, RJ, & Brulle RJ 2011, "The unbearable lightness of politics: Climate change denial and political polarization", *Sociological Inquiry*, vol. 52, pp. 195–202.

Archer, MS 1988, *Culture and agency: The place of culture in social theory*, Cambridge University Press, New York.

Aristotle 2009, *The Nicomachean Ethics*. Edited by Brown L, translated by Ross D (revised ed.). Oxford University Press, Oxford, NY.

Bageant, J 2007, *Deer hunting with Jesus: Guns, votes, debt and delusion in redneck America*, Random House, New York.

Bales, K 2016, *Blood and earth: Modern slavery, ecocide, and the secret to saving the world*, Spiegel & Grau, New York.

Beck, U 2010, "Climate for change, or how to create a green modernity?", *Theory, Culture, & Society*, vol. 27, no. 2, pp. 254–266.

Bellah, RN 2011, *Religion in human evolution: From the Paleolithic to the axial age*, The Belknap Press of Harvard University Press, Cambridge, MA.

Bellah, RN & Joas, H 2012, *The axial age and its consequences*, The Belknap Press of Harvard University Press, Cambridge, MA.

Benford, RD & Snow, DA 2000, "Framing processes and social movements: An overview and assessment", *Annual Review of Sociology*, vol. 26, pp. 611–639.

Berger, PL & Luckmann, T 1967, *The social construction of reality: A treatise in the sociology of knowledge*, Anchor, New York.

Berry, T 1999, *The great work: Our way into the future*, Bell Tower Books, Sharon Hill, PA.

Brown, K 2003, "Integrating conservation and development: A case of institutional misfit," *Frontiers in Ecology and the Environment*, vol.1, no. 9, pp. 479–487.

Brulle, R 2000, *Agency, democracy and nature: The U.S. environmental movement from a critical theory perspective*, MIT Press: Cambridge, MA.

Burke, K 1969a, *A Grammar of motives*, University of California Press, Berkeley, CA.

Burke, K 1969b, *A Rhetoric of motives*, University of California Press, Berkeley, CA.

Burns, TJ 1999, "Rhetoric as a framework for analyzing cultural constraint and change", *Current Perspectives in Social Theory*, vol. 19, pp. 165–185.

Burns, TJ 2009a, "Culture and the natural environment", In P Lopes & A Begossi (eds.), *Current trends in human ecology*, Cambridge Scholars Press, Newcastle upon Tyne, UK, pp. 56–72.

Burns, TJ 2009b, *Building interfaith dialogue in an interconnected world: Approaches, obstacles and prospects*, Paper presented at the Oxford Roundtable, Harris Manchester College, Oxford University, UK.

Burns, TJ 2012, *Canonical texts: Selections from religious wisdom traditions*, Cognella, San Diego, CA.

Burns, TJ 2014, "Reconsidering scripture in late industrial society: Religious traditions and the natural environment", In BS Caniglia, TJ Burns, R Gurney, & EL Bond (eds.), *Rise of environmental consciousness: Voices in pursuit of a sustainable planet*, Cognella, San Diego, CA, pp. 43–60.

Burns, TJ & Boyd, TW 2015, *Global culture, world religions and the new syncretism, with a case study in environmental science*, Paper presented at the Parliament of the World's Religions, Salt Lake City, UT.

Burns, TJ, Boyd, TW & Burns, CM 2014, "Engaging complexity in business and technology: Rethinking old ideas humanistically and ecologically," *International Journal of Business, Humanities and Technology*, vol. 4, no. 1, pp. 1–9.

Burns, TJ, Boyd, TW & Dinger, JR 2003, *Religion and globalization in the twenty-first century: W(h)ither the sacred canopy?* Paper presented at Annual Conference of the Notre Dame Center for Ethics and Culture, South Bend, IN.

Burns, TJ and Caniglia, BS 2017, *Environmental sociology: The ecology of late modernity*, 2nd Ed. Mercury Academic, Norman, OK.

Burns, TJ & Jorgenson, AK 2007, "Technology and the environment", In CD Bryant & DL Peck (eds.), *21st century sociology: A reference handbook*, Sage, Thousand Oaks, CA, pp. 306–312.

Burns, TJ, Kentor, JD & Jorgenson, AK 2003, "Trade dependence, pollution and infant mortality in less developed countries", In WA Dunaway (ed.), *Emerging issues in the 21st century world-system*, Praeger, London, pp. 14–28.

Burns, Thomas J., Kick, Edward L. & Davis, Byron L 2006, "A quantitative, cross-national study of deforestation in the late 20th century: A case of recursive exploitation", In Andrew K. Jorgenson & Edward L. Kick (eds.), *Globalization and the environment*, Brill, Leiden, The Netherlands, pp. 37–60.

Burns, Thomas J., Kick, Edward L., Murray, David A., and Murray, Dixie A 1994, "Demography, Development, and Deforestation in a World-System Perspective." *International Journal of Comparative Sociology*, vol. 32, pp. 221–239.

Burns, TJ & LeMoyne, T 2001, "How environmental movements can be more effective: Prioritizing environmental themes in political discourse", *Human Ecology Review*, vol. 8, no. 1, pp. 26–38.

Burns, TJ & LeMoyne, T 2003, "Epistemology, culture and rhetoric: Some social implications of human cognition," *Current Perspectives in Social Theory*, vol. 22, pp. 71–97.

Burns, TJ & Rudel, TK 2015, "Metatheorizing structural human ecology at the dawn of the third millennium", *Human Ecology Review*, vol. 22, no. 1, pp. 13–33.

Caniglia, BS, Burns, TJ, Gurney, R & Bond, EL 2014, *Rise of environmental consciousness: Voices in pursuit of a sustainable planet*, Cognella, San Diego, CA.

Carson, R 1962/2002, *Silent spring. Mariner books*, Houghton Mifflin, Boston, MA.

Catton, WR, Jr 1994, "What was Malthus really telling us?", *Human Ecology Review*, vol. 1920, pp. 234–236.

Cobb, John B 1971/1995. *Is it too late? A theology of ecology* (Revised ed.). Environmental Ethics Books, Denton, TX.

Colborn, Theo, Dumanoski, Dianne & Myers, John Peterson 1997, *Our stolen future: Are we threatening our fertility, intelligence, and survival? A scientific detective story*, Plume/Penguin, New York.

Coleman, James S 1986, *Individual interests and collective action*, Cambridge University Press, Cambridge, UK.

DeFries, Ruth 2014, *The big ratchet: How humanity thrives in the face of natural crisis*, Basic Books, New York.

Dennett, Daniel 1995, *Darwin's dangerous idea: Evolution and the meanings of life*, Simon & Schuster, New York.

Diamond, Jared 2005, *Collapse: How societies choose to fail or succeed*, Viking, New York.

Dietz, Thomas, Ostrom, Elinor & Stern, Paul C 2003, "The struggle to govern the commons", *Science*, vol. 302, no. 12, pp. 1907–1912.

Dunlap, Riley E & McCright, Aaron M 2015, "Challenging climate change: The denial countermovement", In Riley E. Dunlap and Robert J. Brulle (eds.), *Climate change and society: Sociological perspectives*, Oxford University Press, New York, pp. 300–332.

Findlay, JN 1970, *Axiological ethics*, Macmillan, New York.

Foster, John Bellamy 1999, "Marx's theory of metabolic rift: Classical foundations for environmental sociology", *American Journal of Sociology*, vol. 105, pp. 366–402.

Foster, John Bellamy, York, Richard & Clark, Brett 2011, *The ecological rift: Capitalism's war on the earth*, Monthly Review Press, New York.

Francis, Pope 2015, *Laudato Si': On care for our common home* (encyclical). Vatican website <http://w2.vatican.va/content/francesco/en/encyclicals/documents/papa-francesco_20150524_enciclica-laudato-si.html>.

Frank, Thomas 2005, *What's the matter with Kansas? How conservatives won the heart of America* (Reprint ed.), Holt, New York.

Garrett, Laurie 1994, *The coming plague: Newly emerging diseases in a world out of balance*, Penguin, New York.

Gerth, Hans & Mills, C. Wright. 1964, *Character & social structure*, Mariner Books, New York.

Gilkey, Langdon B 1970, *Religion & scientific future* (Reprint ed.), Mercer University Press, Macon, GA.

Gilkey, Langdon B 2001, Blue twilight: Nature, creationism, and American religion—Kindle edition by Langdon Brown Gilkey, Religion & Spirituality Kindle EBooks @ Amazon.Com, Accessed March 6, 2018 <https://www.amazon.com/Blue-Twilight-Creationism-American-Religion-ebook/dp/B000UK469Q/ref=asap_bc?ie=UTF8>.

Goffman, Erving 1974, *Frame analysis: An essay on the organization of the experience*, Harper Colophon, New York.

Gore, Al 1993, *Earth in the balance: Ecology and the human spirit*, Plume/Penguin, New York.

Gould, Kenneth A. & Lewis, Tammy L 2016. *Green Gentrification: Urban Sustainability and the Struggle for Environmental Justice*, Routledge, New York.

Gould, Kenneth A., Pellow, David & Schnaiberg, Allan 2008, *The treadmill of production: Injustice and unsustainability in the global economy*, Paradigm, Boulder, CO.

Gramsci, Antonio 1971/1989, "The political party and some preliminary points of reference", In Quintin Hoare & Geoffrey Nowell Smith (eds., Reprint, 1989 ed.), *Selections from the prison notebooks*, International Publishers, New York.

Grubler, Arnulf 1991, "Diffusion: Long-term patterns and discontinuities", *Technological Forecasting and Social Change*, vol. 39, pp. 159–180.

Hardin, G 1968, "The tragedy of the commons", *Science*, vol. 162, pp. 1243–1248.

Harvey, David 2015, *Seventeen contradictions and the end of capitalism* (Reprint ed.), Oxford University Press, Oxford.

Hayhoe, Katharine and Andrew Farley 2011, *A climate for change: Global warming facts for faith-based decisions*, FaithWords, Nashville, TN.

Hekmatpour, Peyman, Burns, Thomas J & Boyd, Tom W 2017, "Is Islam pro- or anti-environmental?", *Journal of Asian Research*, vol. 1, no. 1, pp. 92–110.

Hitzhusen, Gregory E 2007, "Judeo-Christian theology and the environment: Moving beyond skepticism to new sources for environmental education in the United States", *Environmental Education Research*, vol. 13, no. 1, pp. 55–74.

Hochschild, Arlie Russell 2016, *Strangers in their own land: Anger and mourning on the American right*, The New Press, New York.

Holling, CS 1973, "Resilience and stability of ecological systems", *Annual Review of Ecology and Systematics*, vol. 4, pp. 2–23.

Horkheimer, Max & Adorno, Theodor W 1944, *Dialectic of enlightenment*, Querido Verlag N.V., Amsterdam.

Hornborg, Alf 2001, *The power of the machine: Global inequalities of economy, technology, and environment*, AltaMira Press, Walnut Creek, CA.

Hornborg, Alf 2016, *Global magic: Technologies of appropriation from ancient Rome to Wall Street*, Palgrave Macmillan, New York.

Hugo, Victor 1877/2001, *Histoire d'un Crime: Déposition d'un témoin (The history of a crime: The testimony of an eye-witness)*, translated by T Joyce & Arthur Locker, University Press of the Pacific, Honolulu, Hawaii.

Hunter, James Davison 1992, *Culture wars: The struggle to define America*, Basic Books, New York.

Hunter, James Davison 2000, *Death of character: Moral education in an age without good or evil*, Basic Books, New York.

Inglehart, Ronald 1990 *Culture shift in advanced industrial society*, Princeton University Press, Princeton, NJ.

Inglehart, Ronald & Baker, Wayne E 2000, "Modernization, cultural change, and the persistence of traditional values", *American Sociological Review*, vol. 65, pp. 19–51.

Isenberg, Nancy 2016, *White trash: The 400-year untold history of class in America*, Viking, New York.

Jaspers, Karl 1953/2011, *The origin and goal of history*, Routledge, Abingdon, Oxon; New York.

Johnson, Bea 2013 *Zero waste home: The ultimate guide to simplifying your life by reducing your waste*, Scribner, New York.

Jorgenson, Andrew K & Burns, Thomas J 2004, "Globalization, the environment, and infant mortality: A cross-national study", *Humboldt Journal of Social Relations*, vol. 28, no. 1, pp. 7–52.

Jorgenson, Andrew K & Burns, Thomas J 2007, "The political-economic causes of change in the ecological footprints of nations, 1991–2001: A quantitative investigation", *Social Science Research*, vol. 36, pp. 834–853.

Kick, Edward L, Burns, Thomas J, Davis, Byron L, Murray, David A, & Murray, Dixie A 1996, "Impacts of domestic population dynamics and foreign wood trade on deforestation: A world-system perspective", *Journal of Developing Societies*, vol. 12, no. 1, pp. 68–87.

Klein, Naomi 2014, *This changes everything: Capitalism vs. the climate*, Simon & Schuster, New York.

LaDuke, Winona 1999, *All our relations: Native struggles for land and life*, South End Press, Boston, MA.

Lerner, Steve 2010, *Sacrifice zones: The front lines of toxic chemical exposure in the United States*, MIT Press, Cambridge, MA.

Maathai, Wangari 2006, *Unbowed: A memoir*, Alfred A. Knopf, Inc, New York.

MacIntyre, Alasdair 2007, *After virtue: A study in moral theory* (3rd ed.). University of Notre Dame Press, Notre Dame, IN.

MacLean, Nancy 2017, *Democracy in chains: The deep history of the radical right's stealth plan for America* (Later Printing ed.), Viking, New York.

Mandel, Ernest 1998, *Late Capitalism*, 2nd ed. Verso Press, London.

Marcuse, Herbert 1964, *One-dimensional man*, Beacon Press, Boston, MA.

Marx, Karl [1867] 1967. *Capital: A Critique of Political Economy*. Vol. 1. International Publishers, New York.

Mayer, Jane 2016, *Dark money: The hidden history of the billionaires behind the rise of the radical right*, Doubleday, New York.

McFague, Sallie 1993, *The body of god: An ecological theology*, Fortress Press, Minneapolis, MN.

McNeill, JR 2000, *Something new under the sun: An environmental history of the twentieth-century world*. W.W. Norton, New York.

Mills, C. Wright 1956, *The power elite*. Oxford University Press, Oxford.

Moore, Jason. 2000. Environmental Crisis and the Metabolic Rift in World-Historical Perspective. *Organization and Environment* 13:123–157.

Moore, Jason. 2003. The Modern World-System as Environmental History? Ecology and the Rise of Capitalism. *Theory and Society* 32:307–377.

Moore, Jason W 2015, *Capitalism in the web of life: Ecology and the accumulation of capital*, Verso, New York.

Mosca, Gaetano 1939, *The ruling class*, In A. Livingston (ed.), translated by H Kahn, McGraw-Hill, New York.

National Oceanic and Atmospheric Administration (NOAA) 2018, *Global Climate Report – Annual 2018*. National Centers for Environmental Information, Washington, DC. (https://www.ncdc.noaa.gov/sotc/global/201813)

National Research Council, Division of Behavioral and Social Sciences and Education, Board on Environmental Change and Society & Committee on the Human Dimensions of Global Change 2002, *The drama of the commons*, In Elke U Weber, Susan Stonich, Paul C. Stern, Nives Dolsak, Thomas Dietz & Elinor Ostrom (eds.), National Academies Press, Washington, DC.

Nesse, Randolph M & Williams, George C 1994, *Why we get sick*, Vintage, New York.

O'Connor, James 1994, "Is sustainable capitalism possible?", In Martin O'Connor (ed.), *Political economy and the politics of ecology*, Guilford, New York, pp. 152–175.

O'Connor, James 1998, *Natural causes: Essays in ecological Marxism*, Guilford, New York.

Olson, Mancur 1965, *The logic of collective action*, Harvard University Press, Cambridge, MA.

Pareto, Vilfredo 1935/1961, "The circulation of elites", In Talcott Parsons, Edward Shils, Kaspar D. Naegele & Jesse R. Pitts (eds.), *Theories of society: Foundations of modern sociological theory*, Free Press, New York, pp. 551–558.

Patel, Raj & Moore, Jason W 2017, *A history of the world in seven cheap things: A guide to capitalism, nature, and the future of the planet*, University of California Press, Oakland, CA.

Peterson, Anna 1999, "Environmental ethics and the social construction of nature", *Environmental Ethics*, vol. 21, no. 4, pp. 339–357. <https://doi.org/10.5840/enviroethics19992142>.

Polanyi, Karl 2001, *The great transformation: The political and economic origins of our time* (2nd ed.), Beacon Press, Boston, MA.

Ponting, Clive 2008, *A new green history of the world: The environment and the collapse of great civilizations* (2nd ed.), Penguin, New York.

Rescher, Nicholas 2005, *Value matters: Studies in axiology*. Ontos Verlag, Frankfurt.

Robinson, William I 2004, *A theory of global capitalism*, Johns Hopkins University Press, Baltimore, MD.

Sandel, Michael J 2013, *What money can't buy: The moral limits of markets* (reprint ed.), Farrar, Straus and Giroux, New York, NY.

Sawyer, Suzana 2004, *Crude chronicles: Indigenous politics, multinational oil, and neoliberalism in Ecuador*, Duke University Press, Durham & London.

Sawyer, Suzana. 2010, "Human energy", *Dialectical Anthropology*, vol. 34, no. 1, pp. 67–77.

Schnaiberg, Allan 1980, *The environment: From surplus to scarcity*, Oxford University Press, New York.

Schnaiberg, Allan, and Gould, Kenneth A 1994, *Environment and Society: The Enduring Conflict*. St. Martin's Press, New York.

Schumacher, EF 1973, *Small is beautiful: Economics as if people mattered*, Hartley and Marks, Point Roberts, WA.

Sikora, Richard I & Barry, Brian M 1978, *Obligations to future generations*, White Horse Press, Winwick, Cambridgeshire, U.K.

Singer, Brent A 1988, "An extension of Rawls' theory of justice to environmental ethics", *Environmental Ethics*, August 1, 1988 <https://doi.org/10.5840/enviroethics198810317>.

Smil, Vaclav 2003, *Energy at the crossroads: Global perspectives and uncertainties*, The MIT Press, Cambridge, MA.

Snow, David A, Burke Rochford, R, Jr, Worden, Steven K & Benford, Robert D 1986, "Frame alignment processes, micromobilization, and movement participation", *American Sociological Review*, vol. 51, pp. 464–81.

Snow, David A & Benford, Robert D 1988, "Ideology, frame resonance, and participant mobilization", *International Social Movement Research*, vol. 1, pp. 197–218.

Stamatellos, G 2015, "Virtue and Hexis in Plotinus", *International Journal of the Platonic Tradition*, vol. 9, no. 2, pp. 129–145.

Stearns, Ami E., & Thomas J. Burns 2011, "About the human condition in the works of Dickens and Marx", *Comparative Literature and Culture*, 13(4).

Steingraber, Sandra 2010, *Living downstream: A scientist's personal investigation of cancer and the environment* (2nd ed.), DeCapo, Philadelphia, PA.

Streeck, Wolfgang 2016, *How will capitalism end? Essays on a failing system*, Verso, London.

Swidler, Ann 2001, "What Anchors Cultural Practices?", In TR Schatzki, KK Cetina & E VonSavigny (eds.), *The practice turn in contemporary theory*, Routledge, New York, pp. 74–92.

Taylor, Charles 2012, *A secular age*, The Belknap Press of Harvard University Press, Cambridge, MA.

Thompson, William Irwin 1971, *At the edge of history and passages about earth: An explanation of the new planetary culture*, Lindisfarne Press, Hudson, NY.

United Nations 2018, Agenda 21: "Earth Summit: The United Nations Programme of Action from Rio: United Nations: 8601400548905: Amazon.Com: Books", accessed March 6, 2018. <https://www.amazon.com/Agenda-21-Summit-Nations-Programme/dp/1482672774/ref=sr_1_1?s=books&ie=UTF8&qid=1520306052&sr=1-1&keywords=UN+agenda+21>.

Vance, JD 2016, *Hillbilly elegy: A memoir of a family and culture in crisis* (reprint ed.), Harper, New York.

Wheeldon, Johannes & Faubert, Jacqueline 2009, "Framing experience: Concept maps, mind maps, and data collection in qualitative research", *International Journal of Qualitative Methods*, vol. 8, no. 3, pp. 68–83.

White, Lynn 1967, "The historical roots of our ecologic crisis", *Science, New Series*, vol. 155, no. 3767, pp. 1203–1207.

Wolf, Eric 1982, *Europe and the people without history*, University of California Press, Berkeley, CA.

Wolff, Richard 2016, *Capitalism's crisis deepens: Essays on the global economic meltdown*, Haymarket Books, Chicago, IL.

8 Regenerative economics[1]

L. Hunter Lovins

Collapse: Is it possible?

Think about it: Total civilizational collapse. The loss of everything that you care about. Impossible?

Climate change and its consequences are existential threats (Ramanathan, Molina & Zaelke 2017; Stockholm Resilience Center 2018). Year 2016 was the hottest ever, the third such year in a row (Greenfieldboyce 2017). Year 2017 only narrowly missed setting a new temperature record (Sevellec & Drijfhout 2018), settling for an American record for most money lost in billion dollar storms. The 2016 annual survey of chief executive officers (CEOs) at the World Economic Forum at Davos named global warming the greatest challenge facing humanity (World Economic Forum n.d.). Bishop Desmond Tutu names it the human rights challenge of our time, responsible for many of the challenges faced by the poorest people, including loss of life, lack of fresh water, the spread of disease, and rising food prices (Tutu 2014).

Religious conflicts and civil wars from Africa to Syria and Iraq to Afghanistan are all worsened by climate change (Fishetti 2015), unleashing a flood of 67 million refugees (Vidal 2009), more people on the move than at any time since World War II (Baker 2015). Threatening the stability of the European Union (Bilefsky & Smale 2016), this flow of humanity is driving xenophobic populism around the world (Woods 2016). Frustrated young men with nowhere to go, no jobs, and no prospects are increasingly easy to radicalize, resulting in attacks for which there is no defense (Aisch, Pearce & Rousseau 2016). The former UN climate chief Christiana Figueres puts it,

> People have lost trust that their lives can get better and that institutions are on their side. This in turn is leading to apathy, depression, despair and in some cases to the development of radical views. This cycle must be stopped, before it consumes our collective future. (Figueres 2016)

Figueres et al. (2015) warn that the world has three years before the worst effects of climate change become inevitable. They urged companies, communities, countries, and citizens that failure to cut carbon emissions means massive insurance costs, fires, floods, droughts, rising sea levels, extreme weather, and agricultural losses. Coupled with inequality and an unsustainable economy this threatens the basis of society.

In 2015, a group of applied mathematicians released the HANDY study (Motesharreia, Rivas & Kalnay 2014). By reviewing five thousand years of history, it was found that, "cases of severe Civilisational disruption due to 'precipitous collapse'—often lasting centuries—have been quite common." The subtitle "Is industrial civilization headed for irreversible collapse," sets forth the thesis. Using a NASA-funded climate model, it explored the history of collapses, warning: "Under conditions [...] closely reflecting the reality of the world today... we find that collapse is difficult to avoid" (Motesharreia, Rivas & Kalnay 2014). Historic collapses, the study argued, were human-induced with two underlying causes (Motesharreia, Rivas & Kalnay 2014):

1 "...the stretching of resources due to the strain placed on the ecological carrying capacity" [emphasis added]; and
2 "...the economic stratification of society into Elites [rich] and Masses (or "Commoners") [poor]"

The study elicited reams of criticism, most posted on ideological websites (Angus 2015). Critics objected that the use of mathematical models made collapse seem unavoidable. To be fair, the authors of the study stated, in terms, that collapse is not inevitable, but it is close. In 2018, much of the world seemed poised for systemic collapse. In late 2017, 15,000 scientists reissued a warning to humanity:

> 'Widespread misery and catastrophic biodiversity loss' unless business-as-usual is changed. By failing to adequately limit population growth, reassess the role of an economy rooted in growth, reduce greenhouse gases, incentivize renewable energy, protect habitat, restore ecosystems, curb pollution, halt defaunation, and constrain invasive alien species, humanity is not taking the urgent steps needed to safeguard our imperiled biosphere. (Ripple et al. 2017)

Expert consensus could not be clearer: Every major ecosystem is threatened (Conservation on Biological Diversity n.d.). Economic inequality has a corrosive effect on health, social cohesion, economic growth, education, and crime (Pickett & Wilkinson 2011). Without narrowing the gap between the richest and poorest in our societies, other attempts to fight poverty and stabilize the environment will fail. Yet the gap grows wider: In 2016, Oxfam (2017) estimated that eight people had as much wealth as the poorest 3.5 billion people on the planet. This number replaced Oxfam's estimate of 62 people being richer than the bottom half just a year before. This means that the richest 1% in the world has more wealth than the other 99% of humanity (Oxfam 2017). Already we see people begging on the streets in major cities (National Alliance to End Homelessness 2015), infrastructure crumbling (Golson 2015), American cities unable to supply clean drinking water to their citizens (Dolan 2016), and companies, communities, and countries who say they cannot afford to solve these crises. In the wake of Hurricane Maria, 3.5 million Americans on Puerto Rico suffered collapse. Months later, thousands of citizens of the wealthiest nation on the earth remained without electricity, fuel, clean water, or reliable supplies of food (Hernández, Leaming & Murphy 2017).

Poor communities hit by violent weather and developing countries without infrastructure to withstand sudden shocks are already experiencing various levels of collapse. Between 2008 and 2016, climate-related disasters displaced 21.5 million people. Millions of Chinese (Mosbergen 2015) and Indians (Reuters 2016) die every year from acute air pollution. These scary statistics seem unendurable. Kids know. They ask if they were going to have a future, knowing that climate change and other environmental harm could cut short their lives (McDonald 2014). Young people now suffer record rates of affective anxiety disorder (fear of the future); some say as high as % of the youth population. Suicide, after years of falling rates, is at its highest level in 50 years, with US homicides being tripled.(Hay 2015). Teen suicide was the second largest cause of death for youths aged 15–24 (Neal 2012). They're not just scared of monsters under the bed. The failure of the nations of the world to curb climate change is now joined by failure to reduce nuclear arsenals. Nuclear brinksmanship creates scares of actual launches. In 2018, a false alarm gave Hawaiian citizens 38 minutes of terror as the Governor scrambled to remember his Twitter password to tell panicked residents that someone had flipped the wrong switch (Gilmer 2018).

Esquire titled a 2015 article, "When the End of Human Civilization is Your Day Job." It mentions: "Among many climate scientists, gloom has set in. Things are worse than we think, but they can't really talk about it." It profiled the emotional trauma, nightmares, and depression felt by climatologists who tracked the indicators showing that climate change is happening far faster, even faster than their most pessimistic models. They have the science. They know just how bad things are going to get. And they have to watch as evermore frightening science fails to rouse a somnolent population to do anything about it (Richardson 2015).

It is a mess

Or if it is not, as Tommy Lee Jones (2007) said: "It will do till the mess arrives." It is important to recognize that the woes of the world are not an accident. Nelson Mandela (2005) said: "Like slavery and Apartheid, poverty is not natural. It is man-made and it can be overcome and eradicated by actions of human beings." The good news is that if we created this mess, we can uncreate it. Buckminster

Fuller said: "You never change things by fighting the existing reality. To change something, build a new model that makes the existing model obsolete." To craft a finer future, institutions around the world must commit to three outcomes (Lovins et al. 2018):

1 Enable all people to achieve a flourishing life within ecological limits.
2 Deliver universal well-being as we meet the basic needs of all humans.
3 Deliver sufficient equality to maintain social stability and provide the basis for genuine security. Security is defined by Webster's Dictionary as "freedom from fear of privation or attack." Privation is defined as "lack of basic necessities or comforts of life."

The threats facing humanity are symptoms of an economy that is degenerative: Liquidating human and natural capital to generate more financial and manufactured capital—money and stuff. This economy is the result of a story, created for purpose by 36 men in 1947 (Raworth n.d.). It is termed neoliberalism, and was built to defend individual creativity from the evils of fascism and communism. The narrative says that humans are at heart greedy bastards. It tells us that money is the only measure of success, and, harking back to the old Calvinist belief, that those who have it are somehow superior to the rest of us. And that is OK, the story goes, because in a perfect free market me against you will somehow aggregate to the greater good. All that matters, says this narrative, is that each of us is free to express our individual liberty in an unfettered market as much as possible. Government, therefore, should be as small as possible. Its role is only to protect private property and access to the market. This story is baked into mental models of mostly everyone in business, academia, and policy (Monbiot 2016). Outside of Scandinavia, it governs global economic policy. And it is impoverishing the planet and most of humanity, driving us over a cliff. As Thomas Berry says,

> We are in trouble just now because we do not have a good story [....] The old story–the account of how the world came to be and how we fit into it [...] sustained us for a long period of time. It shaped our emotional attitudes, provided us with a life purpose, energized action. It consecrated suffering, integrated knowledge, guided education [....] We need a (new) story that will educate man, heal him, guide him. (n.d.)

To supplant the prevailing neoliberal myth, we need a new narrative of a world that works for 100% of humanity, as Buckminster Fuller Institute (n.d.) puts it. The new story we need is emerging. It begins with the three time-tested principles of Natural Capitalism (Lovins et al. 2018). First, as companies and communities begin the journey to becoming more sustainable, they should use all resources dramatically more productively. This is profitable, but more importantly, it buys time by pushing back crises such as climate change. This time should be used to redesign how all goods and services are made and delivered., using such approaches as the circular economy and biomimicry (Ellen MacArthur Foundation 2013). This includes closing loops in material flows to eliminate waste, and the use of renewable energy (Gilding 2015; Leggett n.d.; Lovins 2015; Wijkman 2017). For any system to be truly sustainable, it must be managed so that it is regenerative of all forms of capital, but especially human and natural capital, the forms that our current economic system is liquidating. It has become fashionable to denigrate sustainability because it is not enough. Some people criticize "green" activities as being only "less bad," or "uninspiring." But this circular firing squad confuses and disheartens people facing a world rushing to economic and ecological catastrophe (see http://rio20.net/en/documentos/state-of-the-planet-declaration/).

Around the world activists seek to stop at least some of the destruction facing us: Professors (see http://billmckibben.com/), scientists, businesspeople (Grantham 2014), and young people (see http://www.idlenomore.ca/) get arrested to stop the mining of coal and the pipelines to ship fracked oil. Students and clergy (see http://350.org/) demand that their universities and churches divest from ownership in fossil fuels, and human rights activists fight inhumane conditions in Bangladeshi factories (see http://www.laborrights.org/creating-a-sweatfree-world/sweatshops) or biopiracy in India (see http://www.navdanya.org/). Conservationists save remnants of intact ecosystems (see https://wilderness.org/), and agency personnel enforce pollution regulations (see http://www.eea.europa.eu/). Practitioners of corporate social responsibility drive enhanced profits from cutting business impacts (see http://www.wbcsd.org/Overview/About-us) and ensure that workers and communities are treated decently (see http://www.ilo.org/global/about-the-ilo/lang--en/index.htm). Green developers create less wasteful, more delightful structures to deliver higher productivity because employees do better in cleaner environments (see http://www.usgbc.org/articles/leed-dynamic-plaque-diaries-alliance-center). Organizations such as CDP (https://www.cdp.net/en) and the Global Reporting Initiative (https://www.globalreporting.org/) set standards, measure reduction of impacts, and occupy neighboring rungs—or the lower rungs in a ladder. New accounting systems such as the International Integrated Reporting Committee (http://integratedreporting.org/the-iirc-2/), the Sustainability Accounting Standards Board (http://www.sasb.org/), and other metrics sit on rungs proving the business case for more responsible behavior. These servants of "sustainability" are bringing the system back to neutral—able, unlike now, to endure indefinitely without collapse. As the graph in Figure 8.1 shows, this activism is more than noble; it's essential if we are to preserve life as we know it.

But increasingly, even those who give their lives to this work sense that it is insufficient. As Jo Confino, Executive Editor of Huffington Post puts it, "The status quo is a huge beast with claws sharpened and teeth bared. All the new models that people are pushing are like mice running around bumping into each other" (personal communication).

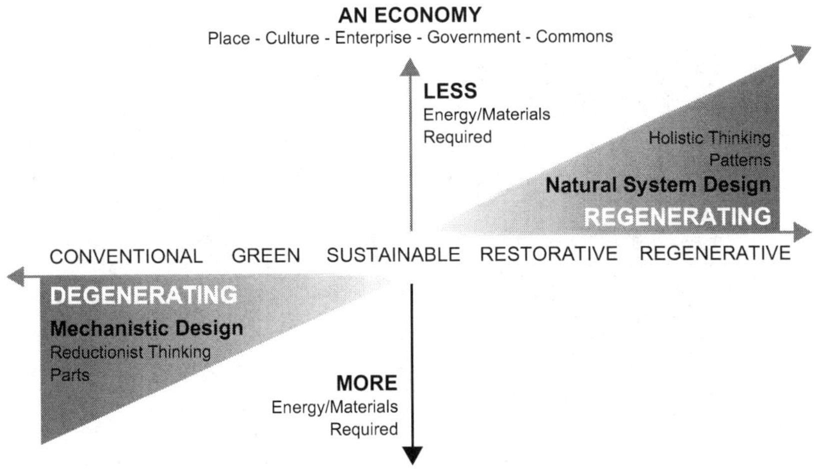

Figure 8.1 Ramp to a finer future, graph from the Capital Institute

So, the real question is, do we have the courage to change our economic system, to create an economy in service to life, not consumption? If we integrate the efforts to create a finer future, seeing them as an arc of transition, we can recognize that all are essential contributions. They form a ramp up away from "he who has the most toys wins" to enable us to achieve equitable distribution of scarce resources and to maximize well-being for everyone within planetary limits.

The regenerative economy

John Fullerton spent 18 years at JP Morgan, leaving as Managing Director. In 2001 he walked away to create a new approach. Realizing that the essence of life and the evolutionary process is regeneration, he realized that sustainability is the result, not the means of getting development right. Nature, he points out, is sustainable *because* it is regenerative. Fullerton outlined what he calls regenerative capitalism (Fullerton 2015), articulating what an economy aligned with natural systems and the laws (not theories) of physics would look like. He pointed out that, according to leading evolutionary theorists, there are patterns and principles that nature (living and nonliving alike) uses to build stable, healthy, and sustainable systems throughout the world (Chaisson 2002). Fullerton set forth eight principles of regenerative capitalism:

1 **Right relationship:** Holding the continuation of life sacred and recognizing that the human economy is embedded in human culture, which is itself embedded in the biosphere. All systems—from molecular scale all the way to cosmic scale—are nested, interconnected, and defined by overarching relationships of mutualism within which any exchange takes place.
2 **Innovative, adaptive, and responsive:** Drawing on the innate ability of human beings to innovate and "create anew" across all sectors of the

society. Humans are innately creative and entrepreneurial. Even in failure, we "begin again."

3 **Views wealth holistically:** True wealth is not money in the bank. It is defined in terms of the well-being of the "whole," achieved through the harmonization of the multiple forms of capital, with system health only as strong as the weakest link. Well-being depends on belonging, on community, and on an array of community-stewarded assets.

4 **Empowered participation:** All participants of the system must be empowered to participate in and contribute to the health of the whole. Therefore, beyond whatever moral beliefs one may hold, financial and non-financial wealth must be equitably (although not necessarily equally) distributed in the context of an expanded understanding of systemic health. We all long to be part of something bigger than ourselves.

5 **Robust circulatory flow:** Like the metabolism of any healthy system, resources (material and nonmaterial) must circulate up and down the system efficiently and effectively. Circular economy concepts for material and energy are one important aspect of this principle at work in a regenerative economy.

6 **"Edge effect" abundance:** In nature the most abundant places are where two ecosystems come together; where a meadow meets a forest, or a river the sea. Creative collaborations across sectors of the economy increase the possibility of value-adding wealth creation through diversity of relationships, exchanges, and resiliency.

7 **Seeks balance:** Balances resilience, the long-run ability to learn and grow stronger from shocks with efficiency, which while more dynamic, creates brittle concentrations of power. Living within planetary boundaries, without collapse, requires economic systems that are designed for a balance of efficiency and resilience and are built on patterns and principles that mirror those found in healthy natural systems.

8 **Honors community and place:** Operating to nurture healthy, stable communities and regions, both real and virtual, in a connected mosaic of place-centered economies. We can have a global exchange of goods and services so long as it ensures the unique integrity of each place.

The principles are not absolutes. They are evolving, and interconnected. Together, they sketch a complex pattern that is beyond linear description, but essential to creating the world we want.

Beginning to be implemented by organizations such as Savory Institute, Natural Capitalism Solutions (www.natcapsolutions.org), Capital Institute, and regenerative agriculture groups across the world (The Nature Conservancy 2013), these principles align economics with the way the rest of the world works. They are a guide to how to create an economy that creates conditions conducive to life (Benyus 2016). Part of a rapidly emerging field of holistic thinking, they draw from the best thinking in ecological economics (Costanza 2010), economic democracy (Peterson 2008), evolutionary biology (Wilson 2013), and the emerging discipline of Humanistic Management (Pirson et al. 2014) to offer a more accurate story of who we are as human beings. They are the essential first step to grounding the new field of regenerative development. They also align with what the world's great religions tell us. Religious leaders of all of the great faiths have joined in acknowledging that climate change hurts the poorest first and worst (Goldenberg 2014). Pope Francis (2015), in his encyclical, *Laudato Si*, called on all humanity to respond to the climate crisis. Moreover he called on all people to address the failure of the current economic system to care for humans or the earth. The Holy Father stated,

> In the face of the emergencies of human-induced climate change, social exclusion and extreme poverty, we join together to declare that: Human induced climate change is a scientific reality, and its decisive mitigation is a moral and religious imperative for humanity.

Transformation ahead

We have the technologies to solve the worst of the crises facing us, and buy the time to deal with the rest. There are many facets to a regenerative economy, far too many to cover in one chapter. But one aspect is preeminent: It will enable us to solve the climate crisis. There are two key transformations needed: The decarbonization of the economy and the shift to regenerative agriculture.

In 2009, a Stanford scientist Dr Mark Jacobson showed that renewable energy could power the entire world by 2030 (see Jacobson & Delucchi 2009). His Solutions Project (http://thesolutionsproject.org/infographic/) calculated how to do this for every state in the United States and in many other countries. More recently, scholars have shown how to do this with photovoltaics alone (Ram et al. 2017). Stanford Professor Tony Seba (2017) argues that not only is the transition to renewables possible, it is also inevitable. In his book *Clean Disruption* (Seba 2014), Seba describes how the convergence of disruptive technologies and business models means that the whole world will be renewably powered by 2030. He ascribes this to four factors: The fall in the cost of solar energy, the fall in the cost of storage (batteries), the rise of the electric vehicles, and the advent of the driverless cars. By delivering renewably powered electric vehicle Transit-as-a-Service, which is ten times cheaper than the current privately owned internal combustion cars, we will move to a completely renewable energy system.

A new solar array goes up in the United States every 150 seconds (Roston 2015). But can the whole world be renewable by 2030? In August 2017, China announced it had already eclipsed its 2020 goal in solar installations (Morgan 2017). It now adds 45 gigawatts (GW) of solar every year (more than the entire installed solar capacity of Germany). California predicted it will hit its declared 2030 target of getting 50% of its power from renewable energy by 2020 (Roselund 2017). The state is now debating resetting the goal to 100% renewable power. Why? Because the cost of solar energy is falling rapidly. When the Kentucky Coal Museum puts solar panels on its roof rather than plug into the coal fired electric grid at its doorstep, you know that the fossil era is over (Watkins 2017). In October, Saudi Arabia announced the new world record low price: 1.7¢/kWh (Dipaola 2017).

So we can run our society on solar energy, but what if the sun isn't shining or the wind blowing? Storage technology to make renewable energy available 24/7 is only in its infancy as an industry, but its prices are falling too, and deployment accelerating (Spector 2017). When the Aliso Canyon natural gas well blew out in 2016, spewing 100,000 metric tons of methane, a potent greenhouse gas (GHG), the local utility feared it would be unable to keep lights on in Southern California (Environmental Defense Fund 2017). Tesla and others brought online more than 70 megawatts (MW) of energy storage at a price cheaper than building a new gas peaking plant, and vastly faster (Pyper 2017). Since then, Tesla built 100 MW battery backup for renewable power in South Australia (Spector 2017), and is now creating a 250 MW distributed power plant made up of solar panels on people's house roofs and batteries in Australian garages (Harmsen 2018).

In late 2016, the Financial Times reported that Fitch Ratings warned: "Widespread adoption of battery-powered vehicles (EVs) is a serious threat to the oil industry (Clark, Ward & Hume 2016)." Transport accounts for 55% of total oil use globally; 71% in the United States (Institute for Energy Research 2012). A significant shift to EVs would rip the guts out of the oil industry, and endanger the banks, as well: A quarter of all corporate debt, perhaps as much as $3.4 trillion, is related to utility and car company bonds that are tied to fossil fuel use (Parkin 2016). Anyone doubting this risk was sobered when China announced in September 2017 that it was going to phase out internal combustion vehicles (LeSage 2017). Given that China represents a quarter of the global automobile market, this—coupled with similar announcements by India, France, the United Kingdom, and Norway—is an existential crisis for both the oil and car industries. Shortly thereafter, General Motors (GM), which had reclaimed its coveted status of the world's leading automobile manufacturer on the strength of its Bolt EV, announced that its future is electric (Davies 2017). Daimler, Volkswagen, and Volvo have also committed to electrifying their product portfolios. In November, Elon released an all-electric long haul truck (Wong 2017), and China announced the launch of the world's first all-electric cargo ship (Kyree 2017).

In Seba's scenario it is the autonomous electric vehicle (AEV) that drives the real reduction in cost; he claims this will make the shift to renewables inevitable. Are AEVs more than just science fiction? Tom Chi, the head of Product Experience at Google X and one of the designers of the self-driving Google car (see https://bsahely.com/2017/11/11/watch-everything-is-connected-heres-how-tom-chi-tedxtaipei-on-youtube/), confirms that cars that drive themselves will be ubiquitous within ten years. Tesla released its driverless vehicle when it was as safe as a human-driven car, noting that hundreds of thousands of people die every year in human-driven car crashes. Teslas have driven more than 5 billion miles in autonomous mode (Lambert 2017), en route to the company's 10 billion mile safety proof point. In fact, all Teslas are now capable of full autonomous mode

(Messner 2017). The Google car has driven four million real miles, and 2.5 billion simulated miles. GM just announced that it is pivoting its business model to offer autonomous electric vehicle Transit-as-a-Service by 2019 (Welch 2017).

If all this is true, it has profound implications for, well, everything. It means that the solution to the climate crisis is already well underway. It also means that we will construct an entirely new economy, like it or not. Failure to manage this transition carefully will mean the dissolution in value of the oil, gas, coal, uranium, nuclear, utility, and auto industries; the banks that hold the loan papers for all of these companies; and the pension funds and insurance companies that are invested in them. It will mean an economic collapse on a scale never before seen coming at us within about ten years. Consider oil. In 2011, Carbon Tracker calculated that at least 80% of the fossil deposits still in the ground would have to stay in the ground if the world is to avoid warming beyond 2C more than pre-industrial levels (Carbon Tracker Initiative 2014). Given that these fossil assets are on the balance sheets of some of the world's wealthiest companies and form the basis of the sovereign wealth funds of many nations, John Fullerton of the Capital Institute predicted that this implied a write-off of at least $20 trillion dollars (Capital Institute 2011). In contrast, Fullerton warned, the 2008 financial collapse was triggered by the stranding of only $2.7 trillion in mortgage assets. If Tony Seba is right, we are looking at the mother of all disruptions.

What can be done? First, it appears unwise to have any of your assets in the industries that will be disrupted. Bevis Longstreth, former Securities and Exchange Commissioner, observed: "It is entirely plausible, even predictable, that continuing to hold equities in fossil fuel companies will be ruled negligence" (Longstreth 2013). Companies such as Change Finance have created truly fossil-free Exchange Traded Funds (Change Finance 2017).

It is also a good idea to look at how you live your life: Where does your energy comes from? Are you dependent on an industry that is at risk? Millions of people, communities, and cities are going 100% renewable. Companies will either become part of the solution, or they will not be a problem because they will not be around. The emerging industries are creating millions of jobs, but millions are at risk. Will we substitute a Universal Basic Income? (Shiller 2017) or a guaranteed jobs program? (Wray 2009) Will we descend into unimaginable darkness? Or will we create a finer future? (Lovins et al. 2018)

Regenerative design will be key to the answer. This is starting to emerge in some cities, as well as in some rural areas. Recognizing that roughly 80% of the world's economic activity came from cities, mayors around the world have begun to join together to take up the jobs that too many national governments are leaving undone (Capital Institute 2016). Mayors are taking seriously the numbers set forth in such reports as Risky Business (Gorden 2014), in which American business leaders Tom Steyer, Hank Paulsen, and Michael Bloomberg show that solving the climate crisis would unleash billions of dollars in investment and create millions of jobs and the New Climate Economy report, which shows that decarbonizing the economy would deliver a net economic gain of $26 trillion to the world economy.

The other half of the solution to the climate crisis, and a key part of creating a regenerative economy is to enhance the ways that nature pulls carbon from the air and restores it to the soil. In nature carbon is not the greatest poison: It is the building block of life. Modern industrial agriculture is energy and chemical intensive. By plowing vast swathes of land, it decarbonizes the soil. Overapplication of nitrogen fertilizers denitrifies the soil and causes emissions of nitrogen oxides, another potent GHGs. Confined animal feeding operations release enormous amounts of methane, a far worse GHG than carbon dioxide which causes about a quarter of the emissions of GHGs, and is itself very vulnerable to climate change. It is also degenerative of rural economic integrity and human health. As a result, there is increasing interest in what Robert Rodale called regenerative agriculture. Regenerative systems, he said, enhance the resources that they use, leaving them more abundant instead of depleted or destroyed. Such systems are holistic, enhancing innovation and delivering environmental, social, economic, and spiritual well-being (see http://regenerationinternational.org/why-regenerative-agriculture/). This aligns well with Fullerton's principles above, and provides a foundation for both rural and urban economies that deliver shared prosperity.

Regenerative agriculture uses nature's wisdom, not human brute force. This approach has been taught for decades by Allan Savory. The Savory Institute (SI) is now the leading proponent of ways to use agriculture to solve the climate crisis. SI (https://www.savory.global/institute/) seeks to impact a billion hectares by 2025 through the teaching and practice of Holistic Management and Holistic Decision-Making. These enable practitioners to turn deserts into thriving grasslands; restore biodiversity; bring streams, rivers, and water sources back to life; and combat poverty and hunger, all while reversing global climate change. Holistically managed grazing animals are one of the best ways to reclaim depleted land. Savory's approach mimics how vast herds of grazing animals coevolved with the world's grasslands: Dense-packed because of predators, moving as a herd, eating everything, fertilizing the land, tilling the manure and seeds into the soil with their hooves, and then moving on, not returning until the grass is lush again. This interaction is one of the more important ways to create healthy communities of soil microorganisms. They, in turn recarbonize the soil, and restore natural nitrogen cycles. Savory, in his many writings and a TED talk viewed by three and a half million people, argues that even achieving zero emissions from fossil fuels would not avert major catastrophe from climate change (for video see: https://www.ted.com/talks/allan_savory_how_to_green_the_world_s_deserts_and_reverse_climate_change). Grassland and savannah burning would continue and desertification would accelerate as soils become increasingly unable to store carbon or water (see http://www.marincarbonproject.org/science/land-management-carbon-sequestration. Averting disaster, Savory says, will require a global strategy to cut carbon emissions, substitute benign energy sources for fossil fuels, and implement effective livestock management practices to put the carbon already in the atmosphere back into the soils. Profitable Holistic Management is the only way, he argues, to reduce biodiversity loss and biomass burning and reverse the desertification that is not caused by atmospheric carbon buildup (see https://www.youtube.com/watch?v=uEAFTsFH_x4). Awarded the 2010 Buckminster Fuller Challenge Prize for decades of work, Savory's Africa Centre for Holistic Management in Zimbabwe and SI shows how to transform degraded grasslands and savannahs into lush pastures with ponds and flowing streams. This award recognized Savory's work to accelerate development and deployment of whole-systems solutions to climate change, and sustainable development.

One of the best examples of regenerative agriculture is the work of Gabe Brown. A commodity corn and soybean farmer, who converted his 2,000 acres near Bismarck, North Dakota to regenerative agriculture to cut costs that were threatening his livelihood. When he began in 1993, his soil quality was poor, requiring annual inputs of fertilizer, pesticides, and herbicides to produce a crop. In 1995, Brown stopped plowing his land. In 1997 he added the use of a wide variety of multispecies cover crops. In 2006, he introduced Savory-style grazing practices, adding different livestock species—now raising cows, sheep, broiler hens, bees, and corn and soybeans. Not needing chemical inputs or fossil energy cut his costs and increased his profitability. In 2014, it cost him $1.35 to produce a bushel of corn, which he sold for more than $3.50. He cannot keep up with demand for his grass-finished beef and lamb, and his fields have never been healthier. The capacity of his acreage to cycle nutrients, including carbon, exceeds that of his neighbors who farm organically—but without animal impact (Brown 2017). It also exceeds two no-till operations that use varying amounts of synthetic fertilizers. Soil samples from his neighbors' operations, show that animal impact increases concentrations of nitrogen (N), phosphorus (P), and potassium (K) dramatically (see https://www.youtube.com/watch?v=9yPjoh9YJMk). The water-extractable organic carbon (WEOC) is, however, the most amazing outcome of Brown agricultural practices. When he bought his farm in 1993, it had shallow soils with 1.3% soil organic matter (soil carbon). By 2013, he had plots with more than 11% soil organic matter. By 2018, it was up to 15%. Brown is rolling climate change backwards, recarbonizing his soils at a profit. As he puts it, if your soil is healthy you will have clean water, clean air, healthy plants, healthy animals, and healthy people. You will have a healthy ecosystem.

Such practices improve more than the soil. Will Harris converted White Oak Pastures (http://www.whiteoakpastures.com/), his family's commodity farm in South Georgia, to a successful regenerative operation by raising, slaughtering, and selling five kinds of poultry, five kinds of red meat (all pasture raised), eggs, and vegetables. The products are sold online, to high-end restaurants as far away as Miami, and to Whole Foods markets across the Eastern United States Will employees 137 residents of the once-decaying town of Bluffton. His commodity farmer neighbor, with the same acreage, employs four (personal communication). White Oak Pastures features agritourism, a restaurant, a general store, and serves as the Savory Hub for the region.

How much carbon can be sequestered in properly managed grasslands and how fast? In some California experiments, manure from dairy and beef operations is blended with green waste that would otherwise go to landfill, impose costs, rot, and release methane. The mix is composted and spread on pastures. Scientists from the University of California, Berkeley, take annual soil cores a meter deep and test whether that soil has soaked up additional carbon. One application of compost to rangeland, for example, doubled grass growth and increased carbon sequestration by up to 70%. Every year the carbon increases. The study found that this can achieve total GHG mitigation rates over a 30-year time frame of more than 18 tons of CO_2-equivalents per acre of land treated with organic amendments (http://www.carboncycle.org/marin-carbon-project/).

Dr David Johnson, Director of the Institute for Sustainable Agricultural Research at New Mexico State University, has developed a similar approach (Johnson, Ellington & Eaton 2015; Kelley 2014; Teague et al. 2016). His research has shown that,

> …promoting beneficial interactions between plants and soil microbes increases farm and rangeland's efficiency for capturing carbon and storing it in soil. These same interactions increase soil microbial carbon-use efficiencies reducing the rate at which soil carbon, as CO_2, is respired from the soil. When this bio-technology is promoted in agro-ecosystems, it is feasible to capture and sequester an average of >11 metric tons of CO_2 per hectare per year in rangeland soils and >36.7 metric tons CO_2 per hectare per year in transitioning farmland soils all for less than one-tenth the cost of EPA's recommended Carbon Capture Utilization and Storage (CCUS) technologies.

The world's permanent pasture and fodder lands amount to roughly 3.4 billion hectares. Back-of-the-envelope calculations with Seth Itzkan of Soil4Climate show that multiplying calculations of carbon sequestration by the global hectares of pastureland gives 10.2 GtC/yr potential soil carbon capture via grazing. That, alone, would offset all human emissions (personal communication).

Clearly, it is a big "if" to say that Holistic Management will be practiced on all of the world pastureland, but coupled with other forms of regenerative agriculture and the reductions in carbon emissions possible and profitable through good energy policy, it is clear that we can solve the climate crisis and do it in ways that are profitable.

Creating a finer future

We have created an economy designed to maximize possession of financial and built capital, but doing this destroys our life support systems. Money and stuff are useful but increasing them by sacrificing human and natural capital is daft, and is just bad capitalism. Intact community and ecosystems are far more valuable forms of capital, because without them there is no social stability and no life, and thus no economy. We need to create what is now being called a well-being economy, one that delivers shared well-being on a healthy planet. The great scientist and lead author of *Limits to Growth* (Meadows, Randers & Meadows 2004), Dana Meadows, believed that we can do it. We *can* avoid collapse. The sustainability vision she outlined looked very like what the world needs to counter the threat of climate change driven collapse (Meadows 1994; Meadows, Meadows & Randers 2016). She observed,

> People don't need enormous cars; they need respect. They don't need closetsful of clothes; they need to feel attractive and they need excitement, variety, and beauty. People need identity, community, challenge, acknowledgement, love, joy. To try to fill these needs with material things is to set up an unquenchable appetite for false solutions to real and never-satisfied problems. The resulting psychological emptiness is one of the major forces behind the desire for material growth. A society that can admit and articulate its nonmaterial needs and find nonmaterial ways to satisfy them would require much lower material and energy throughputs and would provide much higher levels of human fulfillment.

Dana believed that a sustainable world is possible. She believed in human potential, and the power of telling optimistic stories.

Many more such stories are found in the 2018 book: *A Finer Future: Creating an Economy in Service to Life*, (Lovins et al. 2018) that lays out why our current neoliberal narrative is inadequate, and how a new story of a regenerative economy can deliver shared prosperity on a healthy planet.

Conclusion

Humanity stands on a cliff's edge, at risk of total system collapse (Motesharreia, Rivas & Kalnay 2014). As the great environmentalist David Brower once said, when you stand at a precipice, the only progressive move is to turn around (see https://www.youtube.com/watch?v=b-yRWsOBlBg).[2] Regenerative development is the way by which humanity can make this conceptual turn and begin to move forward. It goes beyond older framings of sustainable development to embrace the principles of living systems as the basis for development. It shows how humanity can ascend from the degenerative world in which we live, and how more sustainable practices are rungs on a ladder to reach an economy in service to life.

Note

1. This article is drawn from the book from New Society Publishers: *A Finer Future: Creating An Economy In Service to Life*, by Hunter Lovins, Stewart Wallis, Anders Wijkman and John Fullerton. Published Fall 2018, it outlines the risk facing humanity of total system collapse, and the need for a new narrative upon which to craft a regenerative economy.
2. Hayes, Randy, "The Great U-Turn From Cheater Economics to True-Cost Economics," You tube, 5 February 2014, https://www.youtube.com/watch?v=b-yRWsOBlBg

References

Aisch, G, Pearce, A & Rousseau, B 2016, *How far is Europe swinging to the right?*, viewed 5 July 2018, <http://www.nytimes.com/interactive/2016/05/22/world/europe/europe-right-wing-austria-hungary.html?_r=0>

Angus, I 2015, *What did that 'NASA-funded collapse study really say?*, viewed 31 March, 2018, <http://climateandcapitalism.com/2014/03/31/nasa-collapse-study/>

Baker, A 2015, *How climate change is behind the surge of migrants to Europe*, viewed 7 September 2018, <http://time.com/4024210/climate-change-migrants/>

Benyus, JM 2016, *A Biomimicry Primer, Biomimicry 3.8*, viewed 3 October 2018, <https://biomimicry.net/b38files/A_Biomimicry_Primer_Janine_Benyus.pdf>

Berry, T n.d., *University Story*, viewed 7 September 2018, <http://thomasberry.org/life-and-thought/about-thomas-berry/a-universe-story>

Bilefsky, D & Smale, A 2016, *Dozens of migrants drown as European refugee crisis continues*, viewed 22 January 2018, <http://www.nytimes.com/2016/01/23/world/europe/valls-france-eu-warns.html>

Brown, G 2017, *Can we really regenerate our soils?*, viewed 1 January 2018, <http://www.grazeonline.com/canweregeneratesoils>

Buckminster Fuller Institute n.d., *About Fuller: World game*, viewed 22 January 2018, <https://www.bfi.org/about-fuller/big-ideas/world-game>

Capital Institute 2011, *The big choice*, viewed 3 October 2018, <http://capitalinstitute.org/blog/big-choice-0/>

Capital Institute 2016, *City states rising!*, viewed 3 October 2018, <http://capitalinstitute.org/blog/category/city-states/>

Carbon Tracker Initiative 2014, *Unburnable Carbon – Are the world's financial markets carrying a carbon bubble?*, viewed 3 October 2018, <https://www.carbontracker.org/wp-content/uploads/2014/09/Unburnable-Carbon-Full-rev2-1.pdf>

Chaisson, EJ 2002, *Cosmic evolution, the rise of complexity in nature*, Harvard University Press, UK.

Change Finance 2017, *Change Finance Launches CHGX, Impact Investing ETF*, viewed 10 October 2017, <https://www.prnewswire.com/news-releases/change-finance-launches-chgx-impact-investing-etf-300533895.html>

Clark, P, Ward A & Hume, N 2016, *Oil groups threatened by electric cars*", viewed 18 October 2018, <https://www.ft.com/content/b42a72c6-94ac-11e6-a80e-bcd69f323a8b?ftcamp=crm/email//nbe/CompaniesBySector/product>

Conservation on Biological Diversity n.d., *Global Biodiversity Outlook 3*, viewed 3 October 2018, <https://www.cbd.int/gbo3/>

Costanza R 2010, *Flourishing on earth: Lessons from ecological economics*, viewed 3 October 2018, <https://www.youtube.com/watch?v=PZkTIVPgqG4&feature=relmfu>

Davies, A 2017, *General motors is going all electric*, viewed 2 October, 2017, <https://www.wired.com/story/general-motors-electric-cars-plan-gm/>

Dipaola A 2017, *Saudi Arabia gets cheapest bids for solar power in auction*, viewed 3 October 2018, <https://www.bloomberg.com/news/articles/2017-10-03/saudi-arabia-gets-cheapest-ever-bids-for-solar-power-in-auction>

Dolan, M 2016, *Flint crisis could cost U.S. a $300B lead pipe overhaul, agency warns*, viewed 5 March 2018, <http://www.freep.com/story/news/local/michigan/flint-water-crisis/2016/03/04/flint-crisis-could-cost-us-300b-lead-pipe-overhaul-agency-warns/81316860/>

Ellen MacArthur Foundation 2013, *Towards the circular economy: Economic and business rationals for an accelerated transition*, viewed 23 January, 2018, <https://www.ellenmacarthurfoundation.org/assets/downloads/publications/Ellen-MacArthur-Foundation-Towards-the-Circular-Economy-vol.1.pdf>.

Environmental Defense Fund 2017, *Looking to the states to improve natural gas storage policies*, viewed 24 Feb 2018, <http://breakingenergy.com/2017/02/24/looking-to-the-states-to-improve-natural-gas-storage-policies/?utm_source=hs_email&utm_medium=email&utm_content=43376094&_hsenc=p2ANqtz--CzJBfOlt8Lsc49nhC3xlNEViJBik4w71rKl7TOL44BLv3y-9ASmAlJ1RHY7QX_9pto1s--FwXzK9QJAcc7cTuBFxvk4HnDTpoQtmVZLAfaN0pPUc&_hsmi=43376094>

Figueres, C, Schellnhuber, HJ, Joachim, H, Whiteman, G, Rockström, J, Hobley, A & Rahmstorf, S 2015, *Three years to safeguard our climate*, viewed 2 March 2018, <https://www.nature.com/news/three-years-to-safeguard-our-climate-1.22201>

Figueres, C 2016, *Restoring hope*, viewed 14 July 2018, <http://www.huffingtonpost.com/christiana-figueres/restoring-hope_b_10974734.html>

Fishetti, M 2015, *Climate change hastened Syria's civil war*, viewed 2 March 2018, <http://www.scientificamerican.com/article/climate-change-hastened-the-syrian-war/>

Fullerton, J 2015, *Regenerative Capitalism: How universal principles and patterns will shape our new economy*, viewed 3 October 2018, <http://capitalinstitute.org/wp-content/uploads/2015/04/2015-Regenerative-Capitalism-4-20-15-final.pdf>

Gilding, P 2015, *Fossil fuels are finished – the rest is just detail*, viewed 23 July, 2018, <http://reneweconomy.com.au/2015/fossil-fuels-are-finished-the-rest-is-just-detail-71574>

Gilmer, M 2018, *Forgotten Twitter password comes at a real awkward time for Hawaii's governor*, viewed 23 January, 2018, <https://mashable.com/2018/01/23/hawaii-governor-missile-alert-twitter-password/#wLLPhVPpkmqi>

Goldenberg, S 2014, *Climate change: The poor will suffer most*, viewed 31 March 2018, <http://www.theguardian.com/environment/2014/mar/31/climate-change-poor-suffer-most-un-report>.

Golson, J 2015, *It's time to fix America's infrastructure*, viewed 23 January 2018, <http://www.wired.com/2015/01/time-fix-americas-infrastructure-heres-start/>

Gorden, K 2014, *Risky business, the economic risks of climate change in the United States*, viewed 3 June 2018, <http://riskybusiness.org/site/assets/uploads/2015/09/RiskyBusiness_Report_WEB_09_08_14.pdf>

Grantham, J 2014, "Be persuasive. Be brave. Be arrested (if necessary)", *Nature*, vol. 14, no. 7424, pp. 303.

Greenfieldboyce, N 2017, *2016 was the hottest year yet scientists declare*, viewed 18 January 2018, <http://www.npr.org/sections/thetwo-way/2017/01/18/510405739/2016-was-the-hottest-year-yet-scientists-declare>

Harmsen, N 2018, *Elon Musk's Tesla plans to give thousands of homes batteries: Here's how it would work*, viewed 4 February 2018, <http://www.abc.net.au/news/2018-02-04/how-tesla-sa-labor-free-battery-scheme-would-work/9394728>

Hay, M 2015, *Can big data help us fight rising suicide rates?*, viewed 6 September 2018, <http://magazine.good.is/articles/suicide-prevention-week-data-driven-efforts>

Hernández, AR, Leaming, W & Murphy, Z 2017, *Sin luz: Life without power*, viewed 14 December, 2018, <https://www.washingtonpost.com/graphics/2017/national/puerto-rico-life-without-power/?noredirect=on&utm_term=.aba4c5be0dc2>

Institute for Energy Research 2012, *Petroleum imports and exports: What's the story?*, viewed 05 October 2018, <http://instituteforenergyresearch.org/topics/encyclopedia/petroleum/>

Jacobson, MZ & Delucchi MA 2009, *A plan to power 100 percent of the planet with renewables*, viewed 3 October 2018, <http://www.scientificamerican.com/article/a-path-to-sustainable-energy-by-2030/>

Johnson, D, Ellington, J & Eaton, W 2015, *Development of soil microbial communities for promoting sustainability in agriculture and a global carbon fix*, viewed 3 October 2018, <https://peerj.com/preprints/789/>

Jones, TL 2007, *No Country for Old Men*, viewed 3 October 2018, <http://www.miramax.com/movie/no-country-for-old-men/https://books.google.com/books?id=l4L-YABUSm8C&pg=PA132&lpg=PA132&dq=Or+if+not+it%27ll+do+till+the+mess+arrives&source=bl&ots=x3zTrlYxmA&sig=PJd2BJY1UW2-mI4QCZ_KYfcH_Rg&hl=en&sa=X&ved=0ahUKEwjAk6efxajLAhWEtYMKHUESBFoQ6AEIJjAB#v=onepage&q=Or%20if%20not%20it'll%20do%20till%20the%20mess%20arrives&f=false>

Kelley, EC 2014, *NMSU researcher's carbon sequestration work highlighted in 'The soil will save us'*, viewed 05 October 2018, <https://newscenter.nmsu.edu/Articles/view/10461/nmsu-researcher-s-carbon-sequestration-work-highlighted-in-the-soil-will-save-us>

Kyree, I 2017, *China launches world's first fully electric cargo ship*, viewed 14 November 2018, <https://dailypost.in/news/international/china-launches-worlds-first-fully-electric-cargo-ship/>

Lambert, F 2017, *Tesla's global fleet reaches over 5 billion electric miles driven ahead of model 3 launch*, viewed 12 July 2018, <https://electrek.co/2017/07/12/tesla-global-fleet-electric-miles-model-3-launch/>

Leggett, J n.d., *Climate, energy, tech, and the future of civilization*, viewed 2 October 2018, <http://www.jeremyleggett.net/>

LeSage, J 2017, *China's ban on gas-powered cars could cripple the oil market*, viewed 12 September 2018, <http://www.businessinsider.com/china-ban-gas-powered-cars-effect-on-oil-market-2017-9>

Longstreth, B 2013, *The financial case for divestment of fossil fuel companies by endowment fiduciaries*, viewed 2 January, 2018, <https://www.huffingtonpost.com/bevis-longstreth/the-financial-case-for-di_b_4203910.html>

Lovins, LH, Wallis, S, Wijkman, A & Fullerton, J 2018, *A finer future: Creating an economy in service to life*, New Society.

Lovins, LH 2015, *The triumph of solar in the energy race, unreasonable*, viewed July 2018, <http://unreasonable.is/triumph-of-the-sun/>

Mandela, N 2005, *Speech at Trafalgar Square*, viewed 1 May 2018, <https://www.youtube.com/watch?v=1NennMCLG7A>

McDonald, G 2014, *Youth anxiety on the rise amid changing climate*, viewed 1 May 2018, <http://www.theglobeandmail.com/life/health-and-fitness/health/youth-anxiety-on-the-rise-amid-changing-climate/article18372258/>

Meadows, D, Randers, J & Meadows DL 2004, *Limits to growth*, Chelsea Green Publishing, London, UK.

Meadows, DH, Meadows, DL & Randers, J 2016, *Confronting global collapse, envisioning a sustainable future*, viewed 30 January 2018, <https://natcapsolutions.org/beyond-the-limits-executive-summary/>

Meadows, D 1994, *Envisioning a sustainable world*, viewed 11 October 2018, <http://donellameadows.org/archives/envisioning-a-sustainable-world/>

Messner, W 2017, *The driverless car of tomorrow looks and drives nothing like today's*, viewed 30 January 2018, <http://www.newsweek.com/self-driving-cars-tesla-innovation-550434>

Monbiot, G 2016, *Neoliberalism – the ideology at the root of all our problems*, viewed 15 April 2016, <https://www.theguardian.com/books/2016/apr/15/neoliberalism-ideology-problem-george-monbiot>

Morgan, S 2017, *China eclipses Europe as 2020 solar power target is smashed*, viewed 30 August, 2018, <https://www.euractiv.com/section/energy/news/china-eclipses-europe-as-2020-solar-power-target-is-smashed/>

Mosbergen, D 2015, *Air pollution causes 4,400 deaths in China every single day: Study*, viewed 14, August 2018, <http://www.huffingtonpost.com/entry/air-pollution-china-deaths_us_55cd9a62e4b0ab468d9cefa9.

Motesharreia, S, Rivas, J & Kalnay, E 2014, "Human and nature dynamics (HANDY): Modeling inequality and use of resources in the collapse or sustainability of societies", *Ecological Economics*, vol. 10, pp. 90–102.

National Alliance to End Homelessness 2015, *The state of homelessness in America 2012 – 2016*, viewed April 2018, <https://endhomelessness.org/resource/archived-state-of-homelessness/>

Neal, M 2012, *1 in 12 teens have attempted suicide: Report*, viewed 9 June 2018, <http://www.nydailynews.com/life-style/health/1-12-teens-attempted-suicide-report-article-1.1092622>

Oxfam 2017, *An economy for the 99%*, viewed 1 September 2018, <https://www.oxfam.org/sites/www.oxfam.org/files/file_attachments/bp-economy-for-99-percent-160117-en.pdf>

Parkin, B 2016, *Batteries may trip 'death spiral' in $3.4 trillion credit market*, viewed 18 October 2018, <https://www.bloomberg.com/news/articles/2016-10-18/batteries-may-trip-death-spiral-in-3-4-trillion-credit-market>

Peterson, C 2008, *What is positive psychology and what is it not, psychology today*, viewed 16 May 2018, <https://www.psychologytoday.com/blog/the-good-life/200805/what-is-positive-psychology-and-what-is-it-not>

Pickett, K & Wilkinson, R 2011, *The spirit level: Why greater equality makes societies stronger*, Tantor Media.

Pirson, M, Steinvorth, U, Largacha-Martinez, C & Dierksmeier C 2014, *From capitalistic to humanistic business*, Palgrave Macmillan.

Pope Francis 2015, *Laudato Si*, viewed 3 October 2018, <https://w2.vatican.va/content/dam/francesco/pdf/encyclicals/documents/papa-francesco_20150524_enciclica-laudato-si_en.pdf>

Pyper, J 2017, *Tesla, Greensmith, Aes deploy aliso canyon battery storage in record time: The companies collectively brought on-line more than 70 megawatts of energy storage in less than six month*, viewed 31 January 2018, <https://www.greentechmedia.com/articles/read/aliso-canyon-emergency-batteries-officially-up-and-running-from-tesla-green>

Ram, M, Bogdanov, D, Aghosseini, A, Oyewo, AS, Gulagi A, Child M, Fell HJ & Beyer, C 2017, *Global energy systems based on 100% renewable energy – power sector*, Lapeenrata University of Technology and Energy Watch Group, viewed 3 October 2018, <http://energywatchgroup.org/wp-content/uploads/2017/11/Full-Study-100-Renewable-Energy-Worldwide-Power-Sector.pdf>

Ramanathan, V, Molina, ML & Zaelke, D 2017, *Well under 2 degrees Celsius: Fast action policies to protect people and the planet from extreme climate change report of the committee to prevent extreme climate change*, viewed 1 September 2018, <http://www.igsd.org/wp-content/uploads/2017/09/Well-Under-2-Degrees-Celsius-Report-2017.pdf>

Raworth, K n.d., *Seven ways to think like a 21st-century economist - tell a new story*, viewed 1 September 2018, <https://www.kateraworth.com/animations/>

Reuters 2016, "*More Indians than Chinese will die from air pollution: Researcher*", viewed 18 August 2018, <http://www.financialexpress.com/economy/india-air-pollution-death-rate-to-outpace-china-researcher/351209/>

Richardson, JH 2015, *When the end of human civilization is your day job*, viewed 7 July 2018, <http://www.esquire.com/news-politics/a36228/ballad-of-the-sad-climatologists-0815/>

Ripple, WJ, Wolf, C, Newsome, TM, Galetti, M, Alamgir, M, Crist, E, Mahmoud, MI, Laurance, WF &15,364 scientist signatories from 184 countries 2017, "World scientists' warning to humanity: A second notice", *BioScience*, vol. 67, no. 12.

Roselund, C 2017, *California's big utilities to reach 50% renewable energy in 2020*, viewed 14 November 2018, <https://pv-magazine-usa.com/2017/11/14/californias-big-utilities-to-reach-50-renewable-energy-in-2020/>

Roston, E 2015, *By the time you read this, they've slapped a solar panel on your roof*, viewed 25 February 2018, <https://www.bloomberg.com/news/articles/2015-02-25/in-the-time-it-takes-to-read-this-story-another-solar-project-will-go-up>

Seba, T 2014, *Clean disruption of energy and transportation: How silicon valley will make oil, nuclear, natural gas, coal, electric utilities and conventional cars*, Beta edition.

Seba, T 2017, *Tony Seba: Clean disruption - energy & transportation*, viewed 7 July 2018, <https://www.youtube.com/watch?v=2b3ttqYDwF0>

Sevellec, F & Drijfhout, S 2018, "A novel probabilistic forecast system predicting anomalously warm 2018–2022 reinforcing the long-term global warming trend", *Nature Communications*, vol. 9, no. 3028.

Shiller, B 2017, *A universal basic income would do wonders for the U.S. economy it's not a giveaway; it's a permanent economic stimulus*, viewed 13 September 2018, <https://www.fastcompany.com/40463533/a-universal-basic-income-would-do-wonders-for-the-u-s-economy?utm_source=feedly&utm_medium=webfeeds>

Spector, J 2017a, *Study: We're still underestimating battery cost improvements*, viewed 17 August, 2018, <https://www.greentechmedia.com/articles/read/were-still-underestimating-cost-improvements-for-batteries>

Spector, J 2017b, *Tesla fulfilled its 100-day Australia battery bet. What's that mean for the industry?*, viewed 27 November 2108, <https://www.greentechmedia.com/articles/read/tesla-fulfills-australia-battery-bet-whats-that-mean-industry#gs.Ntm_jj8>

Stockholm Resilience Center 2018, *Planet at risk of heading towards Hothouse earth state*, viewed 1 August 2018, <http://www.stockholmresilience.org/research/research-news/2018-08-06-planet-at-risk-of-heading-towards-hothouse-earth-state.html>

Teague, WR, Apfelbaum, S, Lal, R, Kreuter, UP, Rowntree, J, Davies, CA, Conser, R, Rasmussen, M, Hatfield, J, Wang, T, Wang, F & Byck, P 2016, "The role of ruminants in reducing agriculture's carbon footprint in North America", *Journal of Soil and Water Conservation*, vol. 71, no. 2, pp. 156–164.

The Nature Conservancy 2013, *Wool from sustainably raised sheep play surprising role in Patagonia's grasslands restoration*, viewed 23 January 2018, <https://www.patagonia.com/on/demandware.static/Sites-patagonia-us-Site/Library-Sites-PatagoniaShared/en_US/PDF-US/PATAGONIA_COLLABORATES_WITH_THE_NATURE_CONSERVANCY_ON_SUSTAI.pdf>

Tutu, BD 2014, *Desmond Tutu: We fought apartheid. Now climate change is our global enemy*, viewed 20 September 2018, <https://www.theguardian.com/commentisfree/2014/sep/21/desmond-tutu-climate-change-is-the-global-enemy>

Vidal, J 2009, *Global warming could create 150 million 'climate refugees' by 2050*, viewed 20 September 2018, <https://www.theguardian.com/environment/2009/nov/03/global-warming-climate-refugees>

Watkins, M 2017, *Even the Kentucky coal museum is going solar*, viewed 8 April 2018, <https://www.usatoday.com/story/news/nation-now/2017/04/08/even-kentucky-coal-museum-going-solar/100205662/>

Welch, D 2017, GM's *self-driving cars to be ready for ride-sharing in 2019*, viewed 30 November 2018, <https://www.bloomberg.com/news/articles/2017-11-30/gm-sees-self-driving-ride-share-service-ready-for-roads-in-2019>

Wijkman, A 2017, *Circular economy could bring 70 percent cut in carbon emissions by 2030*, viewed 15 April 2018, <https://www.theguardian.com/sustainable-business/2015/apr/15/circular-economy-jobs-climate-carbon-emissions-eu-taxation>

Wilson, EO 2013, *The social conquest of earth*, Liveright.

Wong, JC 2017, *Elon Musk unveils Tesla electric truck – and a surprise new sports car*, viewed 17 November 2018, <https://www.theguardian.com/technology/2017/nov/17/elon-musk-tesla-electric-truck-sports-car-surprise>

Woods, N 2016, *Populism is spreading: this is what is driving it*, viewed 9 December 2018, https://www.weforum.org/agenda/2016/12/populism-is-spreading-this-is-whats-driving-it

World Economic Forum n.d., *The global risks report, world economic forum 2016*, viewed 13 October 2018, <https://www.weforum.org/reports/the-global-risks-report-2016>

Wray, LR 2009, *Job guarantee*, viewed 23 August, 2009, <https://www.linkedin.com/mynetwork/invitation-manager/>

9 CityCrafting: Evolution of regenerative development and regenerative development in practice

John L. Knott, Jr.

I have believed for many years that the three most important professions in the world are those who teach the human community, those who feed the human community, and those who shelter the human community. Those of us in every profession that shelter the human community are responsible to protect and steward the resources that give us life; clean air and water, healthy food from healthy soils, and durable and resilient shelter. Understanding this is fundamental to our role in our society's journey to a regenerative future. As an ancient Chinese proverb says: "One generation plants the trees. Another gets the shade."

A regenerative future depends first and foremost on the regenerative behavior by each member of our human community. Regenerative development and long-term regenerative management are a by-product of this behavioral change. This future is dependent on understanding our bioregional context, and codependence of our health and well-being on the vitality and health of each living organism in our bioregion (Figure 9.1). If change is to occur, we must recognize how far we are from understanding that our human species' source of life is dependent on the health of our natural world. As Thomas Berry says in *The great work: Our way into the future* (1999, p. 15):

> The indigenous people of the world live in a universe, in a cosmological order, whereas we, the peoples of the industrial world, no longer live in a Universe. We in North America live in a political world, a nation, a business world, an economic order, a cultural tradition, a Disney dream land [...]. We seldom see the stars at night or the planets or the moon. Summer and winter are the same inside the mall [...]. We read books written with a strangely contrived human alphabet. We no longer read the Book of Nature or the Book of the Universe.

This chapter is a reflection on my understanding of the evolution of regenerative development, the leaders and writers who have contributed to our current level of understanding, and the unique contribution our building and development family, rooted in our 110-year master building tradition, have made.

Figure 9.1 Noisette Community Master Plan 2003

There is no specific literature reference that we are sourcing for our conclusions and methods. Our awareness and values, which have been developed over three generations, are core to our own evolution. Wherever a specific book or article has been a significant influence in our learning and evolution those books are referenced. We begin by sharing the significant intellectual contributors to the movement as we have experienced them. We introduce the development of the Sanborn principles as a way of anchoring this work. We will then define the contribution of our family's 110-year history as a master builder, our value system, and my own evolution in this tradition over the last 50 years. The references to energy crisis, material toxicity, and human health are directly related to the decades in which we became aware of these issues through deep experience. Finally, we identify specific developments and methodologies that have formed the CityCraft® mindset and process over time. This chapter concludes with the five platforms that CityCraft® believes are the essential platforms and infrastructure required to enable us, bioregion by bioregion, to reach a regenerative future. The CityCraft® Platform infrastructure has been moved into the CityCraft® Foundation so that we can create an open source and accessible learning environment to allow for these platforms to improve and adapt over time.

Evolution and history of regenerative development—The intellectual contributions

My sense is that regenerative development is a rediscovery of past wisdom much the same as green building was a rediscovery of the wisdom of master building. Our current efforts do not come near the regenerative capacity exhibited in numerous examples of aboriginal cultures across our planet and the many examples in North America before the Agricultural and Industrial Revolutions. The Lakota Medicine Wheel is a classic example of regenerative principles in the United States at the core of the Lakota Culture. The Cistercian Tradition, going back to 900 AD, has many attributes of regenerative design and development. The difference, prior to the Industrial Revolution, was the need to respect and understand our natural resources—excess and limitations, climatic conditions and threats—as well as our dependence on our place and each other to survive as individuals, as a family and as a community. Past cultures understood humans, buildings, and organizations as organic, evolutionary, and integrated ecosystems as a part of the natural world.

There have been a number of intellectual contributors, social and ecological justice advocates and business leaders, who have taken significant risk to move our regenerative future to where we are today. Our ethical foundation for a regenerative future owes a great deal to Aldo Leopold and Thomas Berry. Aldo Leopold's *A sand county almanac* (1949), set forth in his Land Ethic, "a thing is right when it tends to preserve the integrity, stability, and beauty of the biotic community. It is wrong when it tends otherwise [...] We abuse the land because we regard it as a commodity belonging to us. When we see the land as a community to which we belong, we may begin to use it with love and respect" (pp. 224–225). Thomas Berry, in his last and I believe his most important book, *The great work: Our way into the future* (1999), set forth the direction we must follow, moving from a disruptive force on the earth to a benign presence.

Our intellectual foundation for a regenerative economy owes a great deal to the work of Rachel Carson's *Silent spring* (1962), Paul Hawken's *The ecology of commerce* (1993), the work of Amory and Hunter Lovin with the Rocky Mountain Institute and their ground-breaking book, *Natural capitalism: Creating the next industrial revolution* in partnership with Paul Hawken (2007). Storm Cunningham published *The restoration economy* (2002), which highlighted the economic opportunity and conditions for the restoration economy. This helped accelerate the university's and private sectors' interest and understanding of the new paradigm that was emerging, connecting the limitations and challenges presented by our decaying infrastructure, social and environmental break down, as well as climate change demanding a new approach to address the built environment. Paul Hawken's book *Blessed unrest* (2007), "How the greatest movement in the world came into being and why no one saw it coming," was one of the first major modern works that made the connection between carbon and social justice. Hawken revealed this movement by connecting the dots globally of the hundreds of thousands of small and passionate NGOs on every continent fighting with deep passion to heal the social, economic, and environmental crisis following World War II (WWII) along with the increasing scale and anonymity of our global organizations.

From a business leaders' and organizations' standpoint, Alex Wilson—the founder of "Environmental Building News" (EBN)—created one of the leading sources on the impact of equipment and materials in the built environment on human and natural system health. EBN also offered insights to build a pathway to a low energy future. Harry Gordon, Bob Berkibile, and Gail Lindsay, three of the six founders of the AIA Committee on the Environment (see https://network.aia.org/committeeontheenvironment/home), challenged the design profession to a new understanding of their role to lead us to a sustainable future. Ray Anderson, founder of Interface (see https://www.interface.com/US/en-US/about/mission/Our-Mission), responding to the call of Ecology of Commerce, became a leading industrial voice to transform our industrial systems to a sustainable future, which was documented in his book *Mid-course correction: Toward a sustainable enterprise: The interface model* (1998). Stephen Carpenter (http://www.buildingrapport.com), founder of Enermodal Engineering, in Canada, committed his career to engineering a green, energy-efficient, and healthy future for our built environment. Keith Bower, founder of BioHabitats (see https://www.biohabitats.com/), led us to the creation of ecological restoration and design as a vehicle to restore our natural world to a healthy state and reestablish the regenerative capacity of our ecological systems, which our industrial and urban infrastructure have so dramatically changed over the last 100 years. It is their entrepreneuring leadership, applying the work of leading thinkers to their professional and business sectors that have allowed us to progress down this road to our regenerative future.

The Sanborn principles

In 1994, The National Renewable Energy Laboratory (NREL) assembled 22 thought leaders representing numerous disciplines and professions from across North America. This meeting became known as the Sanborn Habitat Conference. I was one of the 22 invitees by NREL to assess the impact of technology on human habitat in the next 50 years. As a group, most of us having never met each other decided that we needed to first define what a healthy human habitat should be and then determine the supportive role of technology in the next 50 years. The final product of this conference was the development of "The Sanborn Principles—Guidelines for Sustainable Development for New and Existing Cities" (see http://www.donaldaitkenassociates.com/sanborn_daa.html). The Sanborn principles are the first set of regenerative principles developed in the United States. The seven Sanborn principles for sustainable cities in current and future times are described below.

1 **Healthy indoor environment for occupants:** Create a living environment that will be healthy for all its occupants. Buildings shall be of appropriate human scale in a non-sterile, aesthetically pleasing environment. Building

design will respond to toxicity of materials, EMF, lighting efficiency and quality, comfort requirement, and attention to the principles of Feng Shui.

2 **Ecologically healthy:** The design of human habitats shall recognize that all resources are limited and will respond to the patterns of the natural ecology. Land plans and building designs will include only those technologies with the least disruptive impact upon the natural ecology of the earth. Density must be most intense near neighborhood centers where facilities are most accessible. Buildings will be organic, integrate art, natural materials, sunlight, green plants, energy efficiency, low noise levels, and water and not cost more than current conventional buildings.
3 **Socially just:** Habitats shall be equally accessible across economic classes.
4 **Culturally creative:** Habitats will allow ethnic groups to maintain individual cultural identities and neighborhoods, while integrating into the larger community. All population groups shall have access to art, theater, and music.
5 **Beautiful**
6 **Physically and economically accessible:** All sites within the habitat shall be accessible and rich in resources to those living within walkable (wheel chair accessible) distance.
7 **Evolutionary:** Habitats design shall include continuous reevaluation of premises and values, shall be demographically responsive and flexible to changes over time to support future user needs.

Knott family master building tradition—110-year journey to a regenerative future

Our families' master building and development tradition, which reaches three generations over 110 years, as well as centuries of master building's legacy across the world, have created a natural orientation to what we today call sustainable building and regenerative design. Our experience and training are based on building and design principles prior to the advent of our advanced technologies that have so radically shifted our focus from human-centered design to technology-centered design. As a member of the profession that shelters the human community, it is important to have a sense of humility about our role and purpose in our society and the larger world. Since 1908, our work has been deeply rooted in historic preservation and urban redevelopment. Our family's culture of master building was defined by my grandfather Henry A. Knott, continued by my father John Knott and our artisans who practiced these values. We were taught that:

- We were in the human habitat and community building business, not the "sticks and brick" or development business.
- We learned from places built more that 250 years ago, that materials were regionally based; buildings responded and respected their context, climate, and place.
- Developments were mixed use and compact, evolved from their core, and grew from their edge.

- Passive solar, daylighting, and natural ventilation were the rule of the day.
- We were responsible for the health of all those we served directly and the larger community we impacted.
- In serving the larger community and the direct users of our environments, we were responsible to serve five basic health needs for a long term: aesthetic, economic, functional, social, and spiritual.
- Ecological health became the sixth health need added by our generation.

The energy crisis was a very public issue in the 1970s as well as sick building syndrome in the 1970s and 1980s. Our master building culture and value system created a different thought process, as we were confronted in the mid-1970s with issues of energy and inefficient buildings, in the 1980s with the growth of sick building syndrome, and in the early 1990s with the discovery of material toxicity in the built environment impacting our natural world and human health. Our commitment to health in all arenas was core to our early drive to learn and begin to address these health concerns for human and natural system impact.

As an early and long-term member of EBN advisory board, as well as our research surrounding Dewees in the early 1990s, we developed a heightened sensitivity causing us to expand our research commitment to screening all materials and equipment for our buildings and developments. This was followed in 2001 by integrating the Sanborn principles into the CityCraft® process—as a foundational framework for the Noisette Community (see http://www.citycraftventures.com).

The CityCraft® process is rooted in our master building tradition, the foundation of our families building and development practice since 1908. Our approach at the building and community scale was guided by the following:

"Values of Place"

- Diversity
- Beauty and aesthetics
- Accidental meeting place
- Surprise and discovery
- Resource efficiency
- Leaving your mark
- Human form emerging naturally from its place

CityCraft® principles for regenerative development and culture

CityCraft® has evolved as a 21st-century master building mindset for new and existing cities to heal the economic, environmental, and social fabric in support of long-term regenerative health for each bioregion. If we are to achieve a regenerative future, we must understand that as humans, we are both a biologic organism and a social being with a spiritual dimension. As a biologic organism, we need adequate clean air and water, healthy soils to provide healthy food, and resilient shelter based on our unique bioregion. As a social being we require a network of others organized and connected to us supporting a healthy and thriving community each with access to the above three resources. A decision that threatens the capacity or health of these required resources is not regenerative.

A state of regeneration ensures that every decision consistently improves the long-term bioregional health and capacity of every social, natural, economic, and existing human built systems. I do believe there is a fundamental reframing of our language and process to properly and consistently address regenerative development versus sustainability. Some of the key terms to be redefined to push a paradigm shift from sustainability to regenerative futures are as follows:

- **Systems scale**—This future will not be achieved unless the systems our people and buildings are connected to are transformed to a regenerative state at the systems scale before we can ever achieve a regenerative future.
- **Cross-sector integration**—This enables benefit transfer between systems just like the human body acts as a set of integrated and interdependent life support systems.
- **Capital mapping**—Community capital is redefined to include all forms of capital, including human, natural, existing physical assets, and financial capital.
- **Capital integration**—The capacity to identify and integrate, in any investment, all forms of capital described in the capital mapping process, which will generate an economic system less dependent on financial capital.
- **Silo destruction**—In every aspect of our society is essential.
- **A bioregional context**—Understanding our unique bioregion is fundamental to a regenerative state. This context will allow us to plan our future and establish our governance and values with an understanding of our natural resource excess and limitations, our cultural context, and our common climatic threats.
- **Social durability**—There are two core principles within social durability: (1) each member of the community holds in common an understanding of the unique history and heritage over the course of time; and (2) each member of the community holds in common a vision for the future to which they contribute over time.
- **Regenerative research and indicator infrastructure**—There is no agreement by the NGO and public and private sectors on what defines a regenerative economy and sustainable culture. A regenerative future is not possible without this.

Now that we have better defined key terms for regenerative future, the core CityCraft® principles are described as follows:

- **Shared vision**—The vision and success for community come from within the community itself.

- **Collaboration**—Large-scale social change requires broad cross-sector coordination and integration.
- **Embedded capital**—The greatest source of capital for urban regeneration already exists within any community.
- **Systems scale**—Take an integrated approach that results in a scale that takes in all interactions.
- **Openness and transparency**—Continuous measurement, research, and accountability.
- **Sustainability**—Success requires an approach that addresses the long-term economic, social, and environmental health of our cities.
- **Resiliency**—Building local capacity and empowerment through targeted career development resulting in wealth creation that is reinvested locally.
- **Regenerative**—Every decision consistently improves the long-term health and grows more capacity across all human, natural, and economic systems.

Other fundamental concepts of **CityCraft®** are:

1. Systems thinking and integration—We cannot arrive at a truly regenerative future unless we address problems at the systems scale.
2. Socially durability is essential for economic durability and a regenerative future.
3. Identify unvalued, undervalued, and overlooked assets in all capital forms.
4. Tap into communities of human capital.
5. Research and independent verification—Set baseline, measure, learn, report, transfer knowledge.
6. Evolution—Who is responsible to ensure the long-term health of the economic, social, and environmental systems of our cities and bioregions, teach the values and principles that build the capacity to grow a sustainable culture and regenerative economy, and see that a plan and value system is lived over time and becomes deeper and richer than the plan itself conceived?

CityCraft® mindset and process for regenerative development

CityCraft® is a mindset that is evidenced by a substantial change in your way of thinking, behavior, and how, why, and what you value. A CityCraft® thinker understands humans, buildings, neighborhoods, and organizations as organic, evolutionary, and integrated ecosystems. A CityCraft® mindset is grounded in process not formulas and is forensic in nature not defined by predetermined outcomes or answers. A CityCraft® leader understands that:

- Every problem or critical issue we are trying to solve must be addressed at the systems scale and solved across sectors with an integrated systems approach.

- Social durability is the foundation for long-term social and economic durability and must be built not only at the community level but also at every scale of the community from neighborhoods to organizations and across every collaborative team.

CityCraft® process of capital mapping, community building, capacity building, and empowerment is grounded in a belief that:

- We must first heal the social, economic, and environmental justice issues of our communities at the most effective unit of each system to achieve a regenerative economy and a sustainable community and culture.
- Cross-sector integrated solutions at systems scale provide the highest value allowing the benefits from one systems to transfer to another creating new forms of capital capacity in our community.
- Capital in a community includes four forums—social, natural, economic, and existing physical assets. Highest capital capacity occurs when solutions value and integrate all forms of capital across NGO and public and private sectors.

CityCrafting is a paradigm shift from short term to long term, anonymity to awareness, aggregation to diversity, extraction to investment, linear to holistic, end state to evolutionary, competitive to collaborative, building things to building community, closed to open source, mistrust to trust, wisdom of experts to wisdom of community, knowing to unknowing, starting with questions not answers. A CityCraft® leader trusts that the process and community will emerge the right solution and ultimate answers by understanding that:

- A current answer is not an end state but only a place on the evolutionary path to a regenerative future. We cannot define this end state and path because it is improving, changing, and evolving over time to a higher state of regenerative and sustainable capacity.
- Empowerment is the removal of obstacles and limitation on resources so that all members of a community can achieve their highest potential and exercise the power that is innately theirs.
- Health of our bioregion, operating at a fully regenerative state, is the only defined result, but we will only know it when we are there. It will mean that we have healed the economic, environmental, and social fabric of our community and bioregion.
- No action by any person or organizations damages another or undermines any of the four forms of capital supporting our community.

The CityCraft® process is composed of three phases:

Phase 1 is a forensic discovery effort wherein data is collected to identify critical issues affecting the community and the uncovering of assets that the community can use to address them. This phase also makes recommendations for the appropriate geographic scale of this effort.

Phase 2 identifies the excess capacity and underutilized capital assets, develops integration ideas to utilize discovered capacity across all sectors, and recommends implementation opportunities.

Phase 3 represents the implementation phase. This phase begins with initial implementation steps including formalizing the implementation partnership through collective impact, completing detailed capital mapping process, testing of integration opportunities, and deeper community engagement.

Strategy for systems scale cross-sector integration

Determining the appropriate geographic scale requires an analysis of the social, economic, natural, and human built life support, communication, and transportation infrastructure systems. The objective is to overlay these systems creating the potential for an integrated approach that results in a scale that takes in all interactions. Each system needs to be at a scale that the system itself can be transformed, which is usually at a subdistrict level. The appropriate geographic scale helps to provide stacked benefits across sectors specific to the needs of the community and builds off existing assets. At the appropriate scale, long-term strategies for sustainable funding and partnering of resources can be achieved. A scale is needed that leverages resources and drives efficiencies across sectors.

The CityCraft® process is distinguished by prioritizing the integration of financial, natural, physical, and social capital at the systems and cross-sector scale as fundamental to create regenerative solutions. Balanced investment in these four areas supports the long-term health of a community. The modern trend toward increasing compartmentalization and specialization manifests itself in narrowly defined interests competing for too few resources. The heart of our approach is to strengthen community assets by increasing the number of groups with a vested interest in any single resource; to combine and streamline public, private, and NGO resources; and to align interests that build broad constituencies. This integration is designed to drive collaboration among a variety of stakeholders and reduce silo-based decision-making. Each community faces a unique combination of critical issues, and the opportunities to meet them vary accordingly. Despite this diversity of critical issues, CityCraft® strategies have a consistent character: they reflect the integration of at least three resources of financial, natural, and physical and social capital; they are adaptive, flexible, and designed to evolve over time. Long-term regenerative local capacity is the goal and measure of success.

The appropriate geographic scale for this effort is selected based on an examination of the relevant systems (such as transportation, geography, hydrology, land use, socioeconomic). In urban areas, many tend to simplify systems that serve single functions. These systems should be viewed as working together and functioning for environmental and human benefit. Not only do the systems themselves require analysis but most importantly, the interactions and interdependence of the systems need to be considered. Determining an appropriate geographic scale helps to provide benefits across sectors specific to the needs of the community and builds off existing assets. At the appropriate scale, long-term strategies for sustainable funding and partnering of resources can be created. A scale is needed that leverages resources and drives efficiencies across sectors (public, private, nonprofit). The pairing of appropriate scale and cross-sector collaboration attracts additional like-minded investment. The appropriate geographic scale should achieve the following goals:

1 Allows for issues to be addressed at a multi-neighborhood scale
2 Provides a large enough area to encourage integration and stimulation of capital
3 Encourages the physical and economic integration of neighborhood efforts
4 Is large enough to affect regenerative change to each system and generate cross-system benefit transfer

CityCraft® Integrated Research Center (CIRC)—Measurement, research, and reporting

One of the most important assets for a regenerative future is the creation of an interdisciplinary university research center to study the evolution toward a regenerative economy and sustainable culture. The main research gaps for a regenerative future are as follows:

- There are currently no agreed-upon metrics that define a sustainable economy or community.
- There are very few cross-disciplinary graduate programs and research efforts.
- Holistic thinking and decision-making are rare as well as training across systems and disciplines.
- Each profession and discipline is characterized by its own unique language and measures for success, which has increased the difficulty of solving long-term problems and avoiding the impact of unintended consequences.
- The aggregation and globalization of financial capital has led to a devaluation of human and natural capital, an increase in anonymity between the investor and investment, and a focus on short-term over long-term investment measures.

CityCraft®'s objective is to identify baseline conditions and metrics that document the success and failures of the evolving health of cities' economic, environmental, and social fabric and their interdependence. Sharing the results in real time, with full transparency, will allow all sectors to challenge existing practice and design new holistic solutions. At the same time, we will be creating new data that becomes the basis for new curriculums at the undergraduate and graduate levels.

The research center is the city itself not a new building or laboratory created on a university campus. Research in this domain should include a renewed focus on all capital assets of the bioregion and how they become valued and counted as part of the cities' complete capital system versus only counting and valuing financial capital. The four capital assets include human, social, natural, and physical and financial. The disciplines required for this research include the social and physical sciences as well as communication professions and design disciplines. As clarified by Thomas Berry (1999, p. 80):

> The universities, however, should have the insight and the freedom to provide the guidance needed by the human community. The universities should also have the critical capacity, the influence over the other professions and the other activities of society. In a special manner, the universities have the contact with the younger generation needed to reorient the human community toward a greater awareness the human exists, survives, and becomes whole only within the single great community of the planet Earth.

Universities must therefore decide whether they will continue training persons for temporary survival in the declining Cenozoic Era or whether they will begin educating students for the emerging Ecozoic Era.

Dewees case study

As the issues of growth increasingly confront those of the environment, we must ask the question: Is man a functional part of the natural system or an interloper that has no place on this earth? This question may sound absurd, but take a few moments to consider the controversies you have either experienced or read about over the last three decades. Dewees answer was guided by Aldo Leopold's belief that human and nature should live in a state of harmony. The philosophy of Dewees was that we are an essential participant in the ecological system that supports our existence. We simply need to rediscover our intuitive base as to how to live in harmony with our environment as well as relearn how to select habitat and nest within as opposed to dominating and destroying more than we need. The underlying principles of Dewees were as follows:

- Development and environment are natural allies.
- All development and building should occur in the context that all resources are limited.
- Communities and buildings can be resource providers not just resource users.
- Land is a stewardship role for future generations.
- It is less extensive short and long term to build in harmony with the environment.

Figure 9.2 Dewees Island aerial

- Communities are planned for people, and technologies should be supportive not dominant.
- Environmental education is an essential "first step" in the rediscovery of our intuitive sense of integrating with the environment.

Dewees Island is a 1,400-acre island community recognized as one of the first successful sustainable communities in the United States (Figure 9.2). Planning and development began in 1991. Development was limited to 150 home sites with more than 65% of the island in permanent conservation status. Homes are restricted to a maximum of 5,000 square feet and may not disturb more than 7,500 square feet of their lot, average 2 acres. The result meant that only 5% of the island would be disturbed and edge habitat would be tripled and in some areas quadrupled. Roads on the island were natural sand base, and vehicle transportation is limited to electric vehicles.

Dewees has long been recognized as one of the early pioneers in the green building movement. Although Dewees has received much recognition for its environmental focus as a leading sustainable development and green building leader, it cannot truly be sustainable without serving the needs of its owners—the human community of Dewees. True sustainability relates not only to the environment and the long-term value of buildings and infrastructure, but also to the ability of the human community to endure and thrive. As a professional community developer and community organizer, I believe that there are principles that are essential to the creation of holistic and sustainable communities that serve its members and maintain their core values over the course of time. It was at Dewees Island that we tested, at the community scale, our belief in these community building principles. We have long seen many green buildings and communities designed and implemented, but rarely have they retained the original commitment to sustainable principles. It was our firm belief that you need to build a sustainable or regenerative culture as well as buildings and infrastructure. Successful communities have at their core a unique heritage, a common vision and a mission for the future, which bonds them together. The key elements that must be present to build an enduring self-sustaining community are education, involvement, control/ownership, and a sense of common values.

Our commitment to research and education was precedent to creating a long-term sustainable culture. The history of great architecture and historic communities throughout the world is a history of designing and building in the context of climatic conditions and threats, natural resource capacity or limitation, and geography. We studied the history of Sea Island Architecture across three Southeast coastal states resulting in the development of design guidelines that defined the forms and elements that characterized Sea Island Architecture for more than 200 years. There is no color or design standard only—rather a siting process and design based on a full inventory of solar orientation, prevailing winds as well as geologic, maritime forest, and neighbor impacts. Our research informed and educated architects, builders, and owners on how to design human habitat in balance with their environment as well as how their home could harvest natural systems and become a resource provider. Daylighting, natural ventilation, passive heating and cooling, as well as water harvesting are just a few examples of how coastal design had to function before the advent of the many technologies we have become dependent on today.

A six-month training program was designed by Harry Gordon of Burt Hill Kosar Rittleman, our master planning and architectural firm to understand the design requirements for green building and energy efficiency standards as well as passive design methods, which were precedent to LEED Home. Engineering, architectural, land, and maritime forest management standards were deeply influenced by a requirement that we meet the threshold of accepting the impact of a category 4 hurricane with the Barrier Island functioning for the benefit of the mainland as if it was not occupied. We created an environmental education program in 1992 prior to development beginning, hired our first full-time environmental educator, and the first building erected was our environmental education center. Our design and development teams were guided by a scientific advisory board to develop our environmental standards for Dewees and later helped coordinate the many scientific research projects at Dewees, with our owners. Our education director was supported by an educational advisory board to design the education programs for owners, their families, as well as many outside groups including elementary and middle school students and their teachers.

In support of ongoing education for Dewees and the surrounding region, we established in 1993 an Environmental Building Product Trade show originally for our architects, builders, and owners and opened to the public two years later. Dewees education programs have been supported long term through a transfer fee at the time of sale paid by the seller and purchaser at the time of settlement, including the developer. The owners at Dewees subsequently established The Dewees Environmental Trust to ensure that education, environmental management, and research would be supported well into the future.

Noisette case study

The Noisette Community is a 3,000-acre urban area with more than 13,000 existing residents. It was one of the most impoverished neighborhoods in South Carolina including numerous brownfield sites. Planning was community-based, rooted in asset-based community development, designed as a regenerative economy for the current residents. Our focus was on healing the social, economic, and environmental justice issues first. This community-based master plan, engaging thousands of residents, resulted in a blueprint for physical regeneration and a new institutional framework supporting a sustainable and regenerative culture. The underlying principles for Noisette were as follows:

- Respect is at the core of every successful human relationship and endeavor.
- Social durability is the foundation for long-term social, economic, and physical resilience.
- All endeavors are approached with a forensic process that starts with observation and inquiry versus a formula orientation.
- Successful problem solving starts upstream with systems thinking.
- Community involvement is essential to great planning and places equal value on the wisdom of the culture and the talent of the planning and development professionals.
- All decisions will be made to serve the long-term health of the economy, ecology, and social fabric of the community being directly developed as well as the larger community in which it participates.
- All planning and decision-making favors collaboration and is based on sustainable partnering of resources.
- All resources are interdependent in natural as well as human communities.

The land north of the city of Charleston, South Carolina, referred to as the north area until incorporation, was rice and indigo plantation land for many years prior to the Civil War. Liberty Hill and Pettigru neighborhoods were established in the 1870s by freedmen and former slaves. In 1901, the US Naval Shipyard was established on the Cooper River and industrial development proliferated in the area replacing a 600-acre park designed and built by the Olmstead brothers. The world's largest asbestos mill was one of the industries that located north of the Shipyard. During WWII, much development occurred as the military base and industries expanded. The city of North Charleston was incorporated in 1972 in an effort to have direct control over future development. Over the next 20 years, North Charleston more than tripled its population and survived Hurricane Hugo. In 1993, the Navy base was closed resulting in the loss of 22,000 jobs (8,000 civilian jobs) for the Tri-County area, and the land was leased out to various public and private entities.

In March 2001, the Noisette Company and the city of North Charleston entered into an agreement that led to an unprecedented public process in this area. In exchange for the right to purchase 340 acres of the old US Navy base property from the city, Noisette created a world-class community-based master plan for the 340 acres as well as the adjacent 2,700 acres of residential, commercial, and municipal property to aid the city's strategic planning efforts. In consultation with an international cast of the leading sustainability experts, including BNIM (architecture and planning), Burt Hill (architecture and engineering), and Anthropogon (landscape architecture and ecology), the Noisette Company created a blueprint to heal long-term economic, environmental, and social fabric of the Noisette Community. The team began their work guided by one of the core principles of CityCraft®'s process—social durability. This requires that every member of the community understand the unique history and heritage and hold in common a vision for the future to which they contribute together. The history informed a unique approach to the sensitivities of the community and helped the team understand how the community had gathered to affect change in the past and what obstacles they had overcome. Based on this preliminary assessment of the community's organizational capacity, the public engagement process was designed to create a forum that would encourage trust and honest communication and would tap all members of the community, including the potential silent majority.

CityCraft® community engagement is premised on the wisdom of the community being seen as equal to or more powerful than the wisdom of the design and development team (Figure 9.3). It is the responsibility of CityCraft® professional teams to honor, respect, and emerge that wisdom. For two years, the team worked to engage as many citizens of North Charleston as possible, in living rooms, church basements, and large community meetings at the convention center. Our engagement process initially was organized across eight neighborhoods and at the "Crayons in the Hand" stage all neighborhoods were brought together. The engagement process touched thousands of community residents and business owners.

This type of process is very time intensive, and one cannot expect to be met with joy by every community member especially those who have suffered injustices at the hands of outsiders and live in a state of poverty. With each step forward, the attention of officials, positive reports from neighbors, and continued invitations to engage in an organized and thoughtful process, more and more community members began to trust and believe that their neighbors were engaging in a meaningful process, and that they would be as well if they become participants. The core group of community participants played a large role in

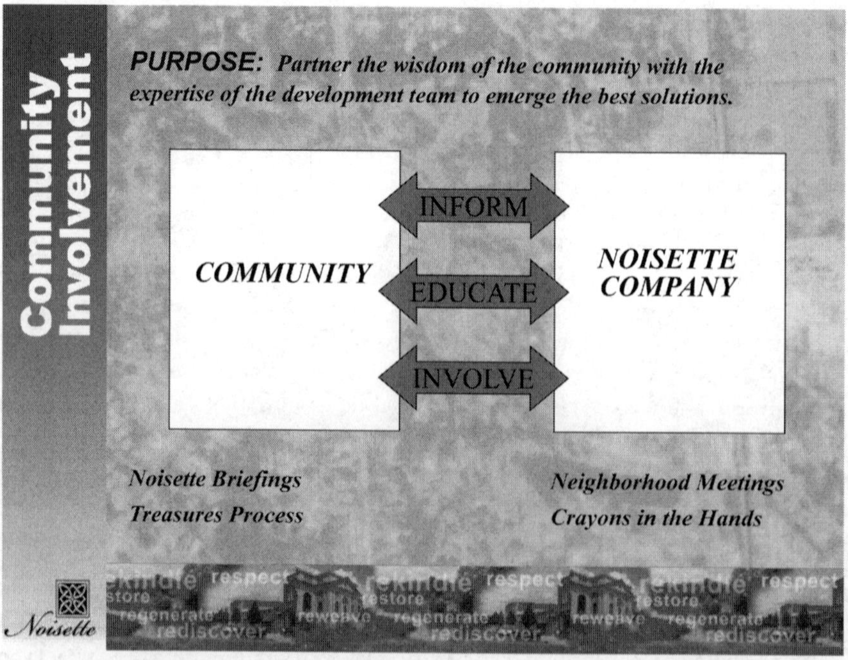

Figure 9.3 Noisette Company presentations

helping to establish that trust. In the best case, they became the advocates and new leadership to sustain this regenerative development process. The group of neighborhood leaders that stepped forward to steward this process have left a legacy of leadership for their subsequent replacements over the last ten years. As the consultant team continued to study the factors of environmental, social, and economic health and to work with the community to set goals, a visualization tool was developed with the community to prioritize and measure progress on achieving these goals. The Noisette Rose became the Noisette Company's guidance for design, tracking, and reporting triple bottom line performance for infrastructure, vertical building, and social infrastructure (Figure 9.4). This tool set the baseline condition of where the community stood in relation to their aspirations for social, environmental, and economic health. Each development scenario was assessed based on its desired outcome and progress over time. Progress was reflected by how full and balanced the "bloom of the rose" was. The Noisette Rose was created to provide a flexible tool to facilitate goal setting and measurement in order to chart the success of this holistic integrated approach with specific measurable criteria. The Rose is a graphic representation that includes both quantitative and qualitative measurements of the components of a planning process or building project within a community. The Rose is composed of groupings of individual metrics to gauge the varying success of specific elements.

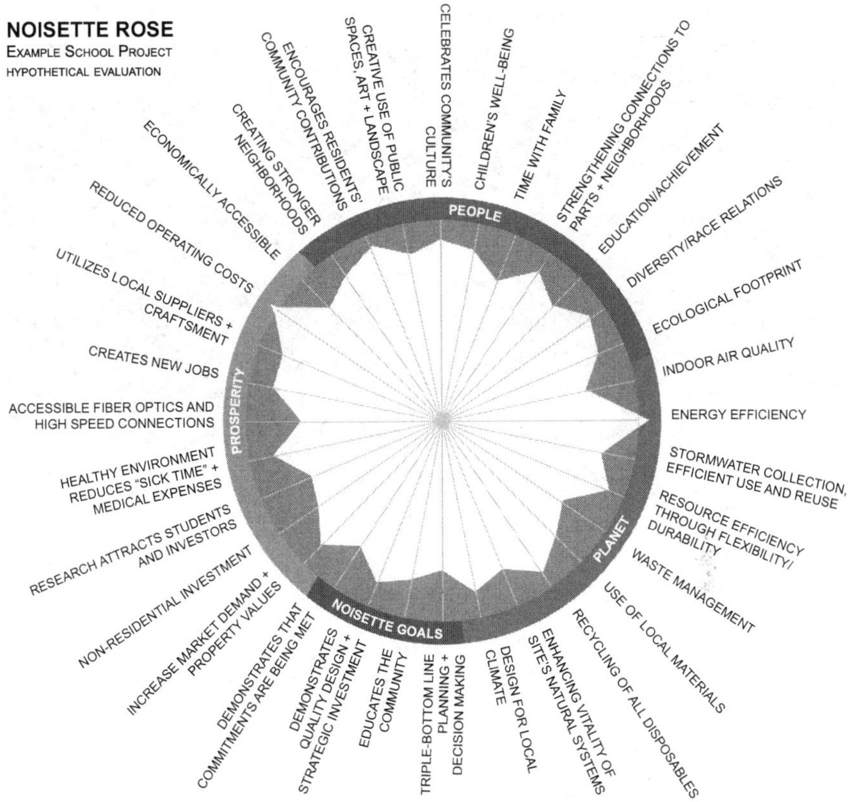

Figure 9.4 Noisette Rose—Noisette Community Master Plan 2003

After two years of work, the master plan was presented to a newly elected City Council of North Charleston (https://issuu.com/citycraftventures/docs/noisette_master_plan). The Council accepted the plan in early 2003, and the final transfer of land occurred in 2006. The Noisette Capital Map and Community Master Plan have been recognized internationally as a new model for urban planning. There were two outcomes of this master plan. The first was a physical blueprint for the regenerative redevelopment of the 3,000-acre urban area representing 90% of the City of North Charleston. The second was a cultural change model and new institutional infrastructure to grow a sustainable culture and create a socially just regenerative future. The community has also been hard at work addressing their priority economic and cultural issues, including reduced recidivism and increased education. With support from the Noisette Foundation many new community organizations have been born to develop relevant solutions to persistent cultural problems present in North Charleston (Figure 9.5). Some of these organizations are the Sustainability Institute, the Michaux Conservancy, Civic Justice Corp, HUB Academy, SC Reentry Initiative (program for recidivism reduction), Energy Conservation Corps, AmeriCorps*VISTA Regional center, SC STRONG, and Lowcountry Local First.

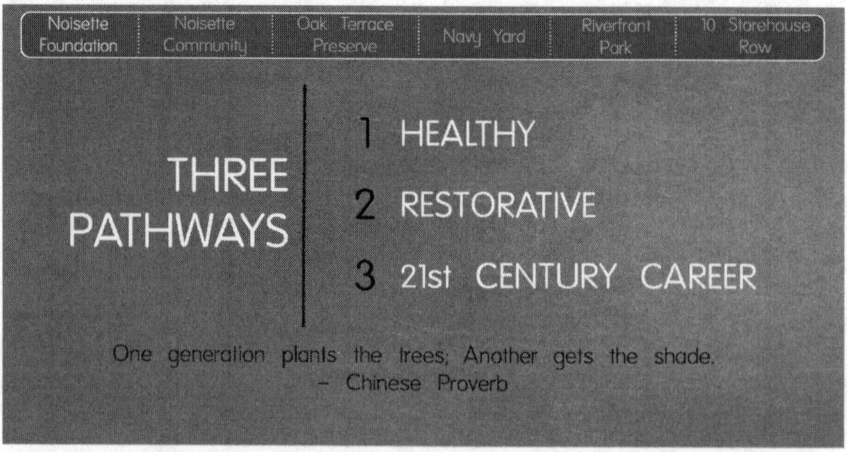

Figure 9.5 Noisette Foundation now CityCraft® Foundation presentations

The community of North Charleston has continued to build their own capacity and increase their health through building more connections between parts of their cultural and natural system. The development of the 15-acre Riverfront Park providing river access to the community for the first time in over 100 years was an early win (Figure 9.6). Today, it is a major regional attraction along with the Navy base memorial.

Figure 9.6 North Charleston Riverfront Park—Noisette Company

Over 8,000 new jobs were created at the Navy yard development including a number of historic buildings restored as technology, economic development, and social justice non-profit and arts incubators. The Noisette Community has seen a major renaissance for the first time in 30 years, with 5 brownfield sites cleaned up, more than 2,000 new housing units planned and developed, millions of dollars invested in new and renovated schools following the centers of community, and green building and arts integration design and building standards adopted as part of the Community Master Plan. Several years after the approval and adoption of the Noisette Community Master Plan with community support and continuous communication with City Council, this plan was adopted and made part of the Comprehensive Plan for this area of the city, guiding the standards for development and approval in the long term.

As we have stated in our CityCraft® principles, evolution is a critical skill to build into our core culture as a regenerative organization as well as building capacity for the community to evolve to a deeper and broader regenerative future. Our experience taught us the critical skills of historic building and neighborhood preservation, social justice issues of affordability and accessibility for some of the most challenged urban communities in the Unied States, and community organizational skills in these same communities. Dewees offered us the first opportunity to apply all of our experimentation in energy efficiency, passive design, natural system land planning, community building, and long-term capacity building through education and research. Dewees Island required us to be responsible for all levels of normal city and county support resources across all levels of infrastructure, police, and fire as well as emergency management and medical emergency. The challenge of building a sustainable culture for our building and design professionals in 1991, as well as our staff and community, was life changing to our understanding of knowledge transfer and leaving behind a governance as well as leadership infrastructure to manage Dewees into the future better than we did at the time we left. The capacity and learning from Dewees enabled us to envision the potential to create a Dewees in the city in one of the most impoverished and socially challenged communities in the Charleston tri-county region. Dewees and Noisette are the foundation for West Denver.

The future of regenerative development—CityCraft® Center West Denver

The West Denver area is composed of 6,400 acres and ten neighborhoods. Despite its proximity to downtown, much of West Denver is isolated physically, psychologically, and economically. Contributing to the isolation is the lack of road connectivity, the South Platte, River (Platte), gulches, and postindustrial sites (Figure 9.7).

Most of the area is zoned residential with many neighborhoods being separated from each other by major roads, gulches, and the Platte. Industrial and commercial use is concentrated on either side of the South Platte River and along a handful of major corridors. Three drainage ways—Lakewood Gulch, Lakewood Dry Gulch, and Weir Gulch—flow from west to east across the area through deep

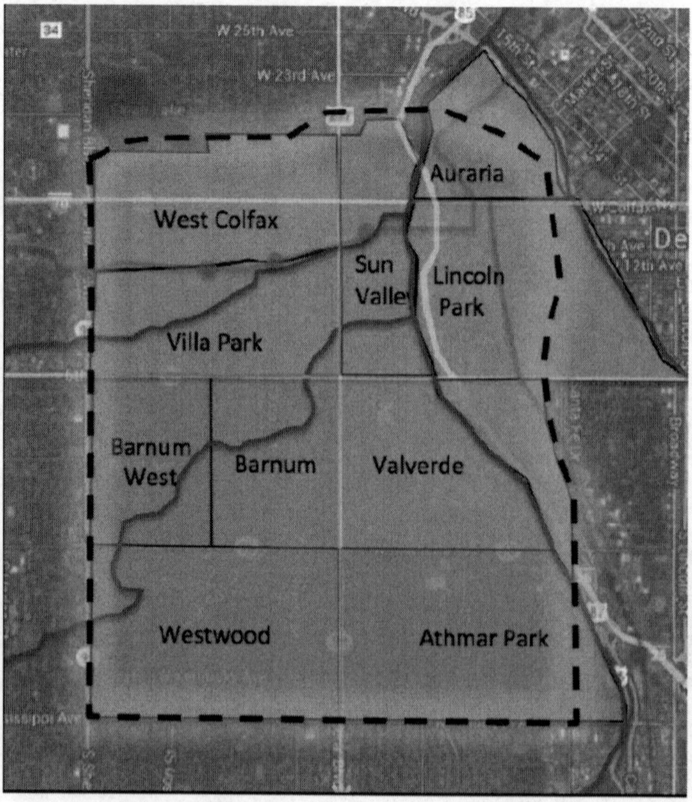

Figure 9.7 West Denver CityCraft® reports

ravines. With a population of 63,000, the area is primarily Hispanic (73%) and has a diversity of other cultures including African-Americans, African, Asian, Anglos, Native American, and others. This mosaic of communities reflects an important part of Denver's cultural and historic heritage. The population is quite young with 33% under the age of 18, and 80% of the population represents the millennials or Gen X generations. Further, the large parking lots surrounding Mile High Stadium have created a void of activity and investment (see https://www.citycraftventures.com/portfolio/west-denver/) (Figure 9.8).

Sun Valley and areas of West Denver share a mix of socioeconomic challenges and opportunities compared to the rest of Denver including the following (Figure 9.9):

- High percentage of population under the age of 18
- Low income levels compared to Denver median
- High levels of unemployment
- High percentage of immigrant population

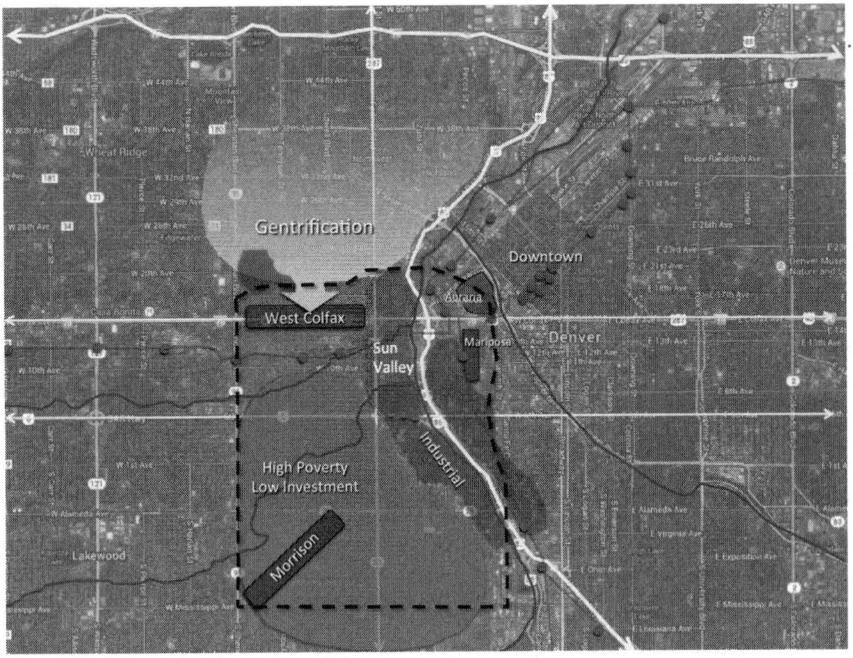

Figure 9.8 Major influence on West Denver neighborhoods—CityCraft® West Denver reports

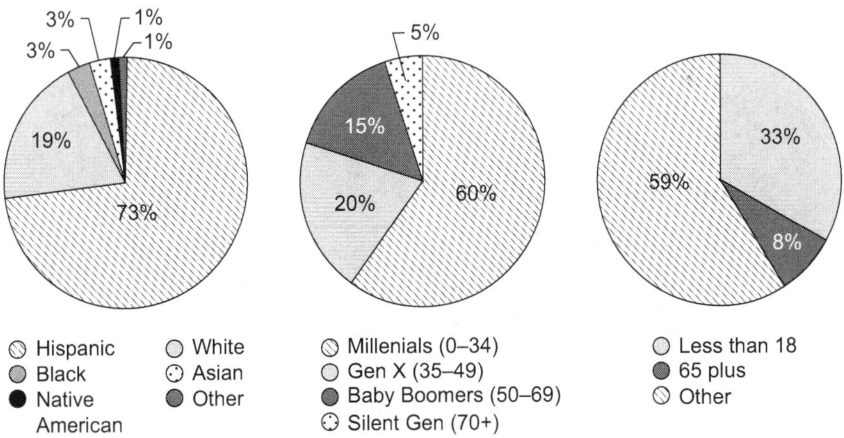

Figure 9.9 West Denver demographics—CityCraft® West Denver reports

The CityCraft® process has been implemented over the last three years in two phases to establish a pathway to achieve a regenerative future. The following regenerative development principles and process are guiding the overall regeneration process:

- Systems scale integration
- Capital mapping
- Capital integration
- Collective impact
- Measurement and research of outcomes—CityCraft® Integrated University Research Center

Critical issues

West Denver is challenged to maintain and improve a place of opportunity for existing residents in the face of uncertain economic and social conditions given the high percentage of population under the age of 18, low income levels compared to Denver median, high levels of unemployment, high percentage of immigrant population, highest concentration of under-performing or failing schools in the city, lack of relative investment, and gentrification encroachment from the north (Figure 9.10). Between phases 1 and 2, there was a transition from

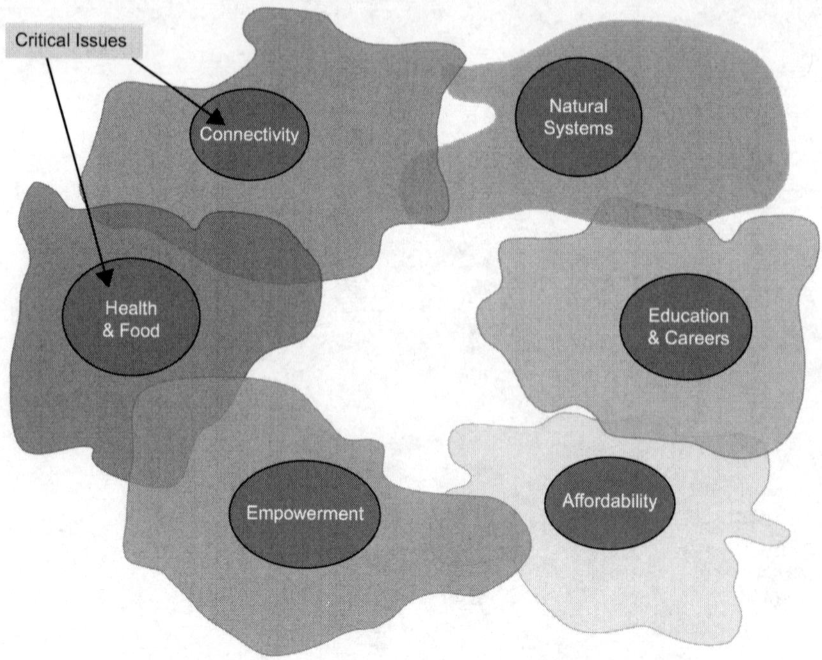

Figure 9.10 Critical issues summary facing West Denver—CityCraft® reports

seven total issues to six because lack of infrastructure funding was a common issue affecting all and became embedded into all critical issues moving forward: connectivity, natural systems, education and careers, affordability, empowerment, and health and food.

Assets

West Denver possesses a variety of existing assets that, if aligned in a long-term integrative manner, could act as a model for urban regeneration for the Denver region and beyond. Phase 1 explored how the development opportunities, catalysts, existing organizations and efforts, the mosaic of cultures, location, young population, and natural systems could be brought together to support improvements that are sustainable and resilient (Figure 9.11). By building upon existing assets and efforts, critical issues and challenges can be addressed to transform West Denver into a model for sustainable community revitalization and urban regeneration.

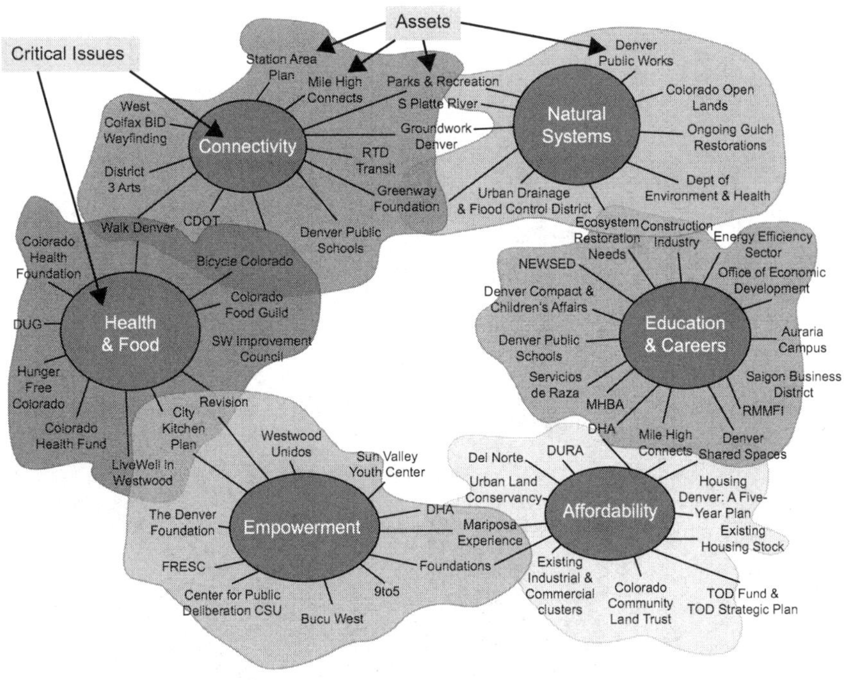

Figure 9.11 West Denver Assets—CityCraft® reports. The above assets are not meant to portray a comprehensive set; provide examples of common assets associated with critical issues

Integration opportunities

Integration opportunities necessary to heal the social, economic, and environmental systems of West Denver emerged during the process of analyzing assets and critical issues (Figure 9.12). These integration opportunities provide recommended specific solutions that address critical issues while building on existing assets. The goal is to always look for integrated solutions first when trying to address a critical issue. These opportunities should be further tested and analyzed but they still represent significant opportunities for West Denver to take a cross-sector integrated approach.

Analysis and synthesis of issues and assets show us that where many see problems or issues, we see opportunities for solutions to overcome those issues. Flipping critical issues from phase 1 into integration opportunities in phase 2 provides the basis for the integration framework (Figure 9.13). The six integration opportunities propose implementation strategies across sectors (public, private, nonprofit) and at the systems scale. The intent is to propose goals that align actions, build local capacity to address issues, and incorporate guiding principles for a 21st-century economy. The goal of phase 2 is to transform the critical issues faced by the community into viable solutions for revitalization and provide an integrated framework for their implementation. Whereas phase 1 identified critical issues,

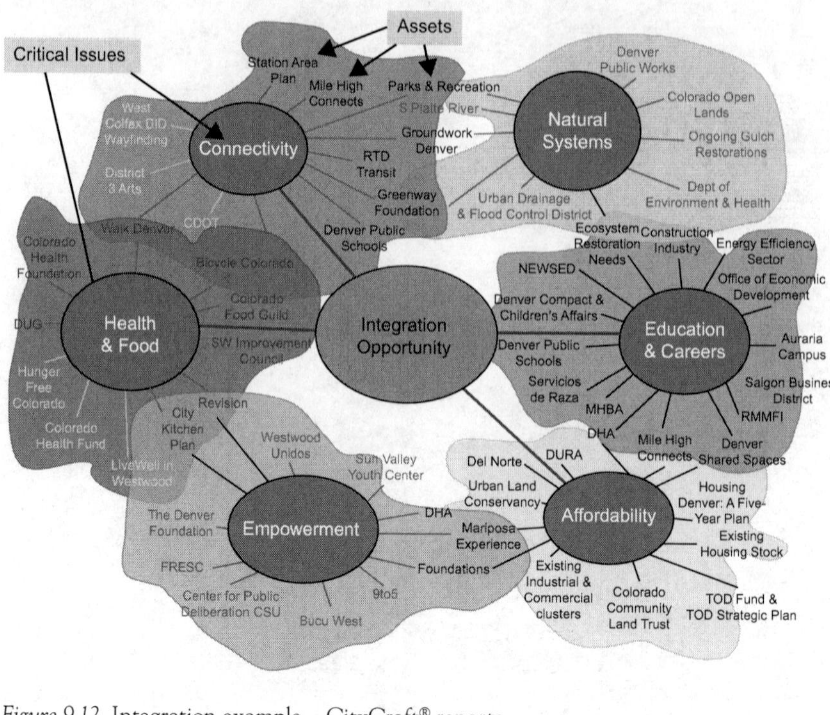

Figure 9.12 Integration example—CityCraft® reports

Phase 1 Critical Issue	➤	**Phase 2 Integration Opportunity**
Connectivity	➤	A Connected Community
Natural Systems	➤	Healing Natural Systems
Education and Careers	➤	A Place of Opportunity: From Cradle to Career
Affordability	➤	Preserving Affordability
Empowerment	➤	An Empowered West Denver
Health and Food	➤	Food for Thought

Figure 9.13 Critical issues transformed to integration opportunities—CityCraft® reports

phase 2 takes the approach that these critical issues can be viewed as opportunities. Because the overarching goal is cross-sector integration at the systems scale, strategies for each opportunity were identified using the following questions:

- Does the strategy benefit more than one type of capital?
- Does the strategy address the systems at scale and across sectors?
- Does the strategy integrate existing assets and build local capacity for future efforts?

Benchmarks for success: Measuring, reporting, and learning from results

The CityCraft® Integrated Research Center (CIRC) represents six university partners and 18 disciplines—CityCraft® and Piton Foundation committed to supporting a regenerative culture and future (Figure 9.14).

CIRC for the Rocky Mountain Front Range Bioregion is a major component of the ongoing cross-sector, systems scale integration strategy for West Denver. The research center is the city itself not a new building or laboratory created on a university campus. Rather, it is a model for cross-disciplinary research and how to conduct applied, long-term, bio-regionally oriented, interdisciplinary, and community-based research. Sharing the results of research in real time, with full transparency, will allow all sectors to challenge existing practice and design new holistic solutions, while at the same time creating new data that becomes the basis for new curriculums at the undergraduate and graduate levels. Partners in CIRC are committed to a new North American learning environment for our regenerative future.

University of Colorado **Denver**
CU IN THE CITY

- College of Architecture and Planning—Dept. of Urban and Regional Planning
- Colorado Center for Sustainable Urbanism

UNIVERSITY of DENVER

- School of Real Estate and Construction Mgmt.
- Dept. of Media, Film and Journalism Studies
- Estlow Center
- Dept. of Anthropology
- SPARK

Colorado State University

- Institute for the Built Environment (IBE)
- Department of Sociology
- Center for Energy and Behavior

Shift RESEARCH LAB | A Program of THE PITON FOUNDATION

METROPOLITAN STATE UNIVERSITY OF DENVER

- Center for Geospatial Research

REGIS UNIVERSITY

- College of Business and Economics
- SEED Center
- Workforce, Employment & Lifelong Learning (WELL)
- The Innovation Center

University of Colorado Boulder

- School of Environmental Design
- Community Engagement, Design and Research (CEDAR)
- Law School + Clinical Programs

Figure 9.14 CityCraft® Integrated Research Center presentation—CityCraft® Foundation

CIRC seeks to support the West Denver project through research, data collection, and translation of research, all embedded in a context of stakeholder engagement. CIRC is committed to a collaborative approach that brings together research and academic professionals from diverse sectors and institutions, while breaking down disciplinary and organizational silos. It is also intended to serve as a bridge between the private, public, NGO, and academic sectors by offering an unbiased forum where questions of scholarly inquiry and best practices can be openly discussed and critiqued. Through its example, we hope to spread this model of engaged urban regenerative research to other cities and research institutions around the world. We also intend that our research findings will form a blueprint for planners and policy makers in other cities to follow, both in the Front Range bioregion and beyond.

Success will be measured based on the area's long-term regenerative capacity and the ability to create bridges of opportunity across neighborhoods, industry sectors, and generations. A key goal is to identify baseline conditions and metrics that document the successes and failures of the evolving health of the city's economic, environmental, and social fabric and their interdependence. To evaluate our success in achieving these goals, we are developing ways of measuring progress, a mechanism for reporting the results, and a method to enable the planners of future projects to learn from the experiences of completed projects. In addition, we have been requested to initiate graduate level classes on our CityCraft® Process at the University of Denver, Daniels College at The Burns School for Real Estate and Construction Management, which have been running for three years. I also serve as an Executive in Residence for the Burns School. We have also become engaged with Regis University in the development of a new specialty master's degree for regenerative development as part of the Anderson College of Business and Economics, where I also serve as Executive in Residence. All of these programs are transferring knowledge of our approach in regenerative development and economics and building a graduate core of regenerative leaders for Denver and the larger Rocky Mountain Front Range Bioregion.

The future of CityCraft®—The evolution continues

Our decision more than ten years ago to move toward establishing replication centers, which transferred our process and methods to 15–20 unique bioregions in the United States and Canada led us to Denver Colorado as our first beta replication site five years ago. Our reason for establishing replication centers is that true sustainability and regenerative economies must be controlled and owned at the bioregional level. National and global enterprises will not take us to that future. As our work evolved in Denver and the integrated research center took hold, we were asked initially by Dr. Barbara Jackson at the University of Denver, Burns School of Real Estate and Construction to establish a graduate level course to teach the CityCraft® process. This then expanded to School of Business and Economics at Regis University, which is focused on establishing a school for Regenerative Business, NGO, and government leadership.

During my first graduate class, one of my graduate students at University of Denver introduced me to Peers Inc: *How people and platforms are inventing the collaborative economy and reinventing capitalism* (2015) a book by Robin Chase, co-founder of Zip Car. The fundamental idea is that you organize platforms that identify, access, and aggregate excess and underutilized capacity. You then integrate and deploy that capacity in a new economic model that gives value to human and other forms of capital, previously seen as having little or no value. What we discovered was that CityCraft® had been developing a platform infrastructure as a new organizing principle to evolve collaborative regenerative economy. We believe the following five platforms are the fundamental infrastructure for attaining a regenerative future that is bio-regionally centered while addressing our economic, environmental, and social justice issues without compromising to either (Figure 9.15).

1. CITYCRAFT® URBAN SYSTEMS SCALE CENTERS: Forensic inventory of assets and critical needs establishes system scale boundary to support cross-sector integration, recommends cross sector systems scale implementation strategies,

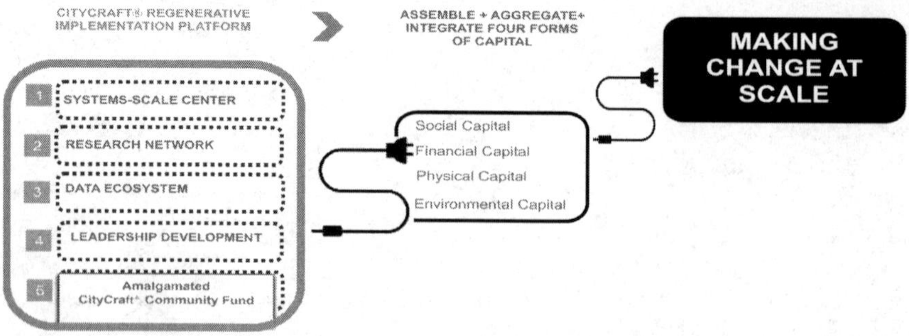

Figure 9.15 CityCraft® Foundation 2018

and becomes the research laboratory for applied learning and multidisciplinary research for that unique bioregion.

2. CITYCRAFT INTEGRATED RESEARCH CENTER NETWORK(CIRC): Multidisciplinary, multi-university collaborative research center focused on solving complex problems through cross-disciplinary education and research in partnership with community residents as part of their research team.

- Builds the research foundation and applied learning center for systems scale urban regeneration
- Studies the evolution of West Denver to a regenerative economy and sustainable culture supporting long-term bioregional health
- Documents the baseline condition for health across all sectors and evolution to a sustainable culture and regenerative economic future in the context of the bioregion

3. DATA ECOSYSTEM AND METRICS FOR REGENERATIVE ECONOMICS:

- Builds data infrastructure defining a state of regenerative capacity for the ecological, economic, and social health for the bioregion
- Develops the baseline condition and reporting framework to measure the community's evolution over time
- Arrives at an agreed-upon set of metrics that define sustainability and a regenerative state, accepted by the NGO and private and public sectors

4. GRADUATE AND PROFESSIONAL LEADERSHIP DEVELOPMENT:

- Development of CityCraft® and integrated project leadership graduate programs and professional certificates with the Burns School of Real Estate and Construction Management

- Development of graduate curriculum and professional certificate programs with Regis University College of Business and Economics
- Continued growth of multidisciplinary graduate courses within and across universities including applied learning opportunities with CityCraft® Integrated Research Center

5. REGENERATIVE CAPITAL ECOSYSTEM: Amalgamated foundation and CityCraft® Foundation Alliance to bridge national philanthropy to local philanthropy and communities to build a regenerative capital ecosystem supporting long-term economic, ecological, and social health through the amalgamated CityCraft® Community Fund (ACCF).

Conclusion

As a learning organization, we are not controlled by predefined outcomes but trust in the process and evolution, wherever it takes us. As West Denver emerged as an urban systems scale center, CIRC expanded to multiple universities and disciplines, graduate-level courses, and degrees and professional certificates for regenerative leaders. It became clear that a knowledge capital and research hub for regenerative outcomes was building in Denver. We recognized that committing our time to Denver as an applied learning laboratory and advanced leadership center would be more effective to growing leadership capacity to create regenerative centers across the United States and provide a continuous learning network as we evolved to our regenerative future.

As our understanding grew, we decided to move the product of our research and applied learning for three generations to the CityCraft® Foundation, which allowed us to focus on capacity building for regenerative leadership with our partner universities. We believe that much of the data, research, and platform infrastructure are replicable to other bioregions as an organizing framework, allowing for the unique attributes and resource conditions to be considered. This also ensures transparency, continuous improvement, and an open source environment to encourage replication to other bioregions and growth of an innovation and learning network across the United States. We have defined ourselves for three generations as the ones, who shelter the human community. It is our choice that we now shift to becoming one of those teachers of the human community and take a position behind a new generation of regenerative professionals, who approach the role of sheltering the human community even more holistically than we have. Fundamental to our understanding is that if you do not count a person, an ecosystem, or an existing physical asset or organization, you have assumed no value and no need for consideration. At the heart of the CityCraft® mindset is a process that ensures that we make every effort to count and value all human, natural, financial, and existing physical capital assets in our community. This discovered excess and underutilized capacity, which forms the foundation for the CityCraft® Platform, is nothing more than counting and valuing all that we have discarded and pushed aside. It is this foundation that creates the new economic capacity for our future regenerative economy.

References

Anderson, R 1998, *Mid-course correction: Toward a sustainable enterprise: The interface model*, Peregrinzilla Press.

Berry, T 1999, *The great work: Our way into the future*, Bell Tower, New York.

Carson, R 1962, *Silent spring*, Houghton Mifflin Harcourt, Boston, MA.

Chase, R 2015, *Peers Inc: How people and platforms are inventing the collaborative economy and reinventing capitalism*, PublicAffairs, New York.

Cunningham, S 2002, *The restoration economy*, Brerrett-Koehler Publishers Inc, San Francisco.

Hawken, P 1993, *The ecology of commerce*, Harper Business, New York.

Hawken, P 2007, *Blessed unrest*, Penguin Books, London, UK.

Hawken, P, Lovins, A & Lovins, HL 2007, *Natural capitalism: Creating the next industrial revolution*, Little, Brown and Company, New York.

Leopold, A 1949, *A sand county almanac*, Oxford University Press, Oxford, USA.

10 Rethinking memorial public spaces as regenerative through a dynamic landscape assessment plan approach

Rebecca Sheehan

While cities have begun to think about the landscape in sustainable and perhaps even regenerative terms, especially regarding agriculture, eco-towns, and industrial complexes (Bayulken & Huisingh 2015; Hemenway 2015; Toensmeier 2016), public spaces and monuments that memorialize and the networks they create tend to lie outside the reach of sustainability and perhaps far outside ideals of regeneration. Whereas regenerative development seeks holistic processes that create feedback loops that generate interlinked physical, natural, economic, and social capital that can be restorative, memorialization is inclined to set in time and space a portion of the public landscape. Often times, these public spaces are quite literally fixed in stone, disregarding the physical landscape as well as restricting the interpretation of the past, limiting the epistemology of memorializing. As Dwyer (2004) explains:

> [Monuments are] authoritative arbiters of the past [that] link them to the city, bringing into question their place in urban redevelopment, real estate speculation and gentrification plans associated with the entrepreneurial city and its selective appropriation of history. (p. 426)

Accordingly, this is a missed opportunity for regenerative development to showcase regenerative thinking that links the physical and sociocultural landscapes to support spatial justice. Spatial justice refers to emphasizing how space is fundamental to achieving justice as we are at all times enmeshed in "physical and natural spatial forms, mentally and materially… to create our biographies and geo-histories" (Soja 2010, p. 17). If memorializing landscapes were conceived as an ongoing dynamic process, then more progressive, inclusive, and just public spaces could emerge that reflect and constitute regenerative sociocultural geohistories.

While I acknowledge the significance of the ecological facets of regenerative design regarding memorializing, including, for example, soil, climate, and material systems as well as potential energy production (Dahmen 2017; Girardet 2018; Shen et al. 2018; Thomson & Newman 2016), in this chapter I focus on the potential of regenerative sociocultural capital, in memorializing. Thus, I first outline the regenerative paradigm and relate it to cultural landscape systems. Using scholarship on Confederate monuments in the US South, I explain how memorializing in the cultural landscape is an act of power, and monuments continue to perform that power to produce understandings of the past with real consequences for the present and future (Brasher, Alderman & Inwood 2017; Dwyer & Alderman 2008; Hoelscher 2003; Leib 2002; Schein 1997, 2003). I connect this literature to ideas of spatial justice, where space is on equal ground with historical and social significance (Soja 2010). Then, I discuss work that evaluates interventions in the landscape that aim to support spatial justice concerning social memory through counter memorializing (Burk 2006; Kobayashi & Peake 2000; Krzyżanowska 2016; Osborne 2017; Stevens, Franck & Fazakerley 2012; Young 1992). However, such interventions do not aim to rework the process of memorializing. Moreover, processes for memorialization in the landscape vary greatly across municipalities, states, and countries, where sensitive sociocultural and political concerns are often addressed only implicitly (Stevens & Sumartojo 2015). Drawing from work concerning impact assessments (Brasher et al. 2017; Clowney 2013; Esteves et al. 2017; Götzmann, VanClay & Seier 2016; Vanclay 1999, 2002), I suggest that dynamic landscape impact assessment (DLIA) plans offer an approach for evaluating proposed and existing memorialization in the landscape that reflects a regenerative paradigm (du Plessis 2012; Hes & du Plessis 2015; Lyle 1994; Mang & Reed 2012). In this regenerative approach, social impacts, human rights, participation from a diverse public, and rethinking memorialization as procedural and iterative are central.

Extending the regenerative paradigm

The power of regenerative development resides in its worldview shift of focus, from a reductionist collection of parts to a holistic synergistic system (Haggard 2002; Mang & Reed 2012). As Mang and Reed (2012, p. 26) explain, "regeneration depends on a developmental process… [that] evokes a set of higher order aims and develops capacities to pursue them." This regenerative paradigm arose from the idea that humans are not separate from nature and a movement away from the idea of sustainability to one of restoration and generation as well as increased potential (du Plessis 2012). The regenerative paradigm requires a change of mind—in fact, an "entirely new mind" (Mang and Reed 2012, p. 26). This means shifting how one understands cities—to see them "as energy systems—webs of interconnected dynamic processes that are continually structuring and restructuring…" (Haggard 2002, p. 25).

This thinking still leaves out problematizing an underlying false mind/body dichotomy, where the mind and body are understood as separate, disconnected realms rather than integral and mutually constitutive. This mindset appears to be at play in regenerative work concerning the built environment thus far, where the mind—which is psychological and spiritual—receives far less attention concerning regeneration through the cultural landscape. Mang and Reed (2012, p. 29) do touch on the idea of understanding the story of place via patterns of relationships as a means to "deepening connection to and growing harmony with place." Their ideas (and others, see for example, Lyle 1994; du Plessis 2012; du Plessis & Brandon 2015; Hes & du Plessis 2015) of spiritual and psychological connection to place, however, is one based in the "natural" environment. The interconnected sociocultural environment remains absent from regenerative work. To begin to address this absence, I consider memorialization in the landscape through a regenerative lens. Du Plessis (2012) outlines four regenerative principles that may be adapted to define regenerative memorialization. Regenerative memorialization:

- Is part of the cultural landscape, and thus an inherent part of embodied sociocultural systems.
- Should contribute positively to the functioning and evolution of sociocultural systems, enabling self-healing.
- Should be rooted in the aspiration of the context.
- Is a dynamic participatory and reflective process.

However, memorials often elude the last three principles. For example as I show below, memorials to the Confederacy in the US South are an inherent part of a white hegemonic sociocultural system. Then, I take up how the last three regenerative principles may begin to be realized through a procedural and iterative approach to memorialization.

Memorialization through the regenerative lens

The materiality of cultural landscapes matters. Cultural landscapes not only reflect power, but they also constitute it (Dwyer & Alderman 2008; Schein 1997; Winberry 1983). In other words, the cultural landscape is never innocent (Schein 1997, 2003), it is always (re)producing power relations especially in terms of belonging and identity, which have real impacts at multiple scales. As such, they are part of embodied sociocultural systems. Soja (2010) explains:

> The geographies in which we live can intensify and sustain our exploitation as workers, support oppressive forms of cultural and political domination based on race, gender, and nationality, and aggravate all forms of discrimination and injustice. (p. 19)

For decades, scholars have shown how remembering and forgetting in the cultural landscape is a form of ongoing domination (e.g., Boyer 1994; Davis & Starn 1989; Dwyer 2004; Fentress & Wickham 1992, Graber 2008; Johnson 1995; Leib & Webster 2012; Nora 1989; Till & Kuusisto-Arponen 2015). As Hoelscher (2003) asserts:

> The interpretation of the past is a salient form of power, and its control carries heavy consequences....No wonder that in times of tension, and in the consolidation of power, so many people have turned to cultural memory or heritage, the means by which the past is domesticated, made familiar, and translated into contemporary language. (p. 660)

Many memorial landscapes, especially in the US South work to maintain a normative social order of white supremacy, for example, where the past valorizes the actions and sacrifices of whites from the Confederacy, especially males, and forgets, minimalizes, and even denigrates others, especially blacks (e.g., Alderman 2012; Dwyer & Alderman 2008; Graber 2008; Hoelscher 2003; Inwood 2009; Leib 2004; Leib & Webster 2012). In fact, a recent study by the Southern Poverty Law Center reveals:

> One thousand seven hundred and forty publicly sponsored symbols honoring Confederate leaders, soldiers or the Confederate States of America in general. These include monuments and statues; flags; holidays and other observances; and the names of schools, highways, parks, bridges, counties, cities, lakes, dams, roads, military bases, and other public works. (Southern Poverty Law Center 2018)

Approximately 700 Confederate monuments, largely made of brass and stone, are littered largely across the U.S. South. Most were erected in the Jim Crow era during the late 19th to mid-20th centuries (Southern Poverty Law Center 2016), but also importantly outside this era even into the 2000s. Additionally, more often the memorials are referential, realistically representing a figure from the Civil War. They are intended to present an immutable version of the past and "a mirror-quality image of the world from which visitors may gather their socio-spatial bearings," obscuring the oppressive processes through which the memorials were emplaced (Dwyer 2004, p. 423). Indeed, critics argue that traditional (e.g., figurative) forms of monumentalizing present a static and closed past based upon dominant powers, leaving little room for change, process, and different interpretations (Dwyer 2004; Nora 1989).

The legacies of atrocities such as enslavement of Africans as well as violence and murder against civil rights activists, for example, have not gained the same traction for memorialization (Young 1992).[1] In the US South, this can be largely attributed to an ongoing reverence that many white Southerners have for their Civil War heritage. Such Southerners feel that their ancestors were valorous in their fight and that the war was not over maintaining a slave economy but instead over state rights (Gallagher & Nolan 2000). However, this *lost cause* myth has been used as a bulwark to threats to white hegemony since the end of Reconstruction (Foster 1987; Gallagher & Nolan 2000; Janney 2008, 2016). Indeed, the emplacement of Confederate commemoration in the landscape is oppressive (Dwyer 2004).

Even when memorialization of non-hegemonic figures are created in the landscape, their messages are often influenced, truncated, or softened, by white hegemony (Alderman 2010; Inwood 2009). For example, Martin Luther King Jr's messages that criticized the US political structure and advocated the "redistribution of wealth and privilege" are missing in Atlanta's Martin Luther King Jr National Park in the service of white hegemony (Inwood 2009, p. 91). As Soja (2010) argues:

> [W]e make our geographies just as it has been said we make our histories, not under conditions of our own choosing but in the material and imagined worlds we collectively have already created—or that have been created for us. (p. 18)

Furthermore, as Burk (2006, p. 949) explains, "if [monuments] challenge [the] dominant social order, [they are] relegated to the colonized rather than the valorized spaces." Alderman and Inwood (2013) also show this to be the case when renaming streets after Martin Luther King Jr, where white business owners and residents fight against street naming in "their" neighborhoods and business districts but "allow" such naming in the so-called black and less prominent areas, further racializing the memorial landscape.

Drawing from Foote (1997, pp. 231–232), Dwyer employs the term symbolic accretion, which "refers to the appending of commemorative elements on to already existing memorials" to argue that a "symbiotic relationship exists between past and contemporary memories-in-the-making as they draw strength from one another" (Dwyer 2004, p. 421). Here, Dwyer is referring to individual memorial sites, but I want to extend this idea to a greater scale. I argue that in the US South a regional system of accretion through Confederate monuments reinforces the sanctity of white domination (Foote 1997). Accordingly, the very act of memorializing much of the cultural landscape in the US South has racialized place.

These issues bring to the fore questions about the epistemology associated with memorializing in the landscapes. Drawing from Nietzsche, Donohoe (2006, p. 11) contends that "memorials can and should function as a call to renewal and critique." This echoes the role that counter and dialogic monuments serve as well as the regenerative paradigm, but Donohoe argues that all monuments should have openness to them where they work to recover and monumentalize while at the same time produce the "imperative to critique" (Donohoe 2006, p. 11). Donohoe (2006, p. 11) makes these assertions in criticizing the rush to memorialize in which doing so "reduces or even eliminates the lapse of time that might give us perspective on and might allow for less politicized transference of its meaning to subsequent generations." She explains that rushing to memorialize "creates imperviousness to anything other than the status quo" (2006, p. 11). However, the status quo cannot be overcome simply through time, as the swath of Confederate monuments built over a 150-year period across the US landscape, for example, illustrates. The hundreds of Confederate monuments alone in the landscape have "overflow[ed] the present, and by their iterative power stretch[ed] out their significance into enduring strands of order" (Allan 1986, p. 200). Additionally, scholars such as Stevens and Sumartojo (2015) and Brasher et al. (2017) show that the inertia of maintaining the status quo is durable. Brasher et al. (2017, p. 8), for example, explore a Yale University report which asserts "there is a strong presumption against renaming a building on the basis of the values associated with its namesake. Such naming should be considered only in exceptional circumstances", showing, by such action, a willingness to perpetuate white supremacy in the landscape or to excuse it—thereby continuing to silence and degrade blacks and other members of vulnerable, marginalized, and underrepresented groups.

In the next section, I argue that counter memorializing may be a constructive intervention into hegemonic memorialization. However, such projects are discrete as has been the case of memorializing in the landscape more broadly and fail to take a systems approach for regenerative memorializing.

Counter memorializing

Some memory-work provides counter narratives to dominant ideologies in the memorialized cultural landscape (Burk 2006; Erőss 2016; Krzyżanowska 2016; Stevens et al. 2012; Till 2005; Young 1992). Particularly since the late 20th century (Erőss 2016; Krzyżanowska 2016; Till 2005), monument forms have also included a variety of appearances, materials, and permanence. Stevens et al. (2012) use "anti-monumental" to explore how such monuments (e.g., Maya Lin's 1982 Vietnam Veterans Memorial) work against the grain of traditional monument forms and messages to invite alternative and multiple interpretations of the commemorated event or person.[2] They also explain that some monuments whether anti-monumental or traditional in form may be "dialogic," which "critiques the purpose and the design of a specific existing monument, in an explicit, contrary, and proximate pairing" (Stevens et al. 2012, p. 952). Anti-monuments and counter-monuments are often used interchangeably, but the use of the term "counter-monument" allows for traditional (i.e., figurative) forms that do not project dominant oppressive narratives of memorializing (see for example, the Memorial for Peach and Justice in Montgomery, Al USA).

On the one hand, counter-monuments have been praised in that they invite multiple meanings and question (traditional) memorializing forms in the landscape. On the other hand, they have been criticized for their sometimes inaccessible and misread messages and meanings (Burk 2006; Stevens et al. 2012). Moreover, often these monuments are temporary and made of short-lived materials, for example, so that their impact may be fleeting or not as grand as the monuments that they aim to counter or to be in dialog. Counter memorializing then should be thought of only as a possible tool within a larger framework of regenerative memorializing. Regenerative memorializing requires a systems process, and in the next section, I explain how building upon impact assessment literature to form dynamic landscape assessment plans may achieve the remaining principles of regenerative memorializing in the landscape, including self-healing landscapes, rooted in the aspiration of the context, and produced through a dynamic participatory and reflective process.

Toward regenerative memorializing

By drawing from landscape, social, and human rights impact assessments, DLIAs may produce regenerative memorializing. For "landscape fairness," Clowney (2013) calls for decision-making regarding the built environment to be integrated procedurally through municipal governing. He argues for landscape impact assessments (LIAs) that would require greater public participation, not require an overhaul to local government, and allow proposed as well as existing memorials to be evaluated. An LIA would work much like environmental impact assessments. Clowney outlines the architecture of LIAs to include a draft impact statement, a citizenry review period of the draft statement, a final report that reflects upon the input of citizenry, and a final decision with justification for acceptance, modification, or denial of the new project. However, Clowney does not specifically draw from social impact assessment (SIA) and human rights impact assessment (HRIA) literatures, which demand inclusion of positive and negative contributions as well as potential human rights violations, respectively, of the project.

In development work, from mining projects to urban renewal, SIAs have been an important avenue to mitigate negative social impacts on communities, especially those that are vulnerable, marginalized, and underrepresented (Vanclay 2002). However, SIAs have not been a part of proposed memorialization in the landscape. While scholars have put forth various notions for SIAs, Vanclay (2002) provides a definition that may be drawn upon when considering the impacts of memorializing:

> Social impact assessment can be defined as the process of assessing or estimating, in advance, the social consequences that are likely to follow from specific … project[s]…. Social impact include all social and cultural consequences to human populations of any public or private actions that alter the ways in which people live, work, play, relate to one another, organize to meet their needs, and generally cope as members of society. Cultural impacts involve changes to the norms, values, and beliefs of individuals that guide and rationalize their cognition of themselves and their society. (p. 190)

HRIA emerged from more cooperative international development (Götzmann et al. 2016). HRIAs are based on three principles including the application of international human rights standards in a human rights focused process that recognizes rights-holders (and their (in)ability to claim rights) as well as duty-bearers' responsibility to uphold human rights (UN Development Group 2003).

More recently, scholars have considered the intersection of SIAs and HRIAs, elaborating on their similarities and differences (Chan 2017; Esteves et al. 2017; Götzmann et al. 2016). SIAs and HRIAs similarities include that both emphasize the need to include public participation; nondiscrimination, especially members of vulnerable, marginalized, and underrepresented groups; and accountability by duty-bearers (in the case of memorialization it would include those that propose the project as well as the appropriate public governing bodies).

Significantly, Vanclay (2003, p. 9) argues that "promoting equity… should be a major driver of development planning, and impacts on the worse-off members of society should be a major consideration in all assessment," and I extend that promoting equity in memorialization in the landscape should also be a major driver. As Götzmann et al. (2016, p. 20) explain, SIAs have typically not attended to this aim and that, "sometimes individuals, groups, or organizations who have more power to influence, or a louder 'voice,' …dominate stakeholder consultation and engagement sessions and consequently … their interests have undue influence in impact mitigation and management decisions." Therefore, focusing on rights-holders and basic human rights defined by the United Nations Universal Declaration of Human Rights (United Nations 1948), as HRIAs do, means procedures should work to include marginalized groups with mechanisms that further their empowerment.

Indeed, the infusion of addressing basic human rights in all matters related to development and the development of memorialization in the landscape would be a fundamental shift in how Clowney's LIA may be imagined. As Götzmann et al. (2016) explain, a human rights based approach (HRBA) does not allow human rights trade-offs. Instead, HRIAs require that all negative impacts be addressed, not offset by positive impacts. Additionally, HRIA acknowledges that human rights are primarily individual, not collective rights; HRIA pays particular attention to the impacts on especially vulnerable or marginalized individuals, and therefore "must be conducted in a way that ensures the identification and assessment of impacts at an individual rights-holder level" (Götzmann et al. 2016, p. 18). This contrasts typical SIAs which look to the "aggregate social welfare" and net benefits, where positive impacts may be calculated to outweigh negative impacts. In such a situation, those included in dominant groups would likely prevail, but Clowney does not address this issue.

Another issue regarding human rights does emerge in an HRBA. The UN's Universal Declaration of Human Rights, Article 19 (United Nations 1948), states that "Everyone has the right to freedom of opinion and expression; this right includes freedom to hold opinions without interference and to seek, receive, and impart information and ideas through any media and regardless of frontiers." Memorialization in the landscape is one form of opinion and expression, and as previously discussed, the memorialized landscape in the United States (especially in the South) is one of white hegemony. This puts forth the need for procedural assessment as opposed to relying (in the United States) on the First and Fourteenth Amendments, which protect rights associated with the freedom of speech and equal protection of rights, respectively. Clowney (2013, p. 34) shows, even "when plaintiffs have summoned evidence against discriminatory symbols that mar the built environment, courts have hesitated to impose the guarantee of equal protection," resulting in few victories with lawsuits based on the First and Fourteenth Amendments. Additionally, work in the public sphere to create, change, remove, and relocate existing names and monuments generally must be undertaken through political power, petitions, and lawsuits that require economic, political, and sociocultural capital to bring forth—capital that vulnerable, marginalized, and underrepresented individuals and groups usually have is far less than what the ruling white classes have (Clowney 2013; Young 1992). Moreover, Clowney argues that the First Amendment "is not a sword to constrain government, but a shield to protect it" (Clowney 2013, p. 37). The "it" here is government speech, which includes monuments approved or built by governments, and thus protected by the First Amendment (Clowney 2013).

Clowney (2013) argues, however, against a ban on racialized spaces for two reasons: (1) Monuments invariably fail to have one coherent meaning and (2) judicial intervention over cultural ideals often results in public backlash, creating even greater obstacles to change. He offers solutions to memorializations that LIAs find unacceptable as proposed:

> For example, in a case where a municipality proposed building a Confederate monument in the town square, several alternatives could be imagined—the local government could place the artwork in a less visible location, erect

prominent signage questioning the underlying valor of the Confederate cause, or choose to honor a hero from a different age. (Clowney 2013, p. 46)

Several issues are involved with these potential solutions. First, Clowney, fails to acknowledge that Confederate monuments generally valorize a cause that aimed to maintain the slavery system in the United States, that the Confederacy symbols have been adopted by white supremacist groups, that signage is almost always less dominant in the landscape than the monument itself and often ignored, and that less visible locations may become more visible by erecting the monument or through the mere development of place. Thus all LIAs should specifically discern whether figures or causes are associated with racialization, division, and subjugation—which should be objectionable by all groups, and if so, should be denied emplacement.

Thus, LIAs should require not only a negative impact assessment as Clowney (2017) explains, but also, as SIAs require, a positive impacts assessment for present and future generations—that is the memorialization in the landscape should be rooted in the aspirations of a city. As ethics and values change, so do aspirations and the sociocultural systems change that are often impossible to predict; therefore, memorialization in the landscape should also be open to such changes through ongoing assessment.

Accordingly, Clowney includes a sunset clause whereby "existing monuments and honorary spaces… would face destruction unless the relevant local governments deliberated over the meaning of the space and voted to reaffirm its value to the landscapes" (Clowney 2013, p. 58). He suggests that governments could deal with the tall task of evaluating existing memorial landscapes by either addressing such landscapes only after a citizen filed a "reassessment request" or giving governments long assessment periods (Clowney 2013, p. 61).

Brasher et al. (2017) question this first suggestion, explaining that memorialization that is at odds with values of a place may go unchallenged. They are also concerned that some places, especially universities, will increasingly rely upon political correctness resulting in color-blind memorializing thereby missing healing memory-work opportunities that the cultural landscape can do by making visible the contributions and tragedies associated with black geographies, and other vulnerable, marginalized, and underrepresented individuals and groups. They reemphasize that memorialization does not exist in a vacuum of simply recording history, refuting ideas against renaming, and removing monuments based upon "erasing history" (Brasher et al. 2017, p. 11). However, Clowney explains that judicial officials are often called upon to interpret the meaning of names and monuments but rarely have the expertise to make such interpretations. Accordingly, those professionals and scholars trained in studying landscape meaning as well as members of a diverse public who experience the memorialization should together contribute to memorialization impacts (Brasher et al. 2017).

I argue that regenerative memorialization includes not only preservation of the past but also being mindful of the past. For example, negative and positive impacts of the memorialization should therefore be reassessed after a specific period of time, as sociocultural values change over time. This requires fundamental changes in the epistemology of memorial landscapes. A participatory iterative process that reexamines, at regular intervals, existing memorial landscapes should be part of any LIA, ensuring dynamism, which is integral to regenerative memorializing. In fact, SIAs and HRIAs do not stop at the assessment of the initial impact projects make but also include follow-up and monitoring impacts of projects (Götzmann et al. 2016). Mindfulness in the memorialized landscape must steer away from simply forgiving cultural values of the past and be open to evolve, that it, to reinterpretation of the past over time. In fact, rather than "preserving" the past, memorialized landscapes ought to be the base for regenerative ideologies that serve to inspire a more just world, emphasizing healing, aspirations, and a dynamic participatory and reflective process.

More specifically, a regenerative approach of memorialization in the landscape should be preceded by and subject to DLIA, where memorialization embraces re-visioning—a regenerative social contract. Some municipalities have provided legislative framework to help with this. In fact, New Orleans relied upon such an ordinance to aid in the controversial removal of four Confederate monuments. The 1993 ordinance (Code of the City of New Orleans 2017) states that the City Council may have monuments removed if the monument (1) praises a subject at odds with the message of equal rights under the law, (2) has been or may become the site of violent demonstrations, or (3) constitutes an expense to maintain that outweighs its historical importance and/or the reason for its display on public property.

The fine print of the ordinance includes that any monument may be removed that "honors, praises or fosters ideologies … in conflict with the requirements of equal protection," would be subject to removal "that participated in the killing of public employees of the city or the state" or anything that lauds any "violent actions" to promote "ethnic, religious or racial supremacy." Nevertheless, after town hall meetings, hearings, lawsuits, appeals, and protesting, it took approximately two years to remove four Confederate monuments.

Clowney proposes such memory-work at the local level; however, a top-down approach has been enacted by several state legislatures and officials to stop localized memory-work. In Alabama, the governor halted the removal of any monument in place for 40 years or more (*Huffington Post* 2017). Similar efforts have been successfully made in Georgia, Mississippi, North Carolina, and Virginia (*Huffington Post* 2017). Swooping legislative acts, however, provide fertile ground for backlash at the local level too. As the mayor of Birmingham, AL had a plywood wall installed around a Confederate monument across from City Hall after the governor's removal ban. This has created ongoing backlash from multiple sides (Barajas 2017; Edgemon 2017).

Moreover, a bottom-up approach could lead to an infinite amount of concerns whereas a top-down approach could leave important concerns unattended (Götzmann et al. 2019). Thus, an either-or scenario is likely not sufficient given the complexity of memorialized landscapes. Thus, construction of DLIAs should be created by DLIA Boards that include diverse stakeholders from local to state levels that have equal power in decision-making, especially vulnerable, marginalized, and underrepresented groups.[3] Such a highly participatory process would include, according to the International Association of Public Participations, informing, consulting, involving, collaborating with, and empowering a diverse public, especially those with traditionally less power (International Association for Public Participation International Federation 2014).

Dynamic landscape impact assessment as regenerative memorializing

Drawing heavily from Clowney's LIA (2013, pp. 45–48) as well as HRIA and SIA literatures (Esteves et al. 2017, pp. 76–79; Götzmann et al. 2016, pp. 18–22; Vanclay 2002, pp. 185–187), I further outline a DLIA framework that supports the regenerative paradigm. At each step the five components of International Association for Public Participation's participatory process should be followed:

1 Describe the purpose of the memorialization in accordance with how it meets the goals of memorialization.
2 Describe the persons, groups, and entities funding and supporting the memorialization.
3 Describe the community: Historical and contemporary groups and their cultural landscapes.
4 Create an assessment report that determines if the memorialization violates human rights of any group or person as designated by the United Nations Universal Declaration of Human Rights.
5 Create assessment reports that determine negative and positive social impacts of the memorialization for present and possible future populations concerning:

    a Multiple groups' way of life:
        i Aspirations for themselves and others
        ii Healing for themselves and others
        iii Empowerment for themselves and others
        iv Interactions with others in and outside their social network in their day-to-day life of living, working, and recreating
        v Perceptions about theirs and others' safety
        vi Perceptions of their place in their community and world
    b Community:
        i Character
        ii Cohesion and stability
        iii Values

6 Relate the social impacts above with broader impacts to any relevant city, region/state, and national and international communities.
7 Make the report available for public review and comment through public hearings at municipal buildings as well as other mechanisms such as social media,

mailings, and meetings at community and neighborhood centers, especially for vulnerable, marginalized, and underrepresented groups who may not have access or feel empowered or safe at public hearings in municipal buildings.

8 Create a revised assessment report that incorporates additional applicable information from the review process for decision makers.
9 The DLIA Board chooses from the following possible decisions:
   a Approve proposed memorialization
   b Deny proposed memorialization based up human rights violations and/or negative social impacts, sending the proposed memorialization back to those who initiated the project with suggestions for revisions.
      i A revised project would then be required to go through the DLIA process again.
10 After approval of memorialization in the landscape, its impacts should be reevaluated by a DLIA Board with every generation—that is every 25 years—as possible negative and positive social impacts for the future will be difficult to assess in the present (Charlesworth 1994).[4]
   a This would require a reevaluation fund dedicated to each memorialization. The responsibility of such fundings should reside with those funding the proposed memorialization.

Existing memorializations in the landscape would be reevaluated through a modified DLIA, with DLIA Boards funded by municipal and state governments. After following steps one through eight, the DLIA Boards would then decide whether the memorialization should be kept or removed.

Clowney (2013, p. 48) includes another option that approves the proposed project as is, even when the project would have "significant adverse impacts on minority communities" because "other specific considerations justify approval." He suggests that adding a "Declaration of Concerns" to such a project to mitigate the impacts. While Clowney does not provide examples of when this might be the case, one can think of recent calls about the "slippery slope" of taking down Confederate monuments, with questions such as "what's next Washington and Jefferson memorials?" (McClendon 2015; Morris 2017; Schuessleraug 2017).

However, this allows certain figures and events to be held beyond reproach and counter to regenerative memorialization. A "Declaration of Concerns" would likely be a plaque or, in the case of toponymic naming, a statement included in official paperwork but not visible on a day-to-day basis. Such efforts are diminutive compared to the monuments and signage that significantly adversely impact members of vulnerable, marginalized, and underrepresented groups. Therefore, I argue that such memorialization must be altered so that human rights today are not violated and that negative social impacts are mitigated. Denying Clowney's proposed forth option means following through with a re-visioning of memorialization that literally stands against white hegemony in the landscape and its perpetuation. Moreover, these processes provide the capacity to pursue a set of higher order (moral) aims in memory-work, a tenant of regenerative thinking (Mang & Reed 2012).

Conclusion

Instituting the DLIA process in the memorialized landscape would entail a significant amount of resources. Considering the power of memorialization that I have outlined in this chapter and that memorialization in the landscape is increasingly part of placemaking in the (re)development of cities for commercial and tourism purposes (Darvil 2014; Leib 2004), such resource expenditure is required in order to establish regenerative memorializing landscapes.

DLIAs form the base of regenerative memorializing. While different groups may well share different views concerning the attributes and impacts of memorializations, diverse DLIA Boards, determining human rights' violations as well as social impacts, should significantly improve the regenerative potential of memorialization. DLIAs provide base information regarding ideologies behind proposed memorialization projects as well as ideologies projected by existing monuments as experienced by diverse stakeholders. DLIAs also ensure positive contributions to and evolutions of sociocultural systems because of initial and repeated evaluation processes. These processes enable healing opportunities through the landscape as social norms, expectations, and aspirations of communities change or are made aware. In this way, DLIAs create the capacity to generate feedback and provide space to restore forgotten contributions and instructive tragedies in place with the hope of being interlinked to a larger regional system of regenerative memory-work. This is the social contract that can be realized in a regenerative memorialized cultural landscape—one that works toward spatial justice.

Notes

1. That said, more recently as Dwyer (2004) points out since the 1990s, an increase in Civil Rights Movement memorializing has been taking place, especially in the US South. Still, funding for and approval of such memorializing has also proven to be difficult (Young 1992; Severson 2012).
2. Such monuments may go by a variety of names, as Stevens, Franck, and Fazakerley (2012, p. 952) point out, including "anti-monument, non-monument, negative-form, deconstructive, non-traditional[,] counter-hegemonic monument[, and]...dialogic."
3. It is beyond the scope of this chapter to detail how a diverse public would constitute DLIA Boards.
4. Mechanisms should also be put in place where grievances may be filed and reevaluation of memorializations may take place sooner that every generation.

References

Alderman, DH 2010, "Surrogation and the politics of remembering slavery in Savannah, Georgia (USA)", *Journal of Historical Geography*, vol. 36, pp. 90–101.

Alderman, DH 2012, "History by the spoonful" in North Carolina. The textual politics of state highway historical markers", *Southeastern Geographer*, vol. 52, no. 4, pp. 355–373.

Alderman, DH & Inwood, J 2013, "Street naming and the politics of belonging: Spatial injustices in the toponymic commemoration of Martin Luther King Jr.", *Social & Cultural Geography*, vol. 14, no. 2, pp. 211–233.

Allan, G 1986, *The importances of the past: A meditation on the authority of tradition*, State University of New York Press, Albany.

Barajas, J 2017, "Alabama attorney general sues Birmingham for partially covering Confederate monument", *PBS News Hour*, viewed 19 October 2018 <https://www.pbs.org/newshour/nation/alabama-attorney-general-sues-birmingham-partially-covering-Confederate-monument>

Bayulken, B & Huisingh, D 2015, "Are lessons from eco-towns helping planners make more effective progress in transforming cities into sustainable urban systems: A literature review (part 2 of 2)", *Journal of Cleaner Production*, vol. 109, pp. 152–165.

Boyer, MC 1994, *The city of collective memory: Its historical imagery and architectural entertainments*, MIT Press, Cambridge, MA.

Brasher, JP, Alderman, DH & Inwood, JFJ 2017, "Applying critical race and memory studies to university place naming controversies: Toward a responsible landscape policy", *Papers in Applied Geography*, vol. 3, no. 3–4, pp. 292–307.

Burk, AL 2006, "Beneath and before: Continuums of publicness in public art", *Social & Cultural Geography*, vol. 7, no. 6, pp. 949–964.

Chan, KW 2017, "Rethinking the mechanism of the social impact assessment with the 'right to the city' concept: A case study of the Blue House Revitalization Project in Hong Kong (2006–2012)", *International Planning Studies*, vol. 22, no. 4, pp. 305–319, DOI: 10.1080/13563475.2016.1273097.

Charlesworth, B 1994, *Evolution in age-structured populations*, (2nd ed.), University of Cambridge Press, Cambridge.

Clowney, S 2013, "Landscape fairness: Removing discrimination from the built environment", *Utah Law Review*, vol. 1, pp. 1–62.

Code of the City of New Orleans 2017, *Article VII – Public monuments, sec. 146–611. – Removal from public property*, viewed November 2017 <https://library.municode.com/la/new_orleans/codes/code_of_ordinances?nodeId=PTIICO_CH146STSIOTPUPL_ARTVIIPUMO_S146-611REPUPR>

Dahmen, J 2017, "Soft futures: Mushrooms and regenerative design", *Journal of Architectural Education*, vol. 71, no. 1, pp. 57–64.

Darvill, T 2014, "Rock and soul: Humanizing heritage, memorializing music and producing places", *World Archaeology*, vol. 46, no. 3, pp. 462–476.

Davis, NZ & Starn, R 1989, "Memory and countermemory", *Representations*, vol. 26, pp. 1–6.

Donohoe, J 2006, "Rushing to memorialize", *Philosophy in the contemporary world*, vol. 13, no. 1, pp. 6–12.

du Plessis, C 2012, "Towards a regenerative paradigm for the built environment", *Building Research & Information*, vol. 40, no. 1, pp. 7–22.

du Plessis, C & Brandon, P 2015, "An ecological worldview as basis for a regenerative sustainability paradigm for the built environment", *Journal of Cleaner Production*, vol. 109, pp. 53–61.

Dwyer, OJ & Alderman, DH 2008, "Memorial landscapes: Analytic questions and metaphors", *GeoJournal*, vol. 73, pp. 165–178.

Dwyer, O 2004, "Symbolic accretion and commemoration", *Social & Cultural Geography*, vol. 5, no. 3, pp. 419–435.

Edgemon, E 2017, *AG files lawsuit against Birmingham over Confederate monument*. *Al.com*, viewed 1 November 2017 <http://www.al.com/news/birmingham/index.ssf/2017/08/defy_state_law_and_remove_conf.html>

Erőss, Á 2016, "In memory of victims": Monument and counter-monument in Liberty Square, Budapest", *Hungarian Geographical Bulletin*, vol. 65, no. 3, pp. 237–254.

Esteves, A, Factor, G, Vanclay, F, Götzmann, N & Moreira, S 2017, "Adapting social impact assessment to address a project's human right impacts and risks", *Environmental Impact Assessment Review*, vol. 67, pp. 73–87.

Fentress, J & Wickham, C 1992, *Social memory*, Blackwell, London.

Foote, K 1997, *Shadowed ground: America's landscapes of violence and tragedy*, University of Texas Press, Austin.

Foster, GM 1987, *Ghosts of the Confederacy: Defeat, the lost cause, and the emergence of the New South*, Oxford University Press, New York.

Gallagher, GW & Nolan, AT 2000/2010, *The myth of the lost cause and civil war history*, Indiana University Press, Bloomington, IN.

Girardet, H 2018, "A call for regeneration: We need to move beyond sustainable development to help heal the Earth's ecosystems", *Resurgence & Ecologist*, vol. Jan/Feb, no. 306, pp. 26–28.

Götzmann, N, VanClay, F & Seier, F 2016, "Social and human rights impact assessments: What can they learn from each other?", *Impact Assessment and Project Appraisal*, vol. 1, no. 34, pp. 14–23.

Graber, S 2008, "'British tribute to Virginia Valor': Unveiling the Stonewall Jackson memorial statue", *American Nineteenth Century History*, vol. 9, no. 2, pp. 141–164.

Haggard, B 2002, "Green to the power of three", *Environmental Design and Construction*, vol. March/April, pp. 24–31.

Hemenway, T 2015, *The permaculture city: Regenerative design for urban, suburban, and town Resilience*, Chelsea Green Publishing, White River Junction, VT.

Hes, D & du Plessis, C 2015, *Designing for hope: Pathways to regenerative sustainability*. Routledge, New York.

Hoelscher, S 2003, "Making place, making race: Performances of whiteness in the Jim Crow South", *Annals of the Association of American Geographers*, vol. 93, no. 3, pp. 657–686.

International Association for Public Participation (IAP2) International Federation 2014, *IAP2's Public Participation Spectrum*, viewed 15 October 2018, <https://www.iap2.org.au/Tenant/C0000004/00000001/files/IAP2_Public_Participation_Spectrum.pdf, accessed 15 October 2018>

Inwood, J 2009, "Contested memory in the birthplace of a king: A case study of Auburn Avenue and the Martin Luther King Jr. National Park", *Cultural Geographies*, vol. 16, no. 1, pp. 87–109.

Janney, CE 2008, *Burying the dead but not the past: Ladies' memorial associations and the lost cause*, University of North Carolina Press, Chapel Hill.

Janney, CE 2016, *The Lost Cause*, viewed 15 November. 2017, <http://www.EncyclopediaVirginia.org/Lost_Cause_The>

Johnson, N 1995, "Cast in stone: Monuments, geography, and nationalism", *Environment and Planning D: Society and Space*, vol. 13, pp. 51–65.

Kobayashi, A & Peake, L 2000, "Racism out of place: Thoughts on whiteness and an antiracist geography in the New Millennium", *Annals of the Association of American Geographers*, vol. 90, no. 2, pp. 392–403.

Krzyżanowska, N 2016, "The discourse of counter-monuments: semiotics of material commemoration in contemporary urban spaces", *Social Semiotics*, vol. 26, no. 5, pp. 465–485.

Leib, J & Webster, GR 2012, "Black, white, or green? The Confederate Battle emblem and the 2001 Mississippi state flag referendum", *Southeastern Geographer*, vol. 52, no. 3, pp. 299–326.

Leib, JI 2002, "Separate times, shared spaces: Arthur Ashe, monument avenue and the politics of Richmond, Virginia's symbolic landscape", *Cultural Geographies*, vol. 9, pp. 286–312.

Leib, JI 2004, "Robert E. Lee, 'race,' representation and redevelopment along Richmond, Virginia's Canal Walk", *Southeastern Geographer*, vol. 44, no. 2, pp. 236–262.

Lyle, JT 1994, *Regenerative design for sustainable development*, Wiley, Hoboken, NJ.

Mang, P & Reed, B 2012, "Designing from place: A regenerative framework and methodology", *Building & Research Information*, vol. 40, no. 1, pp. 23–38.

McClendon, R 2015, *Opposition to removal of Confederate memorials at Lee Circle and elsewhere gains steam*, viewed 15 October 2017, <http://www.nola.com/politics/index.ssf/2015/07/lee_circle_Confederate_monumen_1.html>

Morris, T 2017, *Celebrate Andrew Jackson's 250th birthday. His statue isn't going anywhere*, viewed 15 October 2017, <http://www.nola.com/opinions/index.ssf/2017/03/celebrate_andrew_jacksons_250t.html>

Nora, P 1989, "Between memory and history: Les lieux de memoire", *Representations*, vol. 26, pp. 7–25.

Osborne, JF 2017, "Counter-monumentality and the vulnerability of memory", *Journal of Social Archeology*, vol. 17, no. 2, pp. 163–187.

Schein, RH 1997, "The place of landscape: A conceptual framework for interpreting an American scene" *Annals of the Association of American Geographers*, vol. 87, no. 4, pp. 660–680.

Schein, RH 2003, "Normative dimensions of landscape", In C Wilson & P Groth (eds.), *Everyday America: Cultural landscape studies after J.B. Jackson*, University of California Press, Berkeley, pp. 199–218.

Schuessleraug, J 2017, *Historians question Trump's comments on Confederate Monuments*, viewed 1 November 2017, <https://www.nytimes.com/2017/08/15/arts/design/trump-robert-e-lee-george-washington-thomas-jefferson.html>

Severson, K 2012, *Museums to shine a spotlight on Civil Rights Era*, viewed 1 November 2017, <http://www.nytimes.com/2012/02/20/us/african-american-museums-rising-to-recognize-civil-rights.html>

Shen, W, Zhu, S, Xu, YL & Zhu, H 2018, "Energy regenerative tuned mass damper in high-rise buildings", *Structural Control & Health Monitoring*, vol. 25, no. 2, pp. 1–18.

Soja, E 2010, *Seeking spatial justice*. University of Minnesota Press, Minneapolis.

Southern Poverty Law Center 2018, *Whose heritage? Public symbols of the confederacy*, viewed 10 September 2018, <https://www.splcenter.org/20160421/whose-heritage-public-symbols-confederacy#findings>.

Stevens, Q, Franck, KA & Fazakerley, R 2012, "Counter-monuments: The anti-monumental and the dialogic" *The Journal of Architecture*, vol. 17, no. 6, pp. 951–969.

Stevens, Q & Sumartojo, S 2015,"Memorial planning in London", *Journal of Urban Design*, vol. 20, no. 5, pp. 615–635.

Thomson, G & Newman, P 2016, "Geoengineering in the Anthropocene through regenerative urbanism" *Geosciences*, vol. 6, no. 4, pp. 1–16.

Till, KE 2005, *The new Berlin: Memory, politics, place*, University of Minnesota Press, Minneapolis.

Till, KE & Kuusisto-Arponen, AK 2015, "Towards responsible geographies of memory: Complexities of place and the ethics of remembering", *Erdkunde*, vol. 69, no. 4, pp. 291–306.

Toensmeier, E 2016, *The carbon farming solution: A global toolkit of perennial crops and regenerative agriculture practices for climate change mitigation and food security*. Chelsea Green Publishing, White River Junction, VT.

UN Development Group 2003, *The human rights based approach to development cooperation towards a common understanding among UN agencies*, viewed 27 November 2018, <https://undg.org/wp-content/uploads/2016/09/6959-The_Human_Rights_Based_Approach_to_Development_Cooperation_Towards_a_Common_Understanding_among_UN.pdf>.

United Nations 1948, *The Universal Declaration of Human Rights*, viewed 27 November 2018, <http://www.un.org/en/universal-declaration-human-rights/>.

Vanclay, F 1999, "Social impact assessment", In Petts J (ed.), *Handbook of environmental impact assessment, Vol. 1*. Blackwell, Oxford, pp. 301–326.

Vanclay, F 2002, "Conceptualizing social impacts", *Environmental Impact Assessment Review*, vol. 22, pp. 183–211.

Vanclay, F 2003, "International principles for social impact assessment", *Impact Assess Poj Appraisal*, vol. 21, pp. 5–12.

Winberry, JJ 1983, "'Lest We Forget': The Confederate monument and the southern townscape", *Southeastern Geographer*, vol. 23, no. 2, pp. 107–121.

Young, JE 1992, "The counter-monument: Memory against itself in Germany today", *Critical Inquiry*, vol. 18, no. 2, pp. 267–296.

11 Integrating social science and positive psychology into regenerative development and design processes

Jennifer Eileen Cross and Josette M. Plaut

Regenerative development is made possible by courageous groups of people willing to undergo individual and collective transformation to address local and global challenges. The key processes of regenerative development and regenerative design are largely social in nature and benefit from attention to and integration of a variety of social sciences (Gibbons et al. 2018; Robinson 2004; Robinson & Cole 2015). This chapter explores how insights from social science can inform place-based regenerative development and deepen our understanding to realize the regenerative potential in places (Table 11.1). A fundamental requirement of this work is that people constantly regenerate their thinking, comprehension, and connection to the health of living systems as a whole (Mang, Haggard & Regenesis 2016). Regenerative development is not a time-limited activity or event, but rather a dynamic and iterative process, which strengthens the collective ability to sense what is emergent, what is essential, and where potential exists, thereby enabling regenerative practitioners to evolve themselves, their communities, and all living systems—in the present moment and into the future.

While scholars and practitioners have debated the definition of the terms regenerative development, regenerative design, and regenerative sustainability, we chose the term *regenerative development* for several reasons. First, the term sustainability has long been tied to notions of doing less harm and the regenerative paradigm seeks to move beyond into a mental model that examines ways to develop multidimensional benefits (McDonough & Braungart 2010; Reed 2007). Thus, we depart from Robinson and Cole's (2015) view that regenerative sustainability is the more broadly participatory and constructivist concept. Second, where regenerative design may focus on design efforts like cradle to cradle product design or net positive buildings or urban districts, regenerative development, as it has been conceptualized most recently, includes a larger scope of projects from buildings to neighborhood design to comprehensive community development and environmental justice planning (Mang, Haggard & Regenesis 2016; Wahl 2016). The idea of regenerative development has been and continues to be debated and refined in both scholarly and practice-oriented writings. We will use the term "regenerative development" rather than "regenerative sustainability" because we see regenerative development as more clearly indicating a process that is participatory, evolving, and constructivist.

Table 11.1 Definition of key terms for regenerative development discourses

Key terms	*Definition*
Regenerate	To bring new and more vigorous life. Creating greater capacity for ongoing evolution, resulting in increased vitality and viability (Mang, Haggard & Regenesis 2016).
Regenerative development	The process of building capacity and capability in people, communities, and other natural systems to renew, evolve, and thrive (Center for Living Environments and Regeneration 2017).
Regenerative design	The art and process of planning and creating, based on a deep understanding of place (ecosystem, culture, etc.), using technologies and strategies that result in enduring capability for coevolution and increased vitality and viability (Center for Living Environments and Regeneration 2017).
Regenerative practice	The art and science of realizing regenerative potential in systems.

In this chapter, we explore how insights from social science can help guide regenerative practitioners in the practice of inviting, developing, and guiding networks toward their greatest potential. Regenerative development requires engaging and moving groups of people into higher orders of understanding, capability, and desire to think and act from a regenerative paradigm. We will explore how two unique domains of social science—positive psychology and social network science—can inform and elevate regenerative development projects. We will first discuss the core principles of regenerative development, and then define five core capabilities of regenerative practitioners. We will explore one of those capabilities—Developmental Facilitating—in depth. Of the five capabilities, Developmental Facilitating is most directly connected to social sciences, in that the capability is focused on working with groups and developing capacity through building relationships and capacity for collective thought. Facilitation is as much as a science as an art; when the social sciences are integrated into regenerative development projects, they have the potential to radically transform the power of groups to conceive of new realities and take new action. We make two propositions: First, the principles of human thriving from positive psychology apply equally to the processes of interactions as they do to the outcomes or goals of a regenerative development process; second, regenerative practitioners who are experts in facilitation know how to develop networks and group capacity in ways that overcome many of the common dysfunctions of collective action networks.

Assumptions and key concepts of regenerative development

Drawing on the practice guides of leaders in regenerative work, we have identified six core principles of regenerative development—whole systems approach, being of service, human interdependence with nature, accounting for uniqueness, focus on potential, and intentional network weaving (Center for Living Environments and Regeneration 2017; Gilchrist 2009; Mang, Haggard & Regenesis 2016; Sarkissian et al. 2009). Each of these principles reflects human and natural systems and directs and informs the work of regenerative practitioners. Further, they lay the groundwork for how practitioners can facilitate teams, groups, and communities toward regenerative outcomes.

The first core principle, *whole systems approach*, recognizes that people are embedded in places that constitute complex webs of interdependent ecological, social, material, and economic systems (Bortoft 1996; Liu et al. 2007; Sack 1997). In the past decade, social and natural scientists have been articulating the importance of understanding and modeling the complexities of coupled systems—social, ecological, economic, and technological (Ostrom 2009; Sanford 2017). This interconnectedness within and across systems requires a whole systems approach because the world works as systems of nested wholes, not as a collection of pieces and parts, where the systems are interdependent, have emergent impacts on each other, and produce both long- and short-term impacts on other systems (Liu et al. 2007; Ostrom 2009). Through seeing and working with wholes, practitioners come to understand the interconnections and relationships that are essential to effectively engaging people and organizations to realize the regenerative potential in places (Bortoft 1996; Seamon 2018). Regenerative business leader Carol Sanford (2016) described whole systems in her blog:

> When a seed is dropped into healthy soil, it is nurtured by the whole soil system and the larger ecosystem within which the soil is nurtured. The seed grows into a mature plant, contributing food to the larger system and dropping more seeds into the soil. Looking at the plant in random moments or studying one or another of its phases cuts it into non-living parts. In the same way, looking at organs of the human body independently of their lives as whole beings nested within a human who is nested within a neighborhood within an ecosystem on a living planet is misapprehending them as static, partial objects and thus missing the full reality of their being. This narrow perception of living beings makes it almost impossible to grasp the complexity of living systems in ways that would enable us to make truly regenerative contributions to our communities and the larger wholes within which they are nested.

In recent decades, scholars have been articulating and advocating for a new scientific philosophy, which Dent (1999) called complexity science. This new paradigm is rooted in holism instead of reductionism, mutual causality versus linear causality, constructivism versus objective reality, understanding phenomena within their context rather than as objectively separate, and a focus on relationships between entities versus a focus on discrete entities (Bortoft 1996).

The second core principle, *being of service*, arises from the human need for meaning and to contribute to something larger than the self (Seligman 2012). Being of service has many facets, but at the core, this principle is about being part of something larger than oneself and contributing to the well-being of the others (Center for Living Environments and Regeneration 2017). Seligman (2012) in his book, *Flourish: A Visionary New Understanding of Happiness and Well-Being*, defined meaning as "belonging to and serving something that you believe to be bigger than the self" (p. 17). Similarly, Grant (2013) has argued that giving to others elevates our own success. Grant defined giving as focusing on acting in the interests of others, such as by giving help, providing mentoring, sharing credit, or connecting others to needed resources (Grant 2013). Cross-cultural studies have found that the values associated with giving (helpfulness, responsibility, social justice, and compassion) are the predominant guiding principle across dozens of countries (Grant 2013; Schwartz & Bardi 2001). Being of service is a core principle of regenerative processes not simply because people desire meaning, but also because people are more motivated to act when their work has a deeper meaning and contributes to the welfare of others (Buchanan & Kern 2017).

The third core principle is about *human interdependence with nature*. People are embedded in, and dependent upon natural systems, yet we often create cultural values and world views that deny or ignore this interdependency (Wahl 2016; Whyte, Brewer & Johnson 2016). The dominant cultural paradigm of the 20th century in the United States, and arguably in the developed world, was decidedly anthropocentric, viewing nature as existing for the benefit of human beings, and also purporting that humans are uniquely exempt from the constraints of nature, in contrast to other species (Dunlap & Catton 1994; Sauvé, Bernard & Sloan 2016; Wahl 2016). In recent decades, a contrasting worldview, one that conceptualizes human beings and the well-being of human society as interdependent with nature and environmental conservation, has been growing in the United States and other developed countries (Corral-Verdugo et al. 2008; Dunlap et al. 2000; Sale 1985). By recognizing and appreciating our interdependent relationship with natural systems, we have the opportunity to explore and realize how humans can be positive contributors in natural systems (Center for Living Environments and Regeneration 2017). This journey begins with learning from and understanding the nature of natural systems, and then exploring how we can emulate and enhance nature's processes.

The key principle of *accounting for uniqueness* is related to the local context in which people are embedded. Every culture, place, organization, and community has its own unique history, qualities, and patterns (Sack 1997; Seamon 2018). Anthropologists, cultural geographers, psychologists, and sociologists have long argued for solutions that are not only culturally informed, but are inspired by local wisdom and practices (Agrawal 1995; Barca, McCann & Rodríguez-Pose 2012; Barth et al. 2007; Briggs 2005). A recent effort in Wales found that listening to local citizens and developing messages that reflect local and place-based knowledge and identities were the most effective tools for engaging and encouraging action related to climate change (Marshall 2014). In addition, several fields operate on the recognition that people and their experiences and the meanings they make of those experiences are embedded in a social context (Lin, Cook & Burt 2001; Seligman & Csikszentmihalyi 2000). Not only do local communities have unique knowledge, cultural understandings, and mental frameworks, but engaging people in sharing that knowledge and understanding is a key step in activating authentic, inspired change across natural, social, and economic systems (Mang, Haggard & Regenesis 2016; Sarkissian et al. 2009; Seamon 2018).

Implicit in accounting for uniqueness is the ability to focus on unique potential. The fifth principle is *shifting focus from problems to potential.* Just as each child has a unique potential, so does each place, each project, and each organization. Regenerative development involves shifting our focus from solving problems to focusing on realizing potential. While problem-solving plays a useful role in certain situations, focusing on potential opens up projects and endeavors to a whole world of possibility, and more importantly, relevancy. Across the social sciences, from education to social work and from health care to business management, practitioners have been adopting a strengths-based approach (Kana 'iaupuni 2005; Lopez & Louis 2009; Metcalf & Benn 2013; Whitney & Cooperrider 2011). The strengths-based approach, a philosophy evolving out of social work practice, values the knowledge, skills, capacity, connections, and potential already apparent in individuals, groups, and communities. Realizing potential is accomplished through the integration of the needs and opportunities in the local context and the greater system with the strengths of individuals, groups, and places. The regenerative practitioner works with communities to envision a new future reality and focuses activity toward the desired future (Center for Living Environments and Regeneration 2017).

The final principle is *intentional network weaving.* Not only are human beings embedded in place, but they are also embedded in networks, or patterns of relationships among people. These networks and the patterning of ties, from trust to sharing knowledge to sharing resources, are the key to making significant and lasting change (Gilchrist 2009; Henry & Vollan 2014; Krebs & Holley 2005). It is not the number of people engaged in a change effort that matters, but building the connections that help people develop understanding, access knowledge and resources, and expand their influence (Wenger 2000; Wheatley & Frieze 2006). Intentional network weaving requires some basic understanding of network structures, the ability to map and track the local network, and a capability for building the kinds of ties that will help develop the network into patterns that best facilitate knowledge sharing and resource exchange, called network weaving (Frank 2011; Krebs & Holley 2005). Network weaving, a term that comes out of community development literature, is building a community of practice, where members are committed to not just the needs of the group, but to advancing the practice of regenerative development and sharing their experience and knowledge with a wider audience, thus connecting back to being of service.

Capacities & capabilities of the regenerative practitioner

Operating with the six principles described above as a foundation for action, the regenerative practitioner must also develop a variety of capacities and capabilities to engage in regenerative work. As described in *Becoming a Regenerative Practitioner: A Field Guide* (Plaut & Amedée 2018), regenerative development requires the development of five core capabilities—system actualizing, framework

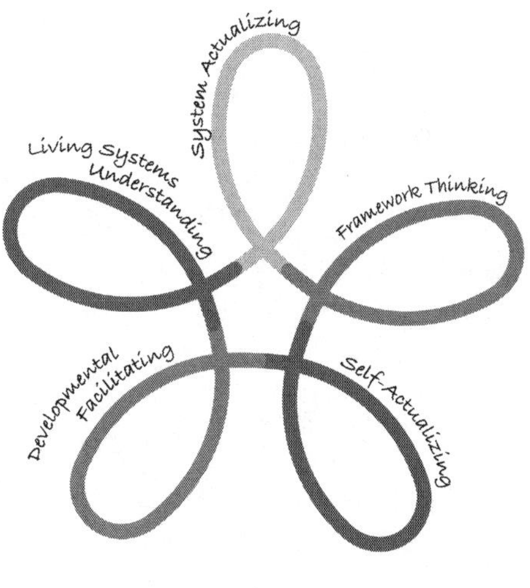

Products of the Five Core Capabilities

System Actualizing – the capacity and capability of a practitioner to engage people in ongoing learning and development, leading to continuing evolution and realization of potential.

Framework Thinking – clear thinking that results in deliberate, effective action, and heightened intelligence for working with whole systems.

Self-Actualizing – personal agency and effectiveness for serving as an agent for developing potential in individuals, teams, and other living systems.

Developmental Facilitating – new thoughts, deeper will, and informed actions that are co-created by the group; increasing levels of capability with the team or community over time.

Living Systems Understanding – actions that align with and support living systems in a reciprocal and mutually beneficial way.

Figure 11.1 Regenerative practitioner framework

thinking, self-actualizing, developmental facilitating, and living systems understanding (see Figure 11.1). "The five capabilities are not buckets of knowledge or disciplines to be studied. Instead, they are a set of capacities and capabilities that are developed over time and used dynamically, in real time application" (Plaut & Amedée 2018, p. 5). The ultimate aim for all those engaged in regenerative development is to become a systems actualizer, to awaken "*the regenerative capability embodied in all living systems to create increasing levels of vitality, viability, and capacity to evolve the systems of which they are a part*" (Plaut & Amedée 2018, p. 10).

The products of the five core capabilities are as follows:

System actualizing: It is the capacity and capability of a practitioner to engage people in ongoing learning and development, leading to continuing evolution and realization of potential.

Framework thinking: It is clear thinking that results in deliberate, effective action, and heightened intelligence for working with whole systems.

Self-actualizing: It is personal agency and effectiveness for serving as an agent for developing potential in individuals, teams, and other living systems.

Developmental facilitating: This core capability has new thoughts, deeper will, and informed actions that are cocreated by the group; increasing levels of capability with the team or community over time.

Living systems understanding: This core capability has actions that align with and support living systems in a reciprocal and mutually beneficial way.

In the remainder of the chapter, we use this framework as an outline to explore how knowledge from various social science disciplines and fields of practice can inform and deepen the understanding of regenerative development. Because Developmental Facilitating focuses on effective group processes and on elevating the collective capacity and capability of groups, the remainder of this chapter will explore how social science informs the practice of Developmental Facilitating, more specifically, we ask the following questions:

1 How can positive psychology and the theory of human thriving inform the way regenerative practitioners work with groups?
   a How can the group processes best be facilitated to help realize potential in regenerative development projects?
   b How do regenerative practitioners engage people in what might be a long and arduous process and how do they keep people engaged?
2 What strategies are used to build robust and productive networks that enable collective action and avoid the pitfalls that block collective action?

Elevating regenerative development through a focus on human thriving

Working on complex systems and creating systems change can be exhausting and daunting. The ultimate aim of Developmental Facilitating is to design and deliver group activities and process that build capacity toward systems actualizing. As Innes and Booher (1999) argued, "processes and outcomes cannot be neatly separated in consensus building because the process matters in and of itself, and because the process and outcome are likely to be tied together" (p. 415). Regenerative development work requires substantive individual and collective growth, therefore, the process of regenerative development must embody the elements of human thriving in order to guide projects that cultivate the capacity to renew, evolve, and thrive. The questions are as follows: How can the group processes of interaction best be facilitated to help realize potential? How do regenerative practitioners engage people in what might be a long and arduous process and how do they keep people engaged?

Positive psychology as a field of study aims to discover and promote the factors that allow individuals and communities to thrive (Seligman & Csikszentmihalyi 2000). Seligman (2012) developed the theory of human well-being comprising five fundamental elements: positive emotion, engagement, meaning, positive relationships, and accomplishment (PERMA). Each of these five elements contributes to well-being, is pursued for its own sake, and is defined and measured separately from the other elements (Seligman 2012). The field of positive psychology presumes that working toward well-being is possible because human beings are assumed to be self-organizing, self-directive, and adaptive entities (Seligman & Csikszentmihalyi 2000). Seligman and Csikszentmihalyi (2000) argued that in order to cultivate well-being, we must look for, address, and cultivate not just opportunities for individuals, but also attend to the creation of positive institutions and communities. This is the work of regenerative development, cultivating groups, institutions, and communities that are expanding their capacity to support the components of human thriving. Understanding the components of human thriving and incorporating them into regenerative development projects sets the foundation for teams and communities to do challenging work.

While most who study regenerative development or sustainable development projects may not have studied positive psychology, ample evidence from a variety of empirical studies illustrates that the most successful processes are incorporating the five aspects of human thriving: (1) they cultivate *positive emotions* within their activities, (2) foster deep *engagement* throughout the process, (3) build *positive relationships* between individuals and groups, (4) generate a sense of *meaning* for all stakeholders, and (5) promote and celebrate *accomplishment* throughout the process.

Positive emotion: Positive emotion encompasses all the usual subjective well-being measures: pleasure, ecstasy, comfort, warmth, and the like. Studies of team effectiveness and creativity have found that positive emotions, having fun, are associated with greater creativity and problem-solving skill (Estrada, Isen & Young 1994;, 2001; Isen, Daubman & Nowicki 1987). In our own study of integrative design teams, we found that those teams that cultivated positive emotions or affect were also described by participants as more creative and successful teams (Cross et al. 2015). When interviewing design teams about their most successful projects and team experiences, one of the first things they comment about is the emotional experience of being on the team. For the most successful teams, participants typically described how fun and creative the process was, how they felt their views were really being listened to, and that they could see specific design choices that came out of group charrettes. In contrast, the teams that designed good, but not great buildings, often made comments that included less positive emotions such as frustration, uncertainty about what happened with information they shared, or lack of emotional safety to share ideas (Cross et al. 2015). The benefits of positive affect do more than boost creativity; they also support other aspects of well-being like increased social interaction, help, generosity, and interpersonal understanding (Isen 2001).

Engagement: Positive psychologists talk about engagement as being in a state of flow, where a person is so engrossed in the present moment and activity that one loses a sense of self and the passage of time (Nakamura & Csikszentmihalyi 2014; Seligman 2012). The flow experience has been found to be associated with all manner of individual human performance from the success of athletes to musical virtuosos; however, only recently has it been examined in team environments (Aubé, Brunelle & Rousseau 2014; Csikszentmihalyi 2014; Lazarovitz 2004). In group processes, cultivating both individual states of flow and team flow can improve team performance. The best team processes give people the opportunity for challenge, deep focus, and collective engagement.

Positive relationships: In group environments and team activities, the quality and effectiveness of teams are influenced by the quality of social relationships on the team. Social trust and positive relationships increase willingness to share information that thus improves team performance (Lee et al. 2010; Wenger 2000). Cross et al. (2015) found that on sustainable design teams, relationship and trust building were consistently described as necessary for information sharing and strong communication loops, which directly impact team creativity, problem-solving, and ultimate performance. In regenerative development projects the cultivation of positive relationships between individuals and groups is both the practice and one of the key objectives.

Meaning: Being a part of something larger than oneself is one of the key components of human well-being, and it has been articulated as the core principle "being of service." Interestingly, giving is associated with greater individual success, higher productivity and salaries, and stronger team outcomes (Grant 2013). In team environments, when teams feel that their work is contributing to a group beyond themselves (society or the company more generally), they are more successful (Bock 2015). Sustainable development teams found motivation to work harder and longer when they knew that their efforts was having a lasting impact (Meyer et al. 2013). Crafting a vision of regenerative work that connects people to something larger than themselves is an essential component of regenerative development projects because the work is hard and the connection to something larger than oneself helps groups maintain motivation in the face of challenges.

Accomplishment: The final component of well-being is accomplishment, the pursuit of achievement, learning, and mastery (Seligman 2012). Just as the positive psychologists have noted, these components of well-being apply equally to individuals, institutions, and communities. Sarkissian et al. (2009) argued that the success of sustainable development efforts hinge on several factors, the development of trust (positive relationships), and accomplishment. When groups are brought together but are not able to take action or develop a sense of accomplishment, they feel that their energy and effort has been wasted (Sarkissian et al. 2009). In the regenerative development process, accomplishment is necessary because without it people lose the willingness to engage in difficult work.

The scientific literature on team performance, creativity in teams, and successful community development has significant overlap with the five components of Seligman's theory of human thriving. Given the synergy with research on teams, the theory of human thriving offers a useful framework for understanding the successful markers of any collaborative process. The degree to which any regenerative development project cultivates and embodies these five elements will directly impact the success of the project.

Cultivating and developing self-evolving networks

The product of Developmental Facilitating is the creation of new thoughts, deeper will, and informed actions that are cocreated by the group, characterized by increasing levels of capability with the team or community over time. How is it possible to realize this potential? Scientists from such diverse fields as systems engineering, natural resource management, and community development have all recognized the importance of one primary strategy in evolving complex systems—using the properties of social networks to advance change, find creative unifying solutions, enable collective action, and anchor change into social habits and institutions (Frank 2011; Krebs & Holley 2005; (Liu & Barabási 2016); Varda, Shoup & Miller 2012).

Networks as framework thinking

Social network analysis (SNA) is a unique tool because it measures and maps the characteristics of social actors (individuals and groups) as well as the quality, frequency, or strength of many types of relational ties and resource exchanges. When tracking a network over time, a regenerative practitioner might examine how trust is expanding in the network or how new clusters of groups have formed strong bonds. SNA provides a powerful framework for understanding how knowledge, information, and other resources are moving through the group. Using network maps to track group development can be a tool to help regenerative development projects self-assess and then adapt more productive patterns of working together.

Network structures

What do regenerative practitioners need to know about social networks in order to create a process that maximizes the potential of the network to do regenerative development work? Networks of people, like those of other living systems, have a variety of universal patterns and traits. Networks are made up of nodes (actors or social agents) that can be individuals, groups, or organizations, and the links or relational ties between them (e.g., trust, kinship, exchange). Regardless of the type of network—financial, biological, electronic, or social—nodes are linked to each other in patterned ways, called a network structure. These patterns or network structures have unique properties; some structures are more efficient at passing information through the network, while others are more resilient to change. Krebs and Holley (2005) argued that "instead of allowing networks to evolve without direction, successful individuals, groups and organizations have found that it pays to actively manage your network" (p. 4).

In place-based, sustainable development work, several idealized types of network structures have been identified that are associated with different levels of capacity to organize, share knowledge, and mobilize action (Bodin 2017; Henry & Vollan 2014; Krebs & Holley 2005; Vance-Borland & Holley 2011). Krebs and Holley (2005) described four stages of community collaboration structures as they develop over time (Figure 11.2). When a community network diagram consists of separated, isolated components, called "scattered fragments," it indicates that some community stakeholders have begun working together in small groups but have not yet become a single network. Without a fully connected structure, the

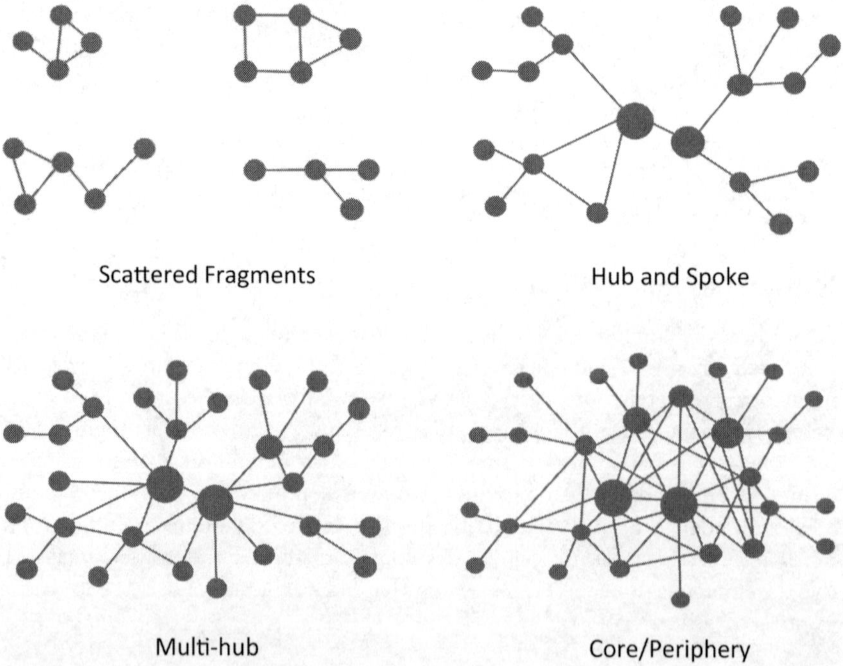

Figure 11.2 Social network structures (Krebs & Holley 2005)

capacity for joint action is severely limited. The Hub and Spoke, or centralized network, indicates that a single actor, often a key community agency, has been acting as the leader or coordinator of community action (Krebs & Holley 2005). This structure is very common at the beginning of a new community development efforts, but does not yet have the variety or density of ties required to really take collective action (Cross et al. 2009; Krebs & Holley 2005). As community stakeholders begin to work together, more ties form in the network, some new hubs and spokes appear as new members join the network, and existing hubs build ties and begin to look more like a web (Cross et al. 2009; Krebs & Holley 2005).

Managing complex problems requires the development of a core/periphery network structure (Cross et al. 2015; Sandström & Rova 2010; Vance-Borland & Holley 2011). This structure has several key features. First, the ties form a web-like or distributed network in the middle, which integrates members with unique knowledge or resources (Henry & Vollan 2014; Vance-Borland & Holley 2011). Second, the core comprises many ties and strong ties in the middle that help to facilitate knowledge and resource flow across the network (Cross et al. 2015; Krebs & Holley 2005). The diversity of the weak ties on the periphery bring in new knowledge and ideas into the network and helps to increase the capacity for creativity and problem-solving (Anklam 2007; Cross et al. 2015; Henry & Vollan 2014). While the structure of collaboration networks impacts the potential of the network to engage in collective action, the dynamics of network formation are recursive. The choices of individuals to form ties influence the structure and the structure then exerts influence on the quality of ties and new tie formation or dissolution. The changing ties and evolving structure of the network alter the capacity of the network to realize its goals or potential.

Building networks that realize potential

Network structures have the potential to either facilitate knowledge sharing, creativity, innovation, and collective action or to thwart them. Is it possible to create self-evolving networks that continually grow their ability to evolve, build capacity, and act collectively to actualize the potential of the system? The scientific evidence suggests that it is possible, but only through intentional network facilitation and the creation of flexible adaptive networks (Bryson, Crosby & Stone 2015; Vance-Borland & Holley 2011). Collaborative groups—policy networks, design teams, and collaborative resource management coalitions—often work off an intuitive understanding of network principles or dynamics. In the last 10–20 years, network science has been increasingly used as a tool for public health, policy networks, and natural resource governance (Cohen, Evans & Mills 2012; Provan et al. 2005; Varda, Shoup & Miller 2011). However, researchers continue to call for the need of professional facilitators, funded by the project to attend to the formation and evolution of relationships and network structures required for regenerative projects (Bryson, Crosby & Stone 2015; Vance-Borland & Holley 2011).

Developmental Facilitating as a capacity requires depth of knowledge in the processes of network formation and the structures that support collective action between a diverse set of actors. It also necessitates depth of knowledge in group activities and processes that have the capacity to evolve positive relationship, systems thinking, and capacity for coordinated action. Networks that form without mindful direction and intervention are more likely to not develop beyond the fragmented state (Figure 11.2), more likely to form structures that are segregated and scale-free and low density, and therefore less likely to succeed in their goals (Bryson, Crosby & Stone 2015; Henry & Vollan 2014). When collaborative networks are mindfully formed and facilitated, they can be shaped in ways that are conducive to resource sharing, collaborative decision-making, or creative thought (Briggs 2005; Bryson, Crosby & Stone 2015; Henry & Vollan 2014).

Regenerative practitioners use a variety of professional facilitation practices that have the potential to cultivate the social relationships and network structures needed for regenerative development and avoid the pathologies that create dysfunctional networks.

Table 11.2 provides an overview of the common professional facilitation practices used in many settings from community development, to adaptive natural resource management, to integrative design (7group & Reed 2009; Butler 2014; Kaner 2014). Paired with each facilitation practice is a short description of how that practice influences the formation of ties, the quality of relationships in the network, and in turn how those relationships help to avoid specific network

Table 11.2 Facilitation practices to evolve a network

Facilitation practice	*Influence on nodes, ties, and network structure*	*Network pathology avoided*	*Network capacity created*
Intentional invitations of diverse stakeholders (Bodin 2017; Vance-Borland & Holley 2011)	Brings diverse membership into the core and periphery, cultivates cross-scale ties, encourages pairwise ties with diverse others	Homophily, segregation, isolates	Group capacity to see and work with the whole system, coordination, group learning
Network weaving: Monitoring the network and suggesting new or strategic ties, and maintaining participation over time (Provan & Milward 2001; Sandström & Rova 2010; Vance-Borland & Holley 2011)	Increases periphery members, builds cross-scale ties, bridges fragmented groups	Segregation, constrained knowledge flow, absence of key roles—like a champion, fragmentation between groups, nodes seeking a position of power or brokerage	New knowledge enhances creativity, adaptive capacity, resource exchange, continuity prevents ideas from getting dropped or forgotten
Group visioning (Hoxie, Berkebile & Todd 2012)	Develops a sense of shared purpose and group identity	Fragmentation, biases assimilation, nodal traits—lack of openness to learning	Capacity for collective action, cooperation, shared mental models
Setting ground rules: Creating norms for even participation, respectful consideration of diverse views, and decision-making based on group vision (Bryson, Crosby & Stone 2015; Cross et al. 2015)	Increases ties across the network, increases trust, increases even participation, reduces opportunities for positions of brokerage, requires participants to consider diverse perspectives	Biased assimilation, uneven participation, segregation, fragmentation, lack of density	Team learning, capacity for collective decision-making, clarity of decision-making for achieving group vision, shared norms improve capacity for collaborative governance
Activities to build social connections (shared meals, fun, social events; Ostrom 1998; Vangen & Huxham 2003)	Increases density of trust ties, cultivates bonding ties in the core, increases reciprocal ties	Selective attachment for individual gain, homophily, segregation	Cooperation, information flow across the network, adaptive capacity

Defined process for facilitating large coalition meetings (Ansell & Gash 2008)	Builds distributed network, increases reciprocal ties, enhances knowledge flow across diverse nodes	Relatively few players control information flow and resources, scale-free networks, high transaction costs for knowledge sharing	Individual learning, team learning and shared mental models, capacity for collective action
Small group work (Cross et al. 2015; Ostrom 1998)	Builds trust, encourages interconnectedness (triadic closure), increases density in the core, increases strength of ties, supports knowledge sharing	Constrained knowledge flow, lower levels of cooperation with diverse others, fragmentation	Individual learning, team learning, capacity for coordinated action
Guided thinking and deliberation: Expert facilitation of divergent, then convergent thinking (Bryson, Crosby & Stone 2015; Kaner 2014)	Increases knowledge sharing across diverse nodes, decreases transaction costs for sharing tacit knowledge and complex knowledge	Limited interactions and knowledge sharing opportunities, ineffective brainstorming, truncated exploration of solutions, using old mental models	Capacity for integrative and creative thinking, discovery of intervention points with the greatest leverage, capacity to see and work with whole system
Clear role definition and flexibility to change roles as conditions change (social, ecological, and political) (Bryson, Crosby & Stone 2006; Innes & Booher 1999; Ostrom 1998)	Improves trust, supports mindful creation of the most appropriate network structure, reveals missing relationships and nodes in the network	Individuals seek to maximize position of power in network, incentives for people to occupy strategic positions, collaborative overload	Cooperation, capacity to develop and adopt innovative solutions, capacity to manage rather than avoid risk, capacity to evolve the network
Shares and discusses network diagrams (Bodin 2017; Sandström & Rova 2010; Vance-Borland & Holley 2011)	Increases group knowledge about effective network structures	Fragmentation, individuals seek to maximize position of power in network, incentives for people to occupy strategic positions	Capacity to evolve the network, adaptive management capacity, shared mental models

pathologies. The fourth column of the table describes how capacities are created in the network through the combination of network structures and quality of relationships. Regenerative practitioners are attending to three primary dynamics that recursively shape the network and its potential: (1) cultivating membership and participation, including diverse stakeholders, and attending to participation over the life span of the project; (2) building social relationships, trust, and group identity; and (3) using a social network framework to track the progress, development, and challenges of the group over time.

Network membership and forming ties

Professional facilitators can prevent many of the dysfunctions of network self-assembly by inviting diverse stakeholders, encouraging interaction between people with diverse knowledge and viewpoints, and ensuring engagement of diverse members across the life span of the project. This attention to network membership and involvement helps to mitigate against the tendencies for individual actors to form and cut ties based on similarity (homophily), to segregate into cliques or clusters with homogenous world views (segregation), and to seek positions of brokerage between subgroups (Henry & Vollan 2014; Vance-Borland & Holley 2011).

Regenerative practitioners prevent homophily and segregation from dominating the network through several activities. Beginning with the intention to invite and mindfully include diverse stakeholders, regenerative practitioners ensure that the process is inclusive of all those whose livelihoods will be impacted by a particular regenerative project. The regenerative practitioner is continually confirming that the project has the necessary sponsors, champions, and facilitators, and that no key roles or groups of stakeholders are missing from the process (Bryson et al. 2015). Finally, the regenerative practitioner is analyzing the structure and membership of the network over time to ensure that key stakeholders are participating throughout the process and have not got marginalized into an isolated subgroup (Cross et al. 2015; Vance-Borland & Holley 2011). Attending to membership, roles, and opportunities for participation are primarily strategies for avoiding segregation, homophily, and fragmentation in the network that reduce opportunities for group learning, cooperation, and knowledge sharing (Phelps, Heidl & Wadhwa 2012).

Cultivating trust and group identity

Diverse membership alone is not enough to enable collaborative groups to realize their potential. They must also share a group vision, develop group norms, and expand levels of trust. Regenerative practitioners accomplish these tasks through several activities. At the beginning of regenerative processes, and periodically throughout, professional facilitators establish ground rules and group norms. Depending on the nature of the collaborative group, sometimes these ground rules are specific to a design charrette or workshop, sometimes they govern an ongoing design process that might last for years (Cross et al. 2015). Ground rules might be formal agreements that include definition of roles, expected resource contributions, and sanctions for violating any part of the agreement (Innes & Booher 1999; Ostrom 1998; Sandström & Rova 2010). Setting expectations for interactions is a key strategy for building trust in collaborative groups.

The second key activity that is used to build trust, which improves the capacity of actors to cooperate and the capacity of the network to take coordinated action, is facilitation of social activities. One architect working on an innovative, high-performance building, said this, "Holding full-day charrettes and eating meals together makes it a social time to come together, where we get to talk to different people and share ideas. Those relationships built during the charrette make it possible for a facilities guy to walk up to the architect and say, 'that is a really dumb idea.'" (Cross et al. 2015, p. 15). This architect described clearly how social events and charrettes build trust and help to integrate the network, creating bridging ties from groups that might typically be isolated in a design network.

The third key activity used by regenerative practitioners and professional facilitators is the guiding of group visioning sessions. These sessions typically explore a variety of topics from the uniqueness of place, to core values of place, to vision and goals (Hoxie, Berkebile & Todd 2012). Facilitated vision sessions improve creativity in groups and can expand what participants see as possible (Vidal & Valqui 2004). The integration of local values and community participation in vision creation, increases the commitment of individuals to the group goal, cooperation, and capacity for collective action (Hoxie, Berkebile & Todd 2012; Innes & Booher 1999; Marshall 2014). Cultivating social relationships while developing a group vision builds trust and expands ties, which improve cooperation and reduce segregation and fragmentation across the network.

Network evolving activities

The last five activities in Table 11.2 work synergistically to evolve the network over time, integrating diverse nodes, and building capacity for knowledge sharing and collective action in the network. Both large and small group meetings create opportunities for building relationships, sharing knowledge, and creating bridging ties across diverse participants (Cross et al. 2015; Sandström & Rova 2010). Regenerative practitioners use a variety of theoretically based deliberative processes (e.g., divergent and convergent thinking, design thinking) to specifically facilitate knowledge sharing across diverse nodes, develop new mental models, and cultivate conditions that increase opportunities for creative and innovative solutions to emerge (Kaner 2014; Vidal & Valqui 2004). Activities that foster collaborative learning become a positive feedback loop in networks; as actors develop shared mental models, their trust in diverse others increases, and as trust increases the ability to share knowledge also increases (Bodin 2017; Ostrom 1998).

Because of the many barriers—nodal qualities, network structure, nature of the knowledge, trust—to sharing knowledge in a diverse network, creating opportunities for group learning and knowledge exchanges are a primary and essential competency of Developmental Facilitating (Bodin 2017; Henry & Vollan 2014; Phelps, Heidl & Wadhwa 2012). Regenerative practitioners improve the outcomes of collaborative networks by attending to the dynamic needs of the network, opportunities to bridge fragmented subgroups, and emergent needs of the network members (Bryson et al. 2015; Hoxie, Berkebile & Todd 2012). The dynamic nature of living systems and social networks require that regenerative practitioners understand the evolving nature of relationships and networks in order to facilitated their evolution to encompass new capacities (Vance-Borland & Holley 2011; Wheatley & Frieze 2006).

Conclusion

Shifting mindsets, paradigms and ways of working is hard work. Those interested in leading regenerative development benefit from understanding two key learnings from social science. First, positive psychology deepens our understanding of the conditions and experiences that cultivate human thriving. Other regenerative practitioners have proposed that "a design process that stimulates regenerative capacity must follow the same rules as the system it seeks to create." Therefore, a regenerative development process that seeks to build opportunities for human thriving must embody those principles throughout the process. Second, decades of research in policy, public health, ecology, and organizational studies have been building the evidence base for understanding how networks (traits of nodes, relationships between nodes, and structures of the whole network) shape the capacity of groups to learn, evolve, and develop capacity for collective action. Yet, there remain a number of gaps in our understanding of how best to engage network science in regenerative development projects.

Current models of regenerative development share a few common propositions: (1) regenerative development requires a process focused on engagement over predetermined goals, (2) the process itself must mirror the dynamics of the system it is trying to generate, (3) participatory processes create the collaborative learning required for creative innovative solutions to emerge, and (4) intentional formation of relationships and network structures is required to evolve collaborative networks (Gibbons et al. 2018; Hoxie, Berkebile & Todd 2012; Mang, Haggard & Regenesis 2016; Plaut & Amedée 2018; Robinson & Cole 2015; Vance-Borland & Holley 2011). These propositions form the foundation for both the practice and study of regenerative development.

Researchers and practitioners both have noted a gap between regenerative development practice and the evidence base. Third-party facilitation has been shown to increase the creativity of teams, improve outcomes of group processes, and evolve the capacity of collaborative groups, yet paid positions for "developmental facilitators" or "network weavers" are not standard practice (Bryson et al. 2015; Cross et al. 2015; Vance-Borland & Holley 2011). The dynamic nature of networks and complexity of the problems facing regenerative practitioners poses challenges for understanding all the causal relationships between network formation, network relationships, network structure, and outcomes (Bryson et al. 2015; Henry & Vollan 2014; Sandström & Rova 2010). The multidisciplinary nature of network science and collaboration networks means that the application of network science to regenerative development is in its nascence. As the field of regenerative development grows, its greatest potential will evolve from the collaboration of scientists and practitioners examining both the practice and science of regenerative processes and how networks evolve within those processes.

References

7group & Reed, B 2009, *The integrative design guide to green building: Redefining the practice of sustainability*, John Wiley & Sons: Hoboken, New Jersey.

Agrawal, A 1995, "Dismantling the divide between indigenous and scientific knowledge", *Development and Change*, vol. 26, no. 3, pp. 413–439.

Anklam, P 2007, *Network: A practical guide to creating and sustaining networks at work and in the world*, Butterworth-Heinemann, Oxford, UK.

Ansell, C & Gash, A 2008, "Collaborative governance in theory and practice", *Journal of Public Administration Research and Theory*, vol. 18. no. 4, pp. 543–571. doi:10.1093/jopart/mum032.

Aubé, C, Eric B, and Vincent R. 2014 Flow experience and team performance: The role of team goal commitment and information exchange. *Motivation and Emotion* 38, no. 1: 120–130.

Barca, F, McCann, P & Rodríguez-Pose, A 2012 "The case for regional development intervention: Place-based versus place-neutral approaches", *Journal of Regional Science*, vol. 52, no. 1, pp. 134–152.

Barth, M, Godemann, J, Rieckmann, M & Stoltenberg, U 2007, "Developing key competencies for sustainable development in higher education", *International Journal of Sustainability in Higher Education*, vol. 8, no. 4, pp. 416–430.

Bock, L, 2015, "*Work rules! : Insights from inside Google that will transform how you live and lead*", Grand Central Publishing, New York.

Bodin, Ö 2017, "Collaborative environmental governance: Achieving collective action in social-ecological systems", *Science*, vol. 357, no. 6352. doi:10.1126/science.aan1114.

Bortoft, H 1996, *The wholeness of nature*, SteinerBooks, Great Barrington, MA.

Briggs, J 2005, "The use of indigenous knowledge in development: Problems and challenges," *Progress in Development Studies*, vol. 5, no. 2, pp. 99–114.

Bryson, JM, Crosby, BC & Stone, MM 2006, "The design and implementation of cross-sector collaborations: Propositions from the literature," *Public Administration Review*, vol. 66, no. s1, pp. 44–55. doi:doi:10.1111/j.1540-6210.2006.00665.x

Bryson, JM, Crosby, BC & Stone, MM 2015, "Designing and implementing cross-sector collaborations: Needed and challenging," *Public Administration Review*, vol. 75, no. 5, pp. 647–663. doi:doi:10.1111/puar.12432

Buchanan, A & Kern, ML 2017, "The benefit mindset: The psychology of contribution and everyday leadership", *International Journal of Wellbeing*, vol. 7, no. 1. doi:http://dx.doi.org/10.5502/ijw.v7i1.538.

Butler, AS 2014, *Mission critical meetings: 81 practical facilitation techniques*, Wheatmark, Tuscon, AZ.

Center for Living Environments and Regeneration 2017, *LENSES facilitator manual: How to create living environments in natural social and economic systems*, CLEAR, Fort Collins, CO.

Cohen, PJ, Evans, LS & Mills, M 2012, "Social networks supporting governance of coastal ecosystems in Solomon Islands", *Conservation Letters*, vol. 5, no. 5, pp. 376–386. doi:10.1111/j.1755-263X.2012.00255.x.

Corral-Verdugo, V, Carrus, G, Bonnes, M, Moser, G & Sinha, JB 2008, "Environmental beliefs and endorsement of sustainable development principles in water conservation: Toward a new human interdependence paradigm scale", *Environment and Behavior*, vol. 40, no. 5, pp. 703–725.

Cross, JE, Barr, SW, Putnam, R, Dunbar, B & Plaut, J 2015. The Social network of integrative design, viewed on 13 February 2019, <http://www.ibe.colostate.edu/news/item.aspx/?ID=131951>.

Cross, JE, Ellyn D, Rebecca NG Jesse MF, 2009, Using mixed-method design and network analysis to measure development of interagency collaboration." *American Journal of Evaluation* 30, no. 3 (2009): 310–329.

Csikszentmihalyi, M., SpringerLink. 2014. *Flow and the foundations of positive psychology: The collected works of Mihaly Csikszentmihalyi* (Springer eBook collection). Springer, Berlin.

Dent, EB 1999, "Complexity science: A worldview shift", *Emergence*, vol. 1, no. 4, pp. 5–19. doi:10.1207/s15327000em0104_2.

Dunlap, RE & Catton, WR 1994, "Struggling with human exemptionalism: The rise, decline and revitalization of environmental sociology", *The American Sociologist*, vol. 25, no. 1, pp. 5–30.

Dunlap, RE, Van Liere, KD, Mertig, AG & Jones, RE 2000, "New trends in measuring environmental attitudes: Measuring endorsement of the new ecological paradigm—A revised NEP scale", *Journal of Social Issues*, vol. 56, no. 3, pp. 425–442.

Estrada, CA., Isen AM, Young MJ. "Positive affect improves creative problem solving and influences reported source of practice satisfaction in physicians." *Motivation and emotion* 18, no. 4 (1994): 285–299.

Frank, KA 2011, "Social network models for natural resource use and extraction", In O Bordin & C Prell, (eds.), *Social networks and natural resource management: Uncovering the social fabric of environmental governance*, Cambridge University Press, Cambridge, UK, pp. 180–205.

Gibbons, L, Cloutier, S, Coseo, P & Barakat, AJS 2018, "Regenerative development as an integrative paradigm and methodology for landscape sustainability", *Sustainability*, vol. 10, no. 6. doi: 10.3390/su10061910.

Gilchrist, A 2009, *The well-connected community: A networking approach to community development*, Policy Press, Bristol, UK.

Grant, A 2013, *Give and take: Why helping others helps drive our success*, Penguin Books, New York, NY.

Henry, AD & Vollan, B 2014, "Networks and the challenge of sustainable development", *Annual Review of Environment and Resources*, vol. 39, pp. 583–610.

Hoxie, C, Berkebile, R & Todd, JA 2012, "Stimulating regenerative development through community dialogue", *Building Research & Information*, vol. 40, no. 1, pp. 65–80.

Innes, JE & Booher, DE 1999, "Consensus building and complex adaptive systems", *Journal of the American Planning Association*, vol. 65, no. 4, pp. 412–423. doi:10.1080/01944369908976071.

Isen, AM 2001. An influence of positive affect on decision making in complex situations: Theoretical issues with practical implications. *Journal of consumer psychology*, 11(2), 75–85.

Isen, AM., Kimberly AD. Gary NP. 1987 "Positive affect facilitates creative problem solving." *Journal of personality and social psychology* 52, no. 6: 1122.

Kana 'iaupuni, SM 2005, "Ka 'akālai Kū Kanaka: A call for strengths-based approaches from a Native Hawaiian perspective", *Educational Researcher*, vol. 34, no. 5, pp. 32–38.

Kaner, S 2014, *Facilitator's guide to participatory decision-making*, John Wiley & Sons, Hoboken, NJ.

Krebs, V & Holley, J 2005 "Building adaptive communities through network weaving", *Nonprofit Quarterly*, pp. 61–67.

Lazarovitz, SM (2004). *Team and individual flow in female ice hockey players: The relationships between flow, group cohesion, and athletic performance* (Doctoral dissertation, ProQuest Information & Learning).

Lee, P, Gillespie, N, Mann, L & Wearing, A 2010, "Leadership and trust: Their effect on knowledge sharing and team performance", *Management Learning*, vol. 41, no. 4, pp. 473–491.

Lin, N, Cook, KS & Burt, RS 2001, *Social capital: Theory and research*, Transaction Publishers, Piscataway, NJ.

Liu, J, Dietz, T, Carpenter, SR, Folke, C, Alberti, M, Redman, CL, Schneider, SH, Ostrom, E, Pell, AN, Lubchenco, J, Taylor, WW, Ouyang, Z, Deadman, P, Kratz, T & Provencher, W 2007, "Coupled human and natural systems", *Ambio*, vol. 36, no. 8, pp. 639–649.

Liu, Yang-Yu, Barabási A. 2016, Control principles of complex systems. *Reviews of Modern Physics* 88, no. 3.

Lopez, SJ & Louis, MC 2009, "The principles of strengths-based education", *Journal of College and Character*, vol. 10, no. 4.

Mang, P, Haggard, B & Regenesis 2016, *Regenerative development: A framework for evolving sustainability*, John Wiley & Sons, Inc.: Hoboken, New Jersey.

Marshall, G 2014, *Hearth and hiraeth: Constructing climate change narratives around national identity*, Climate Outreach Information Network, Oxford.

McDonough, W & Braungart, M 2010, *Cradle to cradle: Remaking the way we make things*, North Point Press.

Metcalf, L & Benn, S 2013, "Leadership for sustainability: An evolution of leadership ability", *Journal of Business Ethics*, vol. 112, no. 3, pp. 369–384.

Meyer, M, Cross, JE, Byrne, ZS, Franzen, WS & Reeve, S 2013, "A Team-approach to green schools: Vision, support, and environmental literacy", In AJ Hoffman & R Henn (eds.), *Constructing green: Sustainability and the places we inhabit*, MIT Press, Cambridge, pp. 381–418.

Nakamura, Jeanne, Csikszentmihalyi, M. 2014. "The concept of flow." In Csikszentmihalyi, M., & SpringerLink. (2014). *Flow and the foundations of positive psychology : The collected works of Mihaly Csikszentmihalyi* (Springer eBook collection). Springer, Berlin.

Ostrom, E 1998, "A behavioral approach to the rational choice theory of collective action: Presidential address, American political science association, 1997", *American Political Science Review*, vol. 92, no. 1, pp. 1–22. doi:10.2307/2585925.

Ostrom, E 2009 "A general framework for analyzing sustainability of social-ecological systems", *Science*, vol. 325, no. 5939, pp. 419–422.

Phelps, C, Heidl, R & Wadhwa, A 2012, "Knowledge, networks, and knowledge networks: A review and research agenda", *Journal of Management*, vol. 38, no. 4, pp. 1115–1166.

Plaut, J & Amedée, E 2018, Becoming a regenerative practitioner: A field guide, viewed 13 February 2019, <https://ibe.colostate.edu/becoming-a-regenerative-practitioner-a-field-guide-2018/≥.

Provan, KG & Milward, HB 2001, "Do networks really work? A framework for evaluating public-sector organizational networks", *Public Administration Review*, vol. 61, no. 4, pp. 414–423. doi:10.1111/0033-3352.00045

Provan, KG, Veazie, MA, Staten, LK & Teufel-Shone, NI 2005, "The use of network analysis to strengthen community partnerships", *Public Administration Review*, vol. 65, no. 5, pp. 603–613. doi:10.1111/j.1540-6210.2005.00487.x.

Reed, B 2007, "Shifting from 'sustainability' to regeneration", *Building Research & Information*, vol. 35, no. 6, pp. 674–680. doi:10.1080/09613210701475753.

Robinson, J 2004 "Squaring the circle? Some thoughts on the idea of sustainable development", vol. 48, no. 4, pp. 369–384.

Robinson, J & Cole, RJ 2015, "Theoretical underpinnings of regenerative sustainability", *Building Reserah & Information*, vol. 43, no. 2, pp. 133–143.

Sack, R 1997, *Homo geographicus*, The Johns Hopkins University Press, Baltimore, MD.

Sale, K 1985, *Dwellers in the land: The bioregional vision*, University of Georgia Press, Athens, GA.

Sandström, A & Rova, C 2010, "Adaptive co-management networks a comparative analysis of two fishery conservation areas in Sweden", *Ecology and Society*, vol. 15, no. 3.

Sanford, C 2016, "What is regeneration? Part 1—A definition and some fundamental ideas", viewed 13 February 2019, <https://carolsanfordinstitute.com/what-is-regeneration-part-1/?utm_content=buffer0e87c&utm_medium=social&utm_source=linkedin.com&utm_campaign=buffer>.

Sanford, C 2017, *The Regenerative business: Redesign work, cultivate human potential, achieve extraordinary outcomes*, Nicholas Brealey, Boston, MA.

Sarkissian, W, Hofer, N, Shore, Y, Vajda, S & Wilkinson, C 2009, *Kitchen table sustainability: Practical recipes for community engagement with sustainability*, Earthscan Press, Sterling, VA.

Sauvé, S, Bernard, S & Sloan, P 2016, "Environmental sciences, sustainable development and circular economy: Alternative concepts for trans-disciplinary research", *Environmental Development*, vol. 17, pp. 48–56.

Schwartz, SH & Bardi, A 2001, Value hierarchies across cultures: Taking a similarities perspective, *Journal of Cross-Cultural Psychology*, vol. 32, pp. 268–290.

Seamon, D 2018, *Life takes place: Phenomenology, lifeworlds, and place making* (1st ed.), Routledge, New York, NY.

Seligman, ME 2012, *Flourish: A visionary new understanding of happiness and well-being*, Simon and Schuster, New York, NY.

Seligman, ME & Csikszentmihalyi, M 2000, "Positive psychology: An introduction", *American Psychologist*, vol. 55, no. 1, pp. 5–14.

Vance-Borland, K & Holley, J 2011, "Conservation stakeholder network mapping, analysis, and weaving", *Conservation Letters*, vol. 4, pp. 278–288.

Vangen, S & Huxham, C 2003, "Nurturing collaborative relations: Building trust in inter-organizational collaboration", *The Journal of Applied Behavioral Science*, vol. 39, no. 1, pp. 5–31. doi:10.1177/0021886303039001001

Varda, D, Shoup, J. A., & Miller, S 2011, "A systematic review of collaboration and network research in the public affairs literature: Implications for public health practice and research", *American Journal of Public Health*, vol. 102, no. 3, pp. 564–571. doi:10.2105/AJPH.2011.300286.

Varda, D., Shoup, JA & Miller, S 2012 "A systematic review of collaboration and network research in the public affairs literature: Implications for public health practice and research", *American Journal of Public Health*, vol. 102, no. 3, pp. 564–571.

Vidal, R & Valqui, V 2004, "The vision conference: Facilitating creative processes", *Systemic Practice and Action Research*, vol. 17, no. 5, pp. 385–405. doi:10.1007/s11213-004-5786-x.

Wahl, D 2016, *Designing regenerative cultures*, Triarchy Press Ltd, Axminster, England.

Wenger, E 2000, "Communities of practice and social learning systems", *Organization*, vol. 7, no. 2, pp. 225–246.

Wheatley, M & Frieze, D 2006, "Using emergence to take social innovation to scale", *The Berkana Institute*, vol. 9.

Whitney, D & Cooperrider, D 2011, Appreciative inquiry: A positive revolution in change, ReadHowYouWant.com, <https://www.researchgate.net/publication/237404587_A_Positive_Revolution_in_Change_Appreciative_Inquiry>.

Whyte, KP, Brewer, JP & Johnson, JT 2016, "Weaving indigenous science, protocols and sustainability science", *Sustainability Science*, vol. 11, no. 1, pp. 25–32.

12 Workforce development: A regenerative perspective

Eugene A. Wilkerson and Allison Dake

The term sustainability has a long and complex history with a foundation grounded in an environmental dimension. George Perkins Marsh, one of America's first environmentalist, states in his book *Man and Nature; or, Physical Geography as Modified by Human Action*, "Sustainability is based on a simple and long-recognized factual premise: Everything that humans require for their survival and well-being depends, directly or indirectly, on the natural environment" (Marsh 1864). It would take over one hundred years before the world began to take action to address a devolving environment. Edwards (2005) suggests that sustainability became a prominent world issue as a result of the 1972 United Nations Conference on the Human Environment. Others assert that the first significant use of the term appeared in the 1980 World Conservation Strategy (IUCN 1980). Despite the consternation as to its beginning, the Brundtland report (1987) provides a working definition that has stood the test of time: "Sustainable development is development that meets the needs of the present without compromising the ability of future generations to meet their own needs." This definition allowed for a gradual movement away from the exclusive focus on the environment and toward a tripartite view of sustainability as a concept that addresses the connection between the environment, social concerns, and the economic issues (Hopwood, Mellor & O'Brien 2005; Littig & Griessler 2005).

In the foreground of this sustainability movement, the regenerative development paradigm has been hedging forward with a whole system perspective with dedicated attention on closing the feedback loops. The priority of regenerative development is to apply holistic processes in a bioregional context to create feedback loops between physical, natural, economic, and social capital that are mutually supportive and contain the capacity to restore equitable, healthy, and prosperous relationships among these forms of capital. Much of the regenerative literature focuses on the built environment (Brandon, Lombardi & Shen 2017; Hes & Plessis 2015; Mang, Haggard & Regenesis 2016) or economics concepts related to business and entrepreneurship (Sanford 2001, 2011, 2014). The work of these scholars demonstrates the evolution of sustainable development to regenerative development within two areas of the traditional tripartite model. The literature surrounding sustainable development continues to struggle to define the social dimension (Boström 2012; Cuthill 2010; Dempsey et al. 2009), which makes it difficult for scholars to discuss the evolution of the social aspect of regenerative development.

The purpose of this chapter is to examine workforce development through the narrow lens of regenerative development. The first section provides an overview of the foundational policies and theories that drive the concept of work and employment in the United States of America (USA). The second section interrogates the concept of social sustainability to tease out the relevant themes that can illuminate a path toward a regenerative workforce development paradigm. The thematic approach to the latter part of this chapter is necessary because of the lack of continuity in the literature surrounding social sustainability and the rapid emergence of regenerative development. A detailed debate of the research associated with social sustainability would devolve the purpose of the chapter into a discussion of definitions and evaluative approaches as opposed to an exploration of this new approach.

Foundational employment policies and theories

Employment in the United States of America: A national focus on employment in the United States of America emerged in the late 1920s around the time of the great Wall Street crash of 1929. Programs such as the New Deal were enacted largely to address the issue of employment and to provide people with a living wage. This thought process was an emphasis in the USA economic policy and was considered important to maintaining our society. Further, employment is a concept for which the government began to take direct responsibility. The 1946 Employment Act was one of the first laws to articulate this direction (Ranney 2003). Section two of the statute states that the United States will

> utilize all its plans, functions, and resources for the purpose of creating and maintaining, in a manner calculated to foster and promote free competitive enterprise and the general welfare, conditions under which there will be afforded useful employment opportunities, including self-employment, for those able, willing, and seeking to work and to promote maximum employment, production and purchasing power. (Employment Act of 1946 2017)

In the late 1950s and early 1960s, organized labor sounded an alarm regarding recent technological innovations that resulted in a decrease in employment for the railroad and steel industries. Large corporations such as General Electric recognized the need to retrain workers and dedicated $40 million toward new workforce education programs. Many politicians became concerned about the USA education system and its ability to keep up with our Cold War opponent the Union of Soviet Socialist Republics. To address the confluence of issues surrounding the US workforce, Congress passed the Manpower Development and Training Act (MDTA) of 1962 that provided federal grants to local agencies working to deliver workforce training (Kremen 1974). The 1960s also brought about the beginning of a shift in the philosophy that employment is a societal goal (Ranney 2003). Noted economist Milton Friedman began to advocate for "free market" policies and saw no reason for a business to pursue goals related to social responsibility. From his perspective, the only role of business is to pursue monetary gains for stockholders (Friedman 1962). Friedman expanded on this view in a 1968 speech to the American Economic Associate where he asserted that the labor unions and minimum wage laws were the reason for rising unemployment. From his perspective, the USA monetary policy could not effectively impact unemployment and attempts to do so were pointless, thus leading to bad policy decisions (Friedman 1968).

Nonetheless, the USA monetary policy continued to focus on full employment as Congress passed the Humphrey-Hawkins Full Employment and Balanced Growth Act of 1978. In practical terms, Humphrey-Hawkins extended the philosophy first articulated in the 1946 Employment Act (Lipford 1999; Ranney 2003). At the same time, there emerged laws that began to lean toward Friedman's philosophy. With the implementation of the 1978 Comprehensive Employment and Training Act (CETA), the federal government began the process of shifting employment and training programs to the states (LaLonde 1995). While CETA retained some public service employment provisions (Holzer 2009), the law also began to codify private sector involving in employment policy through the implementation of the Private Sector Initiatives Program (Heinrich & Lynn 2000). These small policy shifts were the beginning of a shift toward the Friedman philosophy.

The 1980s saw an increase in defense spending, small increases in income tax, and a decline in revenue from corporate taxes among other economic concerns. Unemployment began to rise as the economy experienced a difficult recession (Lipford 1999). As a result, the federal government enacted the Job Training Partnership Act (JTPA) of 1982, which allocated $5 billion to state and local governments to fund employment and training programs. Through JTPA, the next evolution of private sector involvement in employment policy emerged in the form of Private Industry Councils. These organizations, composed primarily of private sector businesses, were responsible for JTPA program delivery and fund distribution (Heinrich & Lynn 2000). For the next ten years, the USA economy fluctuated between periods of growth in the late 80s and then back to a slower economy in the early 90s (Lipford 1999). The evolution of economic theories first espoused by Friedman (1962; 1968) emerged again and were extended when economist Gary Becker won the Nobel Prize for economics in 1992, with his work surrounding human capital. This theory asserts that the ability to gain work and advance in a career is a rational choice that individuals should make in a free market. Individuals can freely select the type of education and training they desire, which provides them the ability to compete in the open market. Social factors such as race, gender, and class were considered human capital investment and their influence on policy would diminish in favor of the free market approach (Ranney 2003). By the late 90s, scholars were calling for a full repeal of the Humphrey-Hawkins Act of 1978 (Lipford 1999). The shift toward a decentralized approach to employment policy driven by the involvement of state and local governments in collaboration with the private sector led to an exploration of the systems surrounding employment or workforce development.

Demographics shifts: The shifts in policies at the national level were affected by the changing demographics of the workforce in the United States of America. In the early 1900s, jobs required manual skills for operating machine-powered mass production assembly lines and manufacturing. With a shift after World War II, the service industry rapidly began changing jobs. In their article Employee Relations Ethics and the Changing Nature of the American Workforce (2001), Kim, Emmet and Sikula note that "in 1945, the service industry accounted for 10% of nonfarm land employment, compared with 38% for manufacturing" as compared to 1996 when "service industry accounted for 29% of nonfarm employment, and manufacturing at 18%" (p. 26). In 1994, it was reported 80% of American workers were in service-oriented jobs.

Each of these shifts in the workforce caused, what is referred to today as technology discrepancy. A major difference in workforce development today as compared to our history is that these technology discrepancies are driving gaps between economic development and business competitiveness at exceptional rates causing economic disenfranchisement in the workforce. As the speed of information increases and product life cycles are reduced to under a year, the net effect is an increase in the rate of knowledge exchange. Overall, this is a key factor in what is considered a relevant and useful product in our society (Shrivastava, Ivanaj & Ivanaj 2016). This rapid firing of technology discrepancies holds the key to our economic future and creates opportunities for business to become a catalyst for change and to create value in a world that is seeking value.

At an organizational level, business, government entities, and nonprofit organizations must be in tune with the talent management needs of their community. According to (Shrivastava et al. 2016):

> The performative function of this core is to create knowledgeable human resources that can function effectively in the knowledge economy of our information society. The core is contextualized by environmental conditions characterized by institutional and social structures of business education. Boundaries between the core and its context are often fuzzy and changing, and to some extent definitional. The important point is to view these elements as part of single organic ecosystem. (p. 3)

The ability to look at demographics shifts at a national, state, local, and across organizational types provides the foundation for a systems thinking approach to workforce development.

Workforce development systems: In the late 1990s and early 2000s, business, government entities, and nonprofit organizations worked to identify the entities at the local and regional levels that encompass workforce development activity and their function as interconnected organizations. The literature began to use the term "workforce development system" consistently as scholars and practitioners sought reforms to address a variety of societal changes. There was a recognition of the limitations associated with an employer-based system designed to meet the fluctuating human capital needs of the private sector; thus, the relationship between communities and employers needed to evolve. Practitioners began to deploy sector-based approaches that would identify high-paying jobs and industries with a need for workers and design program to facilitate entry points for those in low-income communities. Other tactics involved the deployment of a place-based approach, which emphasizes the placement of job training programs in communities. The intended benefit of this approach is to keep the individuals moving into well-paying jobs in the community, thus decreasing unemployment, increasing the tax base, supporting business, and improving housing standards within a community (Bates & Redmann 2002; Fitzgerald 1999; Giloth 2000).

Experts altered their view of the education system and its connection to workforce development. The secondary education environment that had a notable emphasis on college preparation was asked to consider the need for vocational technology. Postsecondary institutions (community colleges and four-year institutions) were forced to increase their coordination at the policy level and to provide career counseling. Education institutions at all levels took notice of cooperative work opportunities that allow students to earn credit while in a workplace setting. Beyond the walls of traditional classrooms, there was great discussion regarding the role of lifelong learning and need for community organizations to assist with job training, job placement, and job retention (Bates & Redmann 2002; Fitzgerald 1999).

Scholars, nonprofit organizations, and government officials advocated for workforce systems to operate holistically as opposed to focusing exclusively on the job training, network connections, and educational alignment. Fitzgerald (1999) advocated for the implementation of transportation, healthcare, and daycare programs. Giloth (2000) stressed the importance of addressing systemic employment inequality. Bates and Redmann (2002) campaigned for work readiness programs to develop individuals in the areas outside the mission of secondary and postsecondary institutions. Skills developed through these programs include relationship building, communication, and interpersonal skills.

Giloth (2000) captures many of the elements mentioned thus far in the following definition of workforce development: "The phrase workforce development implies more than employment training in the narrow sense: It means substantial employer engagement, deep community connections, career advancement, integrative human services support, contextual and industry-driven education and training, and the connective tissue of networks." Scholars quickly recognized that a programmatic focus at the local and regional levels would not be sufficient (Hawley, Sommers & Melendez 2003). Jacobs and Hawley (2009) advanced the notion of a national view workforce development that emphasizes: (1) Globalization, (2) technology, (3) the new economy, (4) political change, and (5) demographic change. Using this paradigm, these scholars define workforce development as: "…the coordination of public and private sector policies and programmes that provides individuals with the opportunity for a sustainable livelihood and helps organizations achieve exemplary goals, consistent with the societal context." In 2017, the Urban Institute, a nonprofit research organization that addresses various social issues, moved the discussion back toward the local and regional levels. Their report uses a logic model to assert that the goals or outcomes for local workforce development should revolve around five areas: (1) collaboration, (2) quality and accessibility, (3) industry engagement, (4) data-driven decision–making, and (5) sustainability (Bernstein & Martin-Caughey 2017).

The concepts of collaboration, quality, and accessibility emphasized by the Urban Institute are similar to the literature from the late 1990s and early 2000s. Their work illuminates two new streams of thought: (1) Data-driven research and (2) sustainability, which are now a major emphasis in conversations surrounding workforce development systems. It is appropriate to argue that workforce development has a connection to both economic sustainability and social sustainability. This connect is the beginning of a regenerative development conversation for the field of workforce development. This chapter will focus specifically on the connection to social sustainability.

Launching a regenerative workforce development paradigm

Peering through the social sustainability lens: The social dimension of sustainability is a broad concept that continues to challenge scholars seeking a clear and concise definition or framework. There are many reasons for this ambiguity. At its core is the presence of two different approaches to the overarching concept of sustainable development. Those espousing the one-pillar approach assert that environmental issues are dominant within the paradigm. Others who advocate for the multi-pillar approach argue that social, economic, and environmental sustainability are equivalent in standing (Bramley & Power 2009; Davidson 2009; Littig & Griessler 2005).

Another confounding issue is that the literature surrounding social sustainability offers a variety of definitions of the term, which differ based on the topic of the text and the views of the authors. In their book, *Understanding the New Sociocultural Dynamics of Cities: Comparative Urban Policy in a Global Context*, Stren and Polèse (2000) define social sustainability as follows:

> development (and/or growth) that is compatible with the harmonious evolution of civil society, fostering an environment conducive to the compatible cohabitation of culturally and socially diverse groups while at the same time encouraging social integration, with improvements in the quality of life for all segments of the population. (pp. 16–17)

In a 2005 article that approaches social sustainability from a sociological and political perspective, Littig and Griessler provide a definition that emphasizes the role of work in society:

> It signifies the nature-society relationships, mediated by work, as well as relationships within society. Social sustainability is given, if work with a society and the related institutional arrangements (1) satisfy and extended set of

> human needs and (2) are shaped in a way that nature and its reproductive capabilities are preserved over long period of time and the normative claims of social justice, human dignity, and participation are fulfilled.

Originally published in the conference proceedings of the Second International Conference on Whole Life Urban Sustainability and its Assessment, Colantonio (2009) lists three additional definitions of social sustainability in addition to those covered in this chapter. It is beyond the scope of this chapter to compare and contrast the different definitions of this concept. Nevertheless, there are overlapping themes such as social equality, social justice, community well-being, social infrastructure, social capital, engaged governance, human service, and human scale that now inform the thinking of scholars about social sustainability definitions (Cuthill 2010; Dempsey et al. 2009; Magis & Shinn 2009).

The limited literature on the space of social sustainability has not evolved to produce specific of sustainable workforce development. Drawing on a variety of literature, Boström (2012) provides a practitioner perspective by outlining 15 goals and actions that drive social sustainability. Four of the 15 have a strong connection to workforce development. These are (1) fair distribution of income, (2) employment issues including the facilitation of small and medium enterprises, (3) opportunity for learning and self-development, and (4) basic needs, which include income opportunity. Using this thin vein of literature as a conceptual foundation the authors iteratively infused the works of Giloth (2000) who outlined a clear definition of workforce development at a regional level, and Jacobs and Hawley (2009) who provide a national perspective along with the Urban Institute's practical workforce development outcomes (Bernstein & Martin-Caughey 2017) to posit a new approach to workforce systems.

Regenerative workforce systems: Regenerative development is a development paradigm designed to push beyond sustainability. While sustainable development focuses on development today that protects the ability of future generations to develop, the priority of regenerative development is to apply holistic processes in a bioregional context to create feedback loops between physical, natural, economic, and social capital that are mutually supportive and contain the capacity to restore equitable, healthy, and prosperous relationships among these forms of capital. Based on the research that informs this chapter, four elements provide an integrated framework for regenerative workforce development: Equity, employability, education, and entrepreneurship. Independently, each aspect of these Essential Elements of Regenerative Workforce Development will have a narrow impact on workforce development. When considered as a foundation for workforce development programming, no matter the industry, sector, or program developer, these elements provide a clear integrated approach with a foundation in scholarly literature. As a working definition, Regenerative Workforce Development Systems are founded upon interactive relationships that build social capital through equitable employment practices within an integrated business educational system that emphasizes the entrepreneurial spirit and develops employable and transferable skills for the residents of a biosphere.

Equity: It is imperative that systems of work and employment provide equal opportunity to all people in a community. The approach to workforce development must accommodate race, gender identity, and disability and many others who struggle with marginalized status. The workforce system must bring those who are undocumented into the traditional economy of the region as opposed to masking the presences of an "under-economy." Addressing these issues will require vigilance on the part of community leaders who must work as advocates for institutional changes in banking and other financial institutions that promote a cycle of poverty. Workforce development must look beyond skills training and address the social factors that "level the playing field" from those from underserved populations. Cuthill (2010) notes that access to transportation, childcare, and elder care can hinder employment by denying access to job skills training. However, perseverance in times of uncertainty has often produced new opportunity out of necessity. In some cases, a new nonprofit organization will step in or a community organization will emerge. While helpful, these organizations often rely on donations and volunteers. Applying concepts surrounding social entrepreneurship can turn a situation born out of a lack of resources into an opportunity that can build the workforce in a bioregion.

Entrepreneurship: To address poverty and to promote the movement of people from an unemployed to employed status many states are using the concept of workforce integration through the deployment of social enterprise (WISE). These organizations create jobs through business ventures that provide job training for the private sector. While there are some benefits, at issue is the migration of the mission of these organizations toward meeting the needs of employers as opposed to meeting the needs of employees (Cooney 2011). While there are issues, these programs are not without merit and deserve further exploration and research. Communities must look beyond "social enterprise and work to support traditional small business growth."

Bygrave and Hofer (1991) define an entrepreneur as a person who "perceives an opportunity and creates an organization to pursue it" (p. 14). Support for local entrepreneurs will create job opportunities within communities and promote positive economic growth. At issue is the realization by scholars and practitioners that local efforts remain challenged. Harper-Anderson and Gooden (2016) found in a recent study that low-income areas are not effectively using community services and programs that promote entrepreneurship. Further, there is evidence that low-income areas may not be an emphasis for these programs.

Entrepreneurship can have an impact beyond economic development. According to Carol Sanford, responsible entrepreneurs seek to transform industries and society itself by challenging cultural assumptions, laws, and regulations, even the process of governance, thus reaching far beyond the traditional scope of business

(Sanford 2014). In addition, a business can serve as a means of cohesion within a community. In the African-American community, black-owned businesses such as barbershops served as a gathering place for socialization and a place where discussions regarding local issues take place. The unique dynamics of barbershops in the African-American community may not appear in a neighborhood with a different ethnic makeup. However, investment in local entrepreneurs and entrepreneurship programs will allow a neighborhood to develop interactions with businesses that fit its culture. Further, as society changes due to dynamic innovation so too will the local entrepreneurs who will bring new opportunities and new ideas to a community. These businesses can serve as a mediator for change and as an opportunity to close the feedback loop through their work with local education systems.

Education: Demographic changes and migration pattern can inadvertently mask deep systemic issues within an education system. For example, various reports routinely rank Colorado as one of the top five most educated states in the United States. Unfortunately, the secondary school systems continue to receive poor marks. Known as the Colorado Paradox, this disparity allows for the perpetuation of poverty in poor neighborhoods while transplants with higher levels of education move to nice suburbs. The tax base in communities of color continues to struggle while the suburbs thrive. When housing prices climb in the suburbs, the net effect is that communities with those from underserved populations begin to experience gentrification as people moving to the area seek for affordable housing.

For communities to thrive, local education systems and the business in communities must work together to provide opportunities for local residents. Cooperative programs that allow high school students to work within a local business can play a major role. Postsecondary institutions can focus on creating apprenticeships and internships for local businesses in addition to the opportunities that come from traditional corporate partners.

Employability: The concept of employability moves beyond the job skills taught in secondary and postsecondary environments. While it is an overused term, concepts such as emotional intelligence and social intelligence are important to individuals who desire long work careers. Pool and Sewell (2007) from the United Kingdom assert a model for employability that focuses on self-efficacy, self-esteem, and self-confidence. They assert that employment is a lifelong pursuit. No one enters the workforce and is "forever" employable based on the skills they acquired upon entry. Business, government entities, and nonprofit organizations must adopt this mindset when working to close the feedback loop associated with workforce development.

Regenerative workforce development in practice

GeoStabilization International (GSI) is an example of a value creation company. Through empirical evidence, their supportive culture offers an example of regenerative workforce development. GSI is a geohazard mitigation firm operating on a passion to develop and install innovative solutions that protect people and infrastructures from the danger of geohazards. The founders Bob Barrett, P.G. and Al Ruckman, P.E. founded the company with their ingenuity and an entrepreneurial spirit. They revolutionized the industry. Today, with their sons Colby Barrett and Tim Ruckman at the helm, their fathers' original ingenuity and entrepreneurial spirit is their compass. These are solid company assets that have secured them as a leader in their industry along with patented tools and a large fleet of equipment. Their entrepreneurial spirit is a strong current at GSI that overlaps into everything from innovation to project management.

GSI is a thriving 30-year-old business, in an industry that demands high levels of safety with enormous amounts of natural resource uncertainty and low employee retention. John Hollander, the Chief People Officer of GeoStabilization International, contributes their success to the simple fact "We all know our mission. We live our mission. We serve our mission and therefore the impact is beyond the four walls. It's a part of our purpose" (J. Hollander, personal communication, 30 October 2017). Their mission is alive and creating a culture of value creation for their employees and their clients by doing the right thing. Culture is at the forefront of GSI's leadership. GSI is on the cutting edge of the implementation of the four essential Es of regenerative workforce development. Each aspect is explained below.

- **Equity:** GSI does not specifically target individuals from underserved communities. Their approach is to look to work with secondary institutions where students are involved in programs that focus on vocational schools. As a by-product of the education system, this lends itself toward working with high schools that have a higher free and reduced lunch rate, a statistic used to assess the economic means of students attending a school.
- **Entrepreneurship:** From the onset of GSI, entrepreneurship has been a major building block of success. Recently in 2016, the founders Bob Barrett and Al Ruckman received the "Excellence in Entrepreneurship" award for the commercial introduction of the Soil Nail Launcher™ Entrepreneurship, at its core, is regarded to as a person who will take risk within a business. This risk shows up in several ways at GSI. First, GSI expects employees not only to think out of the box thinking but never get into the box. Employees are granted resources and freedoms to explore new ideas and technologies. Entrepreneurship is demonstrated by allowing an employee the opportunity to take ownership of their job and to self-manage while being a collaborative team member. This approach is unprecedented in their industry. Second, is the gift of autonomy. GSI has a strong autonomous culture that is practiced at all levels in the business.
- **Education:** GSI thrives and is interested in knowledge-based education; whether it comes from a traditional line of education or is practitioner based. From a traditional sense, they have several avenues in which they are seeking employees. First, in the Denver Colorado area, they collaborate with Arrupe

Jesuit High School. This high school requires its student to participate in a work study/internship to pay for their tuition. As trades have been diminishing from our school system in the United States, this offers an opportunity for students to see trade work from a practitioner perspective while gaining respect for the fortitude these skills have in our society.

GSI also networks with trade schools in the United States of America and Canada. Seeking those who have the certification in the trade skills they require along with the attributes for the job. Often these are younger students who are eager to work. Practitioner experience has weightage as well. When pairing the two together, within their culture, learning becomes the pulse of the business in everything they do. Because of this combination, GSI provides opportunities for their employees to learn and progress vertically not latterly. John Hollander explains, "we make mistakes every day, and we learn together…being a humble and vulnerable employer, make things right, creates a humble employee who wants to learn and grow with the business."

- **Employability:** Employability often refers to set achievements, understandings, or personal attributes that make one more likely to obtain employment. GSI pushes this into a new direction. The industry typically has a 42% job turnover rate in comparison to GSI 22%. According to John Hollander, "determining the success at each level of the business is essential because then you know how to hire and develop jobs. In other words, we find the attributes that align with the job success and hire for those attributes. Then we know we have the right person for the job" (J. Hollander, personal communication, 30 October 2017). Attributes are a large driver of their successful employability equation. GSI hires those who seek autonomy, have an entrepreneurial mentality with an ethical drive, and have an education or practitioner experience while exhibiting an ability to own themselves.

  GSI's work is seasonal. When operating in their season, their workforce is 120% capacity as compared to 30% offseason. Therefore, GSI created a "Talent Network" with other seasonal employers, who are in season during their offseason; their employees can transfer knowledge into another job. This in return allows individuals to remain employed year-round. For example, their heavy and commercial machine drivers and operators will drive snowplows for their counties during the winter months.

Conclusion

The construct of regenerative development ontologically grew from regenerative design, which dwells within the theoretical context of sustainability. This emerging paradigm of a regenerative process serves as the link between the social aspects of sustainability and workforce development. Today, there are many aspects of desirable attributes for a workforce development. Regenerative development defines such desires between the four capitals economic, physical, natural, and social as equitable, healthy, and prosperous relationships. This chapter discussed workforce development from a practitioner lens of regenerative development perspective. Through the thematic approach of this emergent theory framework, the essential Es (equity, employability, education, and entrepreneurship) emerged as a vital ensemble for the future of a prosperous workforce development system.

Fundamentally, our insights and purpose are to extend regenerative development into the business world and workforce development system while serving as a platform for future conversations. We offer this in hopes of stimulating future research on the positive effects of social and workforce regeneration.

References

Bates, RA & Redmann, DH 2002, "Core principles and the planning process of a world-class workforce development system", *Advances in Developing Human Resources*, vol. 4, no. 2, pp. 111–120.

Bernstein, H & Martin-Caughey, A 2017, *Changing workforce systems*, viewed 21 December 2018, <https://www.urban.org/sites/default/files/publication/88296/changing_workforce_systems.pdf>

Boström, M 2012, "A missing pillar? Challenges in theorizing and practicing social sustainability: Introduction to the special issue", *Sustainability: Science, Practice, & Policy*, vol. 8, no. 1.

Bramley, G & Power, S 2009, "Urban form and social sustainability: The role of density and housing type", *Environment and Planning B: Planning and Design*, vol. 36, no. 1, pp. 30–48.

Brandon, PS, Lombardi, P & Shen, GQ 2017, *Future challenges in evaluating and managing sustainable development in the built environment*, John Wiley & Sons, Hoboken, New Jersey.

Brundtland, GH 1987, *Report of the World Commission on environment and development: Our common future*, viewed 21 December 2018, <http://www.un-documents.net/our-common-future.pdf>

Bygrave, WD & Hofer, CW 1991, "Theorizing about entrepreneurship", *Entrepreneurship Theory and Practice*, vol. 16, no. 2, pp. 13–22.

Colantonio, A 2009, "Social sustainability: A review and critique of traditional versus emerging themes and assessment methods", In M Horner, A Price, J Bebbington & R Emmanuel (eds.) *Sue-mot conference 2009: Second international conference on whole life urban sustainability and its assessment*, Conference Proceedings, Loughborough University, Loughborough, pp. 865–885.

Cooney, K 2011, "The business of job creation: An examination of the social enterprise approach to workforce development", *Journal of Poverty*, vol. 15, no. 1, pp. 88–107.

Cuthill, M 2010, "Strengthening the 'social' in sustainable development: Developing a conceptual framework for social sustainability in a rapid urban growth region in Australia", *Sustainable Development*, vol. 18, no. 6, pp. 362–373.

Dacre Pool, L & Sewell, P 2007, "The key to employability: Developing a practical model of graduate employability", *Education+ Training*, vol. 49, no. 4, pp. 277–289.

Davidson, M 2009, "Social sustainability: A potential for politics?" *Local Environment*, vol. 14, no. 7, pp. 607–619.

Dempsey, N, Bramley, G, Power, S & Brown, C 2009, "The social dimension of sustainable development: Defining urban social sustainability", *Sustainable Development*, vol. 19, no. 5, pp. 289–300.

Edwards, AR 2005, *The sustainability revolution: Portrait of a paradigm shift*, New Society Publisher, Canada.

Employment Act of 1946 2017, *Employment Act of 1946*,

Fitzgerald, J 1999, *Principles and practices for creating systems reform in urban workforce development*, viewed 21 December 2018, <http://citeseerx.ist.psu.edu/viewdoc/download?doi=10.1.1.541.5964&rep=rep1&type=pdf>

Friedman, M 1962, *Capitalism and freedom*, University of Chicago Press, Chicago and London.

Friedman, M 1968, "The role of monetary policy", *The American Economic Review*, vol. 58, no. 1, pp. 1–17.

Giloth, R 2000, "Learning from the field: Economic growth and workplace development in the 1990s", *Economic Development Quarterly*, vol. 14, no. 4, pp. 340–359.

Harper-Anderson, EL & Gooden, ST 2016, "Integrating entrepreneurship services into local workforce development systems: Who is doing it and how", *Journal of Poverty*, vol. 20, no. 3, pp. 237–260.

Hawley, J, Sommers, D & Melendez, E 2003, "The impact of institutional collaborations on the achievement of workforce development performance measures in Ohio", *Adult Education Quarterly*, vol. 56, no. 1.

Heinrich, CJ & Lynn, Jr LE 2000, "Governance and performance: The influence of program structure and management on job training partnership act (JTPA) program outcomes", *Governance and Performance: New perspectives*, vol. 68.

Hes, D & Plessis, CD 2015, *Designing for hope: Pathways to regenerative sustainability*, Routledge, New York.

Holzer, H 2009, *Workforce development as an antipoverty strategy, what do we now, what do we do?*, viewed 21 December 2018, <https://pdfs.semanticscholar.org/98ca/a374a2ecdf339d42833c3a82437d5b1d8b58.pdf>

Hopwood, B, Mellor, M & O'Brien, G 2005, "Sustainable development: Mapping different approaches", *Sustainable Development*, vol. 13, no. 1, pp. 38–52.

IUCN 1980, *World conservation strategy: Living resource conservation for sustainable development*, Gland, IUCN, Switzerland.

Jacobs, RL & Hawley, JD 2009, "The emergence of "Workforce Development": Definition, conceptual boundaries and implications", In R Maclean & D Wilson (eds.), *International handbook of education for the changing world of work*, Springer, Dordrecht, pp. 2537–2552.

Kim, CW, Emmett, D & Sikula Sr, A 2001, "Employee relations ethics and the changing nature of the American workforce", *Ethics & Behavior*, vol. 11, no. 1, pp. 23–38.

Kremen, GR 1974, *MDTA: The origins of the Manpower Development and Training Act of 1962*, US Department of Labor, Washington D.C.

LaLonde, RJ 1995, "The promise of public sector-sponsored training programs", *The Journal of Economic Perspectives*, vol. 9, no. 2, pp. 149–168.

Lipford, J 1999, "Twenty years after Humphrey-Hawkins", *Independent Review*, vol. 4, no. 1, p. 41.

Littig, B & Griessler, E 2005, "Social sustainability: A catchword between political pragmatism and social theory", *International Journal of Sustainable Development*, vol. 8, no. 1–2, pp. 65–79.

Magis, K & Shinn, C 2009, "Emergent principles of social sustainability", *Understanding the Social Dimension of Sustainability*, pp. 15–44.

Mang, P, Haggard, B & Regenesis 2016, *Regenerative development and design: A framework for evolving sustainability*, John Wiley & Sons, Hoboken, NJ.

Marsh, GP 1864, *Man and nature: Physical geography as modified by human hand*, Harvard University Press, Cambridge, MA.

Ranney, D 2003, *Global decisions, local collisions: Urban life in the new world order*, Temple University Press, Philadelphia.

Sanford, C 2001, *The responsible business*, Jossey-Bass, San Francisco.

Sanford, C 2011, *The responsible business: Reimagining sustainability and success*, John Wiley & Sons, Hoboken, New Jersey.

Sanford, C 2014, *The responsible entrepreneur*, Jossey-Bass, San Francisco.

Shrivastava, P, Ivanaj, S & Ivanaj, V 2016, "Strategic technological innovation for sustainable development", *International Journal of Technology Management*, vol. 70, no. 1, pp. 76–107.

Stren, R. & Polèse, M 2000, "Understanding the new sociocultural dynamics of cities: comparative urban policy in a global context", *The social sustainability of cities: Diversity and the management of change*, pp. 3–38.

13 Education for regeneration

Kenneth S. Sagendorf and Barbara J. Jackson

A description of the purpose of higher education has not been clearer than that of Wendell Berry. He states (1987, p. 77):

> The thing being made in a university is humanity [...] What universities are mandated to make or help make is human beings in the fullest sense of those words - not just trained workers or knowledgeable citizens but responsible heirs and members of human culture [...] Underlying the idea of a university – the bringing together, the combining into one, of all disciplines – is the idea that good work and good citizenship are the *inevitable by-products* of the making of a good – that is fully developed – human being.

Derek Bok leafs through the history of higher education and describes it like this (2013, p. 167):

> For almost a century, [...] education in the United States has pursued three large, overlapping objectives. The first goal is to equip students for a career either by imparting useful knowledge and skills in a vocational major or by developing general qualities of the mind through a broad liberal arts education that will stand students in good stead in almost any calling. The second aim, with roots extending back to ancient Athens, is to prepare students to be enlightened citizens of a self-governing democracy and active members of their own communities. The third and final objective is to help students live a full and satisfying life by cultivating a wide range of interests and a capacity for self-reflection and self-knowledge.

While Berry's description matches Bok's uncovering of the historical perspective on education in the United States, they both stop short, offering only an allusion to the role of the individual in society (i.e., being "active" or "good" citizens). Clayton Christensen of Harvard Business School fame, author of *The innovative university* (2011) and definer of disruptive innovation, makes a case for the digital future of higher education, quoting Harry Lewis's *Excellence without a soul* (2007, p. 7):

> Universities have forgotten their larger role for college students [...] Rarely will you hear more than bromides about personal strength, integrity, kindness, cooperation, compassion, and how to leave the world a better place than you found it.

George Leef, in describing Bryan Caplan's book *The case against education: Why the education system is a waste of time and money* (2018), takes a pessimistic view of higher education, stating (January 24, 2018) "Education signals three broad traits: intelligence, conscientiousness, and conformity." Indeed, the economists' principle of path dependence, the preference to continue a traditional path upon belief that the historical contexts bind our current competencies, makes the precise case of this chapter. The contrast of education's role as developing conformity (via dependence on what has been done previously) versus leaving the world a better place than you found it (regeneration) is worth exploration. The paradigm shift to individual as an active influencer and influenced part of the industrial and ecological systems—that of regenerative development—requires a change in *how* we offer education. This chapter ponders the questions necessary for offering an education through a regenerative lens.

Current context: Influencers of higher education

Inside institutions of higher education

There is overwhelming agreement about what is fundamental for students to learn to be successful in their future endeavors across different types of colleges and universities and all sectors of companies. That agreement focuses on students' ability to think critically and to evaluate the quality and reliability of information. The same statements are made today by popular media when they describe the hiring desires of global mega companies. And a laser focus has been on critical thinking since the 2010 Arum and Roksa publication *Academically adrift: Limited learning on college campuses* which findings showed no changes in students' critical thinking skills during the first two years of a college education. Since then, many colleges and universities have redoubled their efforts to focus on critical thinking as an institutional outcome.

According to faculty surveyed every couple of years through the Higher Education Research Institute (HERI), additional clarity exists in answering the question, what is the purpose of education? Ninety percent of faculty support self-directed learning, mastering knowledge of a discipline, and the ability to write effectively as necessary for accomplishing those larger purposes of higher education that Bok mentioned earlier (Eagan et al. 2014). There is also agreement among more than two-third of faculty in the importance of preparing students for employment, fostering a tolerance for others' beliefs, developing creative abilities and improving racial understanding, developing appreciation for the liberal arts, and developing moral character (Eagan et al. 2014). Education sometimes focuses on preparation for the next round of education (graduate or professional school) as well. Disciplinary differences in epistemologies and how knowledge is created differently in disciplines often determine what is valued in those areas. For instance, the 2012–2013 HERI survey found that 80% of history faculty frequently assigned work asking for students to weigh the significance and meaning of evidence versus 26% of mathematics and 44.5% of business faculty. Analyzing and interpreting data was reported as frequent in assignments given by scientists (over 76% of physics and biology faculty) but less so by humanities faculty (35%) and education faculty (46.8%) (Eagan et al. 2014).

Without being a mathematician, one can see how the idea of what is in a curriculum becomes a hotly contested nightmare—one that often puts academic disciplines, and schools in competition with one another. And, although there are many examples of models that try to break this down and eliminate the academic silos such as integrative curricula and themed teaching, the silos remain intact in higher education for the most part. These silos cause the most damage to students. Binge and purge learning can easily become the norm as students move through the courses and experiences that make up their education, especially when they find these experiences woefully independent of each other instead of reliant, dependent, and exponentially additive. And this is difficult as we within higher education have been systematically conditioned in these silos for nearly 90 years or more. Hooks describes our conditioning, "given that our educational institutions are so deeply invested in a banking system, teachers are more rewarded when we do not teach against the grain. The choice to work against the grain, to challenge the status quo, often has negative consequences" (1994, p. 203).

Organizations aligned with higher education institutions

Higher education is subject to many influences both internally and externally. Major organizations focused on higher education and access such as the Association of American Colleges and Universities (AACU) and the Lumina Foundation have done mountains of work defining and confirming the purpose of education at the college level by working with constituents and educators. AACU published the Liberal Education and America's Promise (LEAP) Essential Learning Outcomes in 2008 as the culmination of decades of work with colleges, universities, and employers to define four broad categories of focus for liberal education (Association of American Colleges and Universities 2015):

- Knowledge of human cultures and the physical and natural world
- Intellectual and practical skills
- Personal and social responsibility
- Integrative and applied learning

Similarly, The Lumina Foundation published *The degree qualifications profile* (DQP) in 2014 to help define the outcomes of associates, bachelors, and master's degree level work focused in five categories of learning (Adelman et al. 2014):

- Specialized knowledge
- Broad and integrative knowledge
- Intellectual Skills
- Applied and collaborative learning
- Civic and global learning

The AACU LEAP and DQP outcomes are relatively simple targets for the "what" is offered as an education. Simple insofar as there is little disagreement in the value of these foci; their categories and titles have been commonly adopted by institutions as they define the outcomes of their educational offerings. These outcomes have been important guideposts across higher education. However, the outcomes are not as simple as they appear. For example, despite having only four high level outcomes, the AACU LEAP outcomes develop knowledge of human cultures and the physical and natural world through study in sciences, mathematics, social sciences, humanities, histories, languages, and the arts. So, these seven broad disciplinary perspectives are important. Also important are the six sub-categories of intellectual and practical skills, the four sub-bullets of personal and social responsibilities, and the small category of synthesis and advanced accomplishment across general and specialized studies in the integrative and applied learning outcome. In addition, national and regional accreditors also have a large influence on education, often pushing colleges and universities to gain clarity about *what* they are doing as institutions.

External influencers—Drive for market share

Higher education can be seen as a supply chain for employers. Community colleges were (and are) tasked with educating the American populous to prepare them for the work world and continue to be the largest educator of students in the United States. Our departments and academic disciplines often grow when the marketplace has new or different requests or needs. The professional schools certainly take this approach (think Nurse Practitioners who can have a neonatal focus) and concerns regarding the hyperspecialization of knowledge are brewing (Malone et al. 2011). At the author's own school, the development of online curriculum through the hiring of course writers, course reviewers, and course deliverers over 20 years ago was the beginning (but not the end) of a trend to scale out education at low cost to the institution. What has become the overwhelmingly large business and infrastructure we publicly associate with for-profit institutions (versus non-profit and therefore, public good) is, in reality, part of nearly every institution that can manage to get started. These more recent movements in higher education, while often well-meaning (i.e., bringing education to more people or people that need it most), are most often born of desires to increase the bottom line or market share of an institution.

Despite the recent downturn in college enrollments, there were 1,400 more colleges and universities in the United States in 2014 than there were in 1980. This competition exerts pressure on higher education for enrollment. Rankings and raising endowments in our local and globally politically-charged environments can easily shape institutions. Regardless of the intent, the market pressures on higher education most often have the impact of entrenching the existing educational paradigms rather than enhancing or evolving them. Despite some movements and disciplines (e.g., sustainability, green building, environmental science, etc.) moving education away from focusing solely on individuals and the world around themselves toward an external focus on the environment, the core of education (especially undergraduate education) remains relatively unchanged over the last 140 years (Bok 2008). Both academic and popular critiques of higher education continue to be common. In *College: What it was, is, and should be*, Andrew Delbanco (2012) describes the current perspectives on college education warning that "no college is impervious to the larger forces that, depending on one's point of view, promise to transform, or threaten to undermine, it. As these forces bear down on us, neither lamentation nor celebration will do. Instead, they seem to me to compel us to confront some basic questions about the purposes and possibilities of college education" (p. 6).

Current state of higher education

Education cannot focus on critical thinking as the end point if we desire to help develop the next generation of regenerative thinkers. Students experience at least half of an undergraduate degree largely in a fragmented environment. Many consider the academic disciplines to be largely discreet and autonomous. Squires (1992) uses the design of institutions with its distinct colleges, departments, and courses as owing to academic disciplines' own theories, methods, and content. Although there has been some adjustment in higher education (coming later in this chapter), the experience of many undergraduates is still this way. Links between core courses and major courses are often unclear or nonexistent. In this smorgasbord of learning, students often are being introduced to every academic discipline they encounter in a recursive manner—learning and repeating the tenets and foundations of academic areas.

The best-case scenario is that a student experiences their undergraduate education like this and leaves with the depth of critical thinking alongside a specific knowledge from their major field of study as well as the beginnings of understandings of how other academic disciplines encounter the world. Some colleges are designed specifically to make this a reality (e.g., St. Johns College, Colorado College). However, what is more likely is that a student survives their courses outside their desired academic major, often being advised to just get through. If continuing on into graduate school, graduate study most often maintains singular disciplinary focus where depth is honored over an understanding one's field in relationship with other disciplines and approaches.

Sustainability education

As a field, sustainability education is relatively new. With its beginnings with the Tbilisi Declaration in 1977, environmental education moved out of only environmental science and is noted as the precursor to modern sustainability education. The 1987 Brundtland report introduced education for sustainable development. The report, *Our common future*, defined sustainable development as "development that meets the needs of the present without compromising the ability of future generations to meet their own needs" (p. 16) and embraced the triple bottom line of social, economic, and environmental domains, defining the environment as "where we live," and development as "what we all do to improve our lot in our abode" (p. 3).

In 2002, the United Nations declared the UN Decade of Education for Sustainable Development from 2005 to 2014. The results were the UN Sustainable Development Goals and frameworks for K-12 education. Nomenclature of fields (environmental and sustainability education, education for sustainable development) has been a moving target as the fields grow and change. Even Master of Business Administration degrees are offered in sustainability and sustainable development across the United States and abroad.

The work of the Bruntland Commission led to research by Robinson (2004) studying the evolution of sustainability and sustainable development in industrialized nations and drawing a distinction between those who use the related but different terms. He posits that the sustainable development is favored by governments and developers and that sustainability is used more frequently by academics and environmentalists; the former maintaining an anthropocentric focus on incremental change, while the latter maintains the view that the interaction of humans in the natural context requires behavioral changes in order to address existing constraints. The anthropocentric versus biocentric perspectives require different kinds of thinking. El-Haram (2007) describes the thinking of sustainability to be that of integration between policies, programs, plans, and projects with an eye on issues and impacts across space and time. He equates the thinking of sustainability to a decision-making process. Regenerative development is yet another furthering iteration in the growth of the field. Figure 13.1 shows the evolving nature of relationships between sustainable and regenerative design thinking as well as environmentally responsible design (Guzowski 2011). The upper right sections of both the figures showcase the move from sustainability to regeneration, focusing on the move to interactions, co-evolutions, and systems thinking.

Another perspective on the move from sustainability to regeneration comes from Hauk's 2007 paper "The new 'Three Rs' in an age of climate change: Reclamation, resilience, and regeneration as possible approaches for climate-responsive environmental and sustainability education." Hauk pays special attention to the growth and evolution of fields, describing the extension of sustainability education as a move from reclamation to resilience to regeneration, borrowing from Pelling's overlay of climate change (2011) as a move from mitigation to adaptation to transformation. She lays out the three modes side by side in her depiction of the climate responsive environmental and sustainability education graph (Figure 13.2).

Hauk's description of transformative and regenerative sustainability education is spot on for what needs to be done. And the long list of in-depth knowledge and skills in the right hand column describe the next levels of education. But how does the experience of an undergraduate degree, focused on a breadth of knowledge and singular disciplinary ways of knowing that results in critical thinking, prepare anyone for graduate level work that encompasses what Hauk describes? The gap between a critical thinking undergraduate degree and a largely systems-thinking graduate degree seems an unreasonable expectation, especially given the complexity

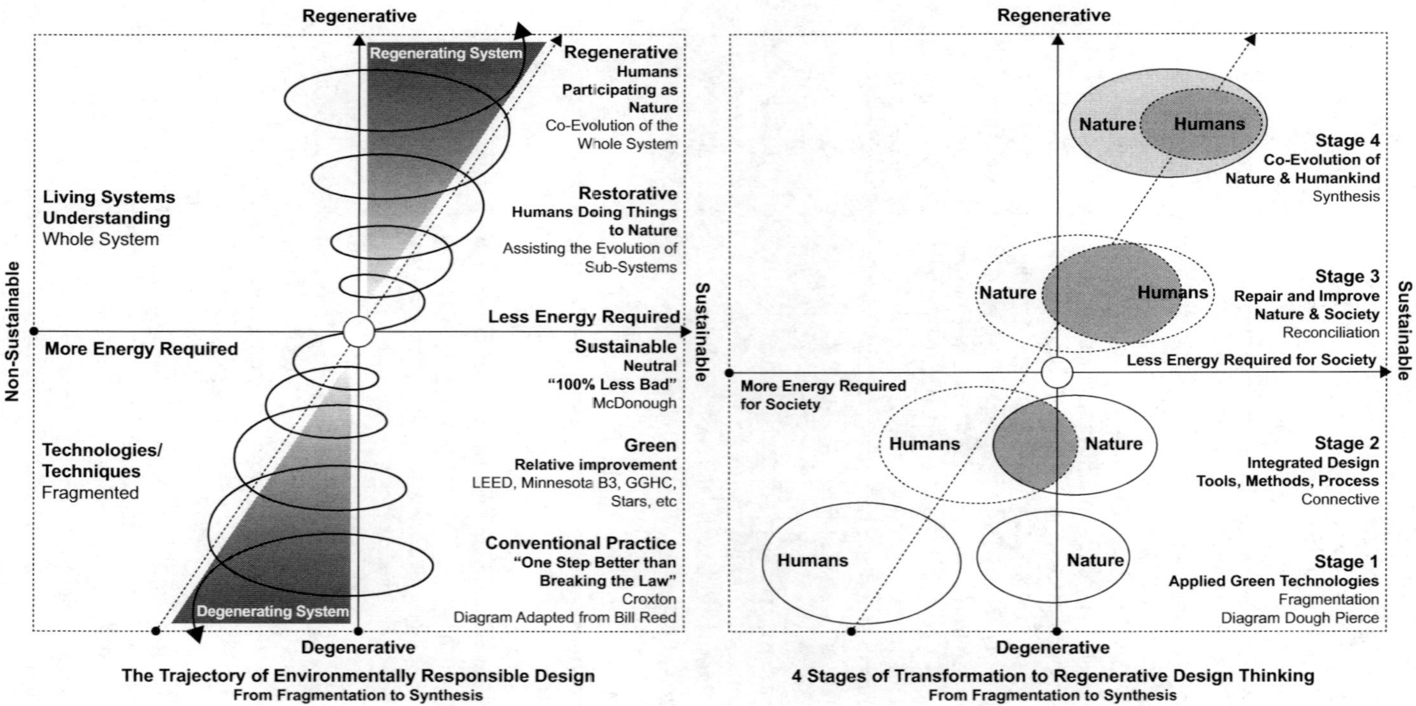

Figure 13.1 The growth from sustainability to regeneration in both design (figure on the left) and design thinking (figure on the right). Both figures reprinted from Guzowski (2011) and used with permission from Bill Reed (as adopted by Doug Pierce) (left) and Doug Pierce (right).

Climate Responsive Environmental and Sustainability Education (CRESE)

CRESE Mode	Parallel Climate Change Strategy	Approach	Skills
Reclamation	Mitigation	Libraries, Sanctuaries, Collective Samplings	Skills of Conservation, Historical Systems Understandings, Recording and Preserving
Resilience	Adaptation	Networks of Adaptive Capacity; Transition	Skills of Scenario-Building, Network-Mapping and Extension; Capacity Building
Regeneration	Transformation	Emergent, Transformative Living Systems	Skills of Systems Evolution, Multi-Scale Metacognition, Regenerative Design; Catalyzing Emergence

Figure 13.2 Hauk's comparison of reclamation, resilience, and regeneration, outlining the skills needed in each area. Table used with permission.

of systems thinking (see Caniglia in this book). As we stated earlier, critical thinking is necessary but not sufficient for educating regenerative thinkers.

How can we change education?

Mulgan (2016) lays out the following fundamental expectations that current education must accept in order to create the future of education:

- Students want to be useful in the world and make a difference. They want to do good.
- Students want to be better prepared for work and life and be readily employable.
- Engagement with others must be enabled.
- There is a pull from society and communities to help solve real problems.

To meet these expectations, critical thinking is the launching point but cannot be the destination of education. Retrofitting traditional disciplines or adding more programs is a temporary, superficial approach that may or may not result in the type and depth of regenerative thinkers that we must create. To change this, some radical thinking is necessary. Williams describes the need like this (2012, p. 363): "to achieve regenerative places will require a generation of practitioners with a new set of attributes. They will need to be skilled technically and understand a portfolio of solutions around ecological and human systems." We must rethink subject structures, foster interdisciplinary work, and find a balance of specialization with application that drives students to effectively influence positive change in a dynamic ever-changing society. This is the blueprint to designing an education for regenerative developers. And all of these components must focus on nature, people, policies, the built environment, and their interconnected and changing impacts over time.

Some of the potential answers for what a regenerative development education should be lie within this book. In Chapter 9 of this text, John Knott lays out the CityCraft principles for regenerative development, which are systems scale, cross sector integration, capital mapping, silo destruction, a bioregional context, and social durability. These principles and their intersections provide a design framework for the how we can adjust education to develop the next generation of regenerative developers. These six areas of focus allow for and invite both disciplinary and interdisciplinary approaches all while seeking a new frontier of understanding. And because these principles are bound by and tied to a place, they require an applied approach. These concepts are a solid place to begin when thinking about the redesign of education. There have also been other approaches that may provide models to learn from. Specifically, experiential learning and integrative education make excellent progress toward bringing this approach into reality.

Experiential learning

The roots of experiential learning have long been known in education. Dewey described this in the 1930s when he wrote about education as understanding the connection points and influences on what is learned. He posed many questions about whether a student is truly learning from place or learning in place. Roberts (2013) raises these questions in "Experiencing sustainability: Thinking deeper about experiential education in higher education." Roberts claims that learning by doing is an insufficient way to define experiential learning, and he draws upon Dewey's work (1916, p. 80) advocating a "continuous reconstruction of experience" when he states that the amount of truly experiential learning depends upon the degree to which learning is actually integrated and made continuous. These are the key points of Roberts's argument—that learning must actually be integrated and made continuous. It is not merely about having the experience, but rather sense-making from the experience (repeatedly). Common methods of experiential learning often affect students as project-based learning and service learning. And while these methods of experiential learning are often highlighting experiences in a higher education setting, it should be noted that one or two of these activities in a college career often constitute fulfillment of experiential learning, thus creating a structure that is neither continuous nor integrated into the curriculum. It needs to become the norm, not the exception.

Connelly and Clandinin (1988) called experiential education the "commonplaces"—the dynamic roles of the learner, teacher, subject matter, and sociopolitical milieu. This would certainly describe a way to educate students in Knott's principles of regenerative development. Biesta and Burbules (2003, p. 9) note that experiential learning is different from other education in that it deals with "questions of knowledge and the acquisition of knowledge within the framework of a philosophy of action." Regenerative education takes this and expands the philosophy of action into interconnected systems.

Integrative Education

Integrative learning (Huber & Hutchings 2004) may be the most recent model to continue to carry the torch of Dewey's work. As a phenomena more closely studied in the K-12 area rather than higher education, integrative learning has gained some attention through the partnership of the Association of American Colleges & Universities and the Carnegie Foundation for the Advancement of Teaching in 2004–2005. In their statement on integrative learning, they lay out a construct that asks for connecting knowledge and skills from multiple sources and experiences, utilizing diverse and contradictory perspectives and understanding issues and positions contextually. However, Huber, Hutchings, and Gale (2005) point out that this was long the goal of liberal education that their organizations promoted and the responsibility remained with the students to do the true integration. At one of the authors' own schools, the liberal arts school had done yeoman's work in developing an integrative core and training faculty to teach, but with only four courses taken in the final two years of school, the messaging to students is that this should remain secondary to the way they learn in a more distributive core and major courses. Kemp (2006) uses the work of Huber, Hutchings, and Gale to lay out a concept map (Figure 13.3) of the linkages between and among the constructs of integrative learning. This is a systems thinking approach to education. Flood (2001) describes the impact of systems thinking as driving

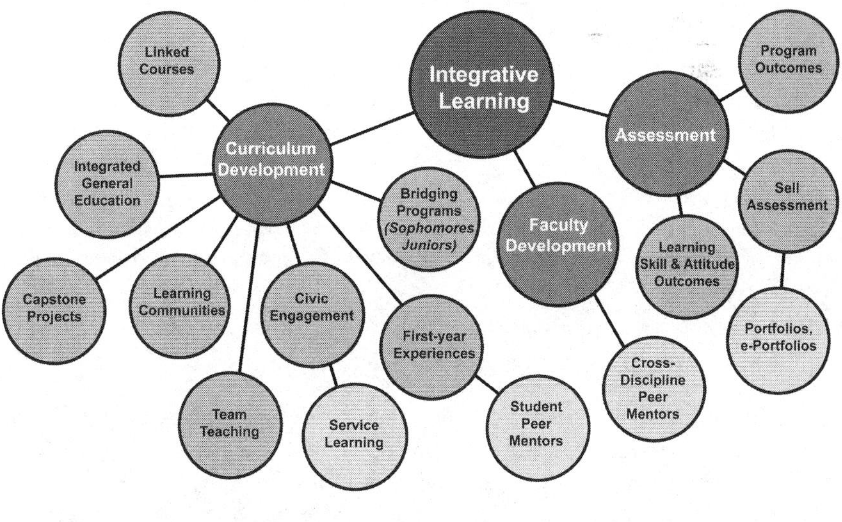

Figure 13.3 Integrative learning concept map.

Source: Courtesy of Dr. Jeremy W. Kemp (2006)

knowledge and meaningful understanding "from building up whole pictures of phenomenon, not breaking them up into parts" (p. 133).

Although the curriculum development side of the concept map speaks to how integrative learning can be developed, the other components must be addressed simultaneously. Also important to note is the distinction between truly integrative learning (different perspectives, their impacts, and relationships) and interdisciplinary learning (where different approaches are presented). Commitment to the principles of integrative education would ideally drive the behaviors of administrators and faculty. However, higher education spends an inordinate amount of time distributing resources to new modes of education (i.e., distance and online education) and new markets (e.g., non- and post-traditional learners). This makes commitment to adjusting traditional educational models and the engaged examination of what and how learning should take place virtually impossible. More than likely, programs or even just courses may take this approach, falling short of the ideal.

The integrative learning concept map (Figure 13.3) gives us a framework to consider when considering educating for regenerative development as the approach is not solely on the learner, but rather on the learning. An example of focusing on the learning can be seen in Sanford's book, *The regenerative business* (2017). Sanford (2017) focuses on mindset when she ponders how we actually learn to think anew—the requirement for regenerative thinking. She defines regeneration as "a process by which people, institutions, and materials evolve the capacity to fulfill their inherent potential in a world that is constantly changing around them" (p. 2). Sanford continues on to describe her themes of regeneration: wholeness, potential, reciprocity, authenticity, nestedness, nodal intervention, and development of capability. In his Institute for Cultural Action (IDAC) interview, Paulo Freire (n.d., p. 7) argued that "men and women are human beings because they are historically constituted as beings of praxis, and in the process they have become capable of transforming the world – of giving it meaning." The good news is that examples of this work are being created and instituted currently. We share an example of educating regenerative thinkers in real estate education below as a model of a successful approach. This approach has been at the center of the vision for a new curriculum in the Franklin L. Burns School of Real Estate and Construction Management at the University of Denver and work in progress for the past few years. It is presented as a way to understand the context and outcomes of change in the way students are receiving education in the real estate development area.

The future of real estate education: A case study

The ownership of land and real property has always been a measure of wealth in the United States. And yet, the discipline of real estate has never held either the professional or academic esteem that other business disciplines have enjoyed such as finance, accounting, or even marketing. That is more than likely because of the familiarity that the average consumer has with their personal residential real estate transactions. In so many of these cases, the average consumers' experience with real estate is limited to buying or selling a home. However, the depth of knowledge and education required to achieve this important, but basic function, of sales does little to demonstrate the deep and extensive knowledge needed to research, analyze, examine, valuate, forecast, and mitigate the many risks associated with making sound business decisions regarding the acquisition, investment, finance, and development of real estate. These skill sets position those educated in real estate play a significant role in regenerative development. This deep knowledge and expertise has somehow not been recognized by the general consuming public as the significant business acumen that it is. And likewise, real estate education has never received the thoughtful consideration as a specialized academic discipline that it deserves.

Moving beyond the transactional perspective

It is time to consider real estate and its significant impacts in the broader perspective—way beyond just a transactional perspective to a regenerative one. The fact is, more than any other asset, real estate has a transformational impact on all of civilized society. Real estate and real estate development impact people's lives in far more ways than just providing shelter. Commerce, in all forms requires some kind of real property influence. In addition to housing, most, if not all of the goods and services that are available to society, require some type of manufacturing, retail, office, medical, educational, recreational, or industrial type of facility in order to exist and do business. But as true as this is, it's rarely recognized by that society as playing a significant part in the quality of life for a community's residents. Real estate is still simply seen as primarily a financial transactional function and not as the major positive community growth and development influence that it is, or can be. The fact is, real estate and real estate development in particular has unfortunately been more broadly seen in a negative light—in some cases a very negative light. And there are many reasons for this negative perception—some of them real (e.g., gentrification) and some of them simply perceived to be real (e.g., change in community dynamics). However, those of us in education have the chance to begin to influence the negative beliefs about the real estate business into a more positive viewpoint about the profession. One way to do that is to begin to redefine real estate beyond its transactional nature. Although we do an excellent job of teaching all of the transactional aspects of real estate, today's students are looking for more. Today's students want to learn about how real estate and real estate development can be transformative and deliver triple-bottom-line results and truly sustainable and regenerative solutions for neighborhoods and communities. It may be time to expand the traditional understanding and definitions of real estate and rebrand real estate education.

Rebranding real estate education

Students who are attracted to real estate are no longer in it just for the money. They want to make a difference and unlike how many of us were educated and trained in the pure transactional aspects of the field, it seems that students today see real estate as an opportunity to impact community engagement and connectedness, health and wellness, and many other aspects of the built environment that do not typically translate into our traditional understanding of return on investment or ROI. We are seeing more and more students today who are seeking educational programs that focus on integrated, holistic solutions for our buildings and neighborhoods. Whereas real estate development of the past was often seen as hurting communities, current students expect real estate solutions to heal communities and the social, economic, and environmental fabric of our cities and address the most challenging economic, environmental, and social issues of our time. As educators, we really have an opportunity to turn the negative perceptions about real estate around, and begin to express and engage in cross-sector integration strategies that not only address the important transactional elements of real estate, but also address the important transformational elements which can restore the economic, environmental, and social health and wellness of our cities.

New Course Development

Some of the new course titles that have emerged in the last several years that indicate a shift in the teaching and learning perspective are:

- Business of the built environment (University of Denver—The Burns School)
- City-crafting (University of Denver—The Burns School)
- Sustainable development
- Green investment and finance
- Re-generative development (University of Denver—The Burns School)

There are also instances where universities have renamed their real estate programs or degree titles to reflect this new perspective in an effort to begin to redefine the real estate profession. For example, the Burns School at the University of Denver has changed its degree name from Real Estate and Construction Management to Real Estate and the Built Environment—to indicate a bigger perspective. And, universities are working together bringing varied expertise closer to reality such that schools without built environment and architecture programs can now bring those theories and systems into exchange with business, sociology, and other expertise.

Systems Thinking

Beth Caniglia's chapter on systems theory provides crucial foundations for the future of education in this arena. In order for us to educate the next generation of regenerative developers, we must focus on systems thinking as we design our education anew. Center and institutes within colleges and universities, often seen as external focused think tanks or actors, may present the best opportunity to create this kind of curriculum because they are less bound by the confines of the ways that curriculum is designed and offered. Examples include the Center for the Study of the Built Environment (CISBE) at the University of Denver that is redesigning real estate education and the Sustainable Economic and Enterprise Development (SEED) Institute at Regis University, both participating in cross-institutional research on regenerative development and adding and proposing new curricula.

As higher education has been slow to engage fully in organizing around systems thinking, alternative methods of educating for systems thinking have emerged outside of academe. The work of Ken Wilbur on integral psychology and education, and Dr. Leyla Acaroglu and her Disruptive Design Method from her aptly named UnSchool of Disruptive Design are just two examples (see https://unschools.co. And although integral education seems to relate more easily to higher education, I think the simplicity of Acaroglu's work and focus on sustainability make her work more applicable to regenerative development. She states that she calls it the UnSchool because it "undoes the damage that mainstream education has done" and seeks to develop systems thinkers who are proficient in understanding and dissecting the dynamic systems that are at play in the world around them.

Her work focuses on the interconnectivity of three systems—the ecological, the social, and the industrial systems that affect humans, the planet, and the connections between the two. More specifically, her work outlines and works on distinguishing the contrasts of parts vs wholes and objects vs relationships. This version of a systems thinking approach to education is full of personal examples professionals bring when they enroll in the UnSchool.

According to their website, the UnSchool models their curriculum around systems thinking, gamification and game theory, ethics and empathy, life cycle thinking, research strategies, applied systems intervention, cognitive and behavioral sciences, ideation and creative problem solving, and sustainable production. If you have been reading this text, the alignment of the tools of the system thinker from the UnSchool with the CityCraft principles mentioned earlier should be obvious, the main distinction being a grounding in location (the bioregional context) (Table 13.1).

The UnSchool and Citycraft agreement in the integration of different components of systems thinking shows there to be alignment in the approach to

Table 13.1 Comparison of the approaches/areas of study used by the UnSchool for Disruptive Design and the CityCraft method as foundations for understanding systems thinking in regenerative development

UnSchool's tools of a system thinker	*CityCraft principles for regenerative development*
Interconnectedness	Systems scale
Synthesis	Cross-sector integration
Systems mapping	Capital mapping
Feedback loops	Silo destruction
Emergence	A bioregional context
Causality	Social durability

educating systems thinkers as these two entities arrived at their methods separately and independently. Furthermore, both align with Hauk's work mentioned earlier in this chapter in her category of regeneration. All three point to systems, emergence, metacognition, connectedness, and, ultimately, regenerative development. The blueprint for how to educate regenerative development experts exists in these spaces outside of higher education. Even the name, the UnSchool, calls out that school does not accomplish the same things. Higher education must take these ideals into serious consideration if it wants to keep pace. Our current practices, while having some one-off examples of success, do not go far enough to bring this kind of thinking to all of our students. We need to move away from relying on a recursive education where we ask students to reproduce knowledge we already know and into a discursive approach that asks students to look for sustainable solutions for questions where we do not know the answers. And anchoring this work in the context of the bioregion is necessary for the application and depth of learning for regenerative development. Our current system that relies upon students to cross disciplinary backgrounds and make sense of the relationships between and applications of those disciplines on their own is not developing the regenerative thinkers that we need for our collective futures. As the first regenerative developers emerge out of higher education, the task of educating for regeneration will really be about developing leaders in the field. It is a new enough paradigm that the first graduates of such programs will need to create, reengineer, and resituate traditional ways that disciplines are taught and experienced. Therefore, developing regenerative thinkers will be tied to leadership development.

Developing regenerative thinkers is leadership development

The concept of regeneration layers a whole new set of standards, values, and metrics on the environmental, social, and economic landscapes of business, education, and society in general. This is where the real challenge lies in regard to developing the leaders we need. Traditional leadership perspectives and approaches simply will not produce the kinds of outcomes and results that are needed and missing as we continue to develop neighborhoods and communities.

When leadership is a part of the curriculum in business, planning, real estate, or AEC (architects, engineers, construction) education, it is usually modeled after the traditional leadership theories which often focus on the leader's traits, or style, or may reference the importance of situational factors. And most recently, leadership is often taught or presented in the context of the volatile business environment characterized by change and disruption, and how well the leader can cope with and manage through such change. The acronym VUCA—volatile, uncertain, complex, and ambiguous—is often used to describe the business environment today, and the state of the world in general. And given this state of VUCA, it is becoming more evident that traditional leadership education is inadequate and thus resulting in the gap between the leaders we have and the leaders we need across most businesses and social enterprises today.

The leadership literature is mountainous. Hundreds of thousands of books and articles have been written about leaders and becoming one. With nearly 350,000 books and articles written with "leader" or "leadership" in the title over the decade between 2005 and 2015, it can be difficult to identify which ones apply to regenerative education. There are several scholars and authors who have written about a new paradigm for leadership that applies here. For example, Russell Ackoff (1998) was both a pioneer in systems thinking and a highly recognized organizational theorist. He believed that leadership was poorly understood largely because it is primarily an aesthetic function and aesthetics are also poorly understood. Science, technology, and economics focus on efficiency, but not effectiveness. The difference between efficiency and effectiveness is important to an understanding of transformational leadership. Efficiency is a measure of how well resources are used to achieve ends; it is value-free. Effectiveness is efficiency weighted by the values of the ends achieved; it is value-full. In *Leading from the emerging future: From ego-system to eco-system economies* (2013), authors Scharmer and Kaufer focus on two different sources of learning—learning from the experiences of the past and learning from the future as it emerges. They refer to this type of operating (or learning) from the future as "presencing." Senge, Scharmer, Jaworski, and Flowers introduced the idea of "presence"—a concept borrowed from the natural world that the whole is entirely present in any of its parts—to the worlds of business, education, government, and leadership in their 2004 book *Presence: Human purpose and the field of the future.* Too often, the authors found, we remain stuck in old patterns of seeing and acting. By encouraging deeper levels of learning, we create an awareness of the larger whole, leading to actions that can help to shape its evolution and our future.

Other leadership work espouses different "types" of leader: the servant leader, the transformational leader, the transactional leader, etc. In *The regenerative leadership handbook*, Hardman (2017) defines regenerative leadership as the capacity to restore the damage caused by human activity on natural, social, and economic systems, while at the same time securing lasting, desirable futures for all living beings through the design of integrated approaches that lead to resilient, thriving and life-affirming organizations, communities, regions, and the world. The core content of the handbook focuses on the evolving theory associated with regenerative leadership, and the regenerative capacity index that emerged as a tool designed to provide a concrete understanding of what Hardman calls the state of readiness or maturity of organizations to engage in regenerative practice.

Hardman's work is foundational for regenerative leadership as it focuses on both the self and the collective; it draws the reader to consider the type of leadership needed in regenerative development as both personal and applied. Leadership in regenerative development must focus on both the subjects and the systems and the place but must also consider the deeper consciousness development of individuals. As mentioned earlier, the work of Freire in understanding ourselves as beings of praxis and the jedi-like mindset of reflection in, and on, our actions are necessary thinking patterns in regenerative leaders.

C. Otto Scharmer, author of *Theory U: Leading from the future as it emerges* (2016), identifies a significant flaw in leadership today that he refers to as the "blind spot." The blind spot is the place from which our attention and intention originate. The blind spot refers to the interior condition (or source) of the leader. It is the place from which we operate when we do something. We are blind to it because it is an invisible dimension of our habitual social field, of our everyday experience in social interactions. Much of the leadership work arises from or focuses on knowing oneself, and identifying these blind spots is imperative.

The work of Chris Lowney on heroic leadership may provide a blueprint for educating regenerative leaders. Lowney's heroic leadership is a holistic approach to leadership focused not on techniques and tactics but rather who to be and how to live. This is why this construct is important to educating for regenerative development. Lowney's experience is born of becoming a manager for JP Morgan and being put through leadership training. But he had the rare background of previously being a Jesuit priest and experiencing a 450-year-old history of world-affecting work. As he reflected upon how the worldwide network of Jesuits proceeded, he uncovered much that is useful for educating for regeneration.

Just as critical thinking is necessary but not sufficient for regenerative development, self-awareness is merely the beginning of leadership development required in regenerative development. McCallum, Connor, and Horian (2012), whose work is based upon Chris Lowney's *Heroic leadership* (2003), describe self-awareness as the essential starting point and "the cornerstone of emotional and leadership intelligence." They further describe self-awareness as inner freedom and a precondition for the "agility required by leaders in the face of challenges and opportunities" (2012, p. 8). Reflection is paramount in designing education with the intent of developing self-awareness. Putting students in places to receive large amounts of feedback through 360° evaluation, utilizing inventories and instruments that help develop interior insights, and applied projects with goal setting and review can all be useful tools in developing students' self-awareness.

Lowney's heroic leadership model, while stemming out of self-awareness, also includes ingenuity, love, and heroism. Ingenuity is where innovation is born from imagination, adaptability, creativity, flexibility, and the ability to respond rapidly. Lowney's description of ingenuity can be summarized as new and different information requires new and different actions. It is the human characteristic of not only desiring to seek new, different, and a multitude of perspectives but believing in the continued validity of that information in the world. This is the human side of systems thinking that results in action. Ingenuity can be developed through the use of unstructured projects where conditions and actions change, thus requiring students to seek new information and focus on solutions rather than answers. It is the seeking and reaction to new information that drives the curiosity and inquisitive nature necessary for regenerative development.

The trust required of heroic leaders manifests through the concept of love (which is really what regenerative development is). Love is vision, passion, commitment, courage, and loyalty. It is the vision to see talent and potential inherent in the dignity of others (including the personal, social, and environmental) and the passion and commitment to unlock that potential. It is about the development and practice of emotional intelligence, of both empathy and affect. It is about the constant uncovering of biases and assumptions in self and others and yet not being swayed. And finally, understanding the mutually and increasingly beneficial results of regenerative development requires the loyalty to continuously support interactions.

But the greatest of the leadership traits in regenerative leaders is not love but rather heroism. Heroism is the ability to identify and challenge the status quo. Failure is not a deterrent because the possibility of a future state is clear. Heroism is active in that it is not mere thought but rather comes with a vision and tactics and is enacted through tireless implementation, evaluation, and learning. It requires leaving our comfort zones to accomplish. Problems without solutions, not merely in capstone situations but worked on throughout one's education so that the idealism and realism are thrown together, would be ideal for students to wrestle with during their education.

Leadership education is lacking in higher education, often relegated to a field or a course, but leadership is cornerstone to regenerative development. Although we have new models for regenerative leadership emerging through the work of Hardman, the amount of conceptual understanding around leadership is vast and the task of educators is to make the journey through these concepts, theories, and applications concrete enough to matter to students and their lives. Current education often focuses on finding oneself and one's place in the world at an undergraduate level. This is only the beginning if we want to educate students in regenerative development. If CityCrafting is the *what* we must educate students in to be regenerative developers, self-awareness, ingenuity, love, and heroism are the *how* we must accomplish that type of education. It is the mix of knowledge with skills but also developing one's affect for a truly holistic educational approach. There is a way ahead.

Conclusion

With some exceptions, higher education is not capable of developing the needed future generation of regenerative developers in its current iteration. Much of this arises out of a lack of agreement as to the purposes of education. And the traditional thinking of viewing the world through distinct disciplinary lenses and epistemologies only adds to the confusion for students despite their straight forward and reasonable expectations from their higher education experiences to be useful and do good in the world, to be prepared for work and life, to engage with the diversity around us, and to be drawn into society and communities to solve real problems. Arguably, these are the traits of regeneration.

There is a long history in education that is influenced by constructs both inside and beyond the walls of academe. There have been wonderful descriptions of offering education differently and studies of what makes education impactful. Many scholars have laid out the specific skills needed in regeneration, especially those already working in sustainability and related fields. This chapter, while introducing some of this work, focuses on the larger vision of what is necessary to educate for regeneration. We have taken the position that two constructs need to be fundamentally changed.

First, critical thinking cannot be the outcome of an education for a regenerative developer. Regeneration is foundationally about systems. Systems thinking is not only the basis of education for regeneration, providing the areas of focus for this work, but also a tool for examining how we might construct the best experiences for our students.

Second, educating for regeneration is really leadership development; leadership development in an increasingly complex, uncertain, volatile world. There is a growing body of knowledge on regenerative leadership. This work focuses on self as a component and the greater good as the outcome. While sharing some of that work here, we focused on heroic leadership as a construct, only reversing the order and coming from the lens of heroism and love to truly guide the development of leadership in this context.

The chapter also provided an example of curricular changes and approaches in real estate, sharing the work presently being done to change an entire field within the university structure. This work is slow and tedious and necessary. And the wise work in that program brings applied expertise to bear on the curriculum. Some processes born outside of higher education provide models to be invited into (such as CityCraft into real estate education) and adapted for higher education, not merely wandered upon if we truly want to educate the next group of regenerative developers.

References

Ackoff, R 1998, A *systemic view of transformational leadership*, viewed 3 December 2018, <https://pdfs.semanticscholar.org/dd7e/f924e44861877b76d65a865754d741d07b0f.pdf>

Adelman, C, Ewell, P, Gaston, P & Schneider, C 2014, *The degree qualifications profile*, Lumina Foundation, Indianapolis, IN, viewed 3 December 2018, <https://www.luminafoundation.org/files/resources/dqp.pdf>

Arum, R & Roksa, J 2011, *Academically adrift: Limited learning on college campuses*, University of Chicago Press, Chicago, IL.

Association of American Colleges and Universities 2015, *The LEAP challenge: Education for a world of unscripted problems*, viewed 3 December 2018, <https://www.aacu.org/sites/default/files/files/LEAP/LEAPChallengeBrochure.pdf>

Berry, W 1987, *The Loss of the university*, North Point Press, San Francisco, CA.

Biesta, G & Burbules, NC 2003, *Pragmatism and educational research*, Rowman & Littlefield Publishers.

Bok, DC 2008, *Our underachieving colleges: A candid look at how much students learn and why they should be learning more*, Princeton University Press, Princeton, NJ.

Brundtland, G 1987, *Report of the World Commission on Environment and Development: Our Common Future*. United Nations General Assembly document A/42/427.

Caplan, BD 2018, *The case against education: Why the education system is a waste of time and money*, Princeton University Press, Princeton, NJ.

Christensen, CM, Burkholder, JP & Eyring, HJ 2011, *Innovative university: Changing the DNA of higher education from the inside out*, Jossey-Bass, San Francisco, CA.

Connelly, M & Clandinin, J 1988, *Teachers as curriculum planners: Narratives of experience*, Teachers College Press, New York.

Construction, Next Generation Green Buildings Conference, viewed 3 December 2018, <http://www.integrativedesign.net/images/ShiftingOurMentalModel.pdf>

Delbanco, A. 2014, *College: What it was, is, and should be*, Princeton University Press, Princeton, NJ.

Dewey, J. 1916, *Democracy and education: An introduction to the philosophy of education*, Macmillan, New York.

Dewey, J. 1938, *Experience and education*, Macmillan, New York.

Eagan, MK, Stolzenberg, EB, Berdan Lozano, J, Aragon, MC, Suchard, MR & Hurtado, S 2014, *Undergraduate teaching faculty: The 2013–2014 HERI faculty survey*, Higher Education Research Institute, UCLA, Los Angeles, CA.

El-Haram, M, Walton, J, Horner, M, Hardcastle, C, Proce, A, Bebbington, J, Thomson, C & Atkin-Wright, T 2007, "Development of an integrated sustainability assessment toolkit" in M Horner, C Hardcastle, A Price & J Bebbington (eds), Proceedings: SUE-MOT conference 2007, *International Conference on Whole Life, Urban Sustainability and its Assessment*, 27–29 June 2007, Glasgow UK, pp. 30–44.

Flood, RL 2001, "*The relationship between 'systems thinking' to action research*", In P Reason & H Bradbury (eds), *The handbook of action research*, Sage, London.

Freire, P 1972, *Pedagogy of the oppressed*, Herder and Herder, New York.

Graff, G 1991, "Colleges are depriving students of a connected view of scholarship", *Chronicle of Higher Education*, February 13, 1991.

Guzowski, M 2011, The next generation of architectural education: Integrating a regenerative approach to sustainable design. In 40th ASES National Solar Conference 2011.

Hardman, J 2017, *The regenerative leadership handbook: A conscious, practical guide for creating purposeful, prosperous, resilient solutions to sustainability challenges in business, community, and education*, Kindle edition, Retrieved from Amazon.com.

Hauk, M 2017, "The new 'Three Rs' in an age of climate change: Reclamation, resilience, and regeneration as possible approaches for climate-responsive environmental and sustainability education", *The Journal of Sustainability Education*, vol. 12, <http://www.susted.com/wordpress/content/the-new-three-rs-in-an-age-of-climate-change-reclamation-resilience-and-regeneration-as-possible-approaches-for-climate-responsive-environmental-and-sustainability-education_2017_02/>

Hooks, B 1994, *Teaching to transgress: Education as the practice of freedom*, Routledge, New York.

Huber, M & Hutchings, P 2004, *Integrative learning: Mapping the terrain*, Association of American Colleges and Universities and the Carnegie Foundation for the Advancement of Teaching, Washington, DC.

Huber, M, Hutchings, P & Gale, G 2005, *Integrative learning for liberal education*, viewed 3 December 2018, <https://www.aacu.org/publications-research/periodicals/integrative-learning-liberal-education>

Institute for Cultural Action (IDAC) (n.d.), *Conscientisation and liberation: A conversation with Paulo Freire*, viewed 3 December 2018, <https://archive.org/details/ConscientisationAndLiberation-AConversationOfPauloFreire/page/n0>

Johns, TW, Laubacher, R & Johns, T 2014, *The big idea: The age of hyperspecialization*, viewed 3 December 2018, <https://hbr.org/2011/07/the-big-idea-the-age-of-hyperspecialization>

Kemp, MC 2006, "Integrative learning concept map", In PJ Palmer, M Scribner & A Zajonc (eds), *The heart of higher education: A call to reneval: Transforming the academy trough collegial conversations*, Jossey-Bass, San Francisco, CA.

Kuh, GD 2008, *High-impact educational practices: What they are, who has access to them, and why they matter*, AAC&U, Washington, DC, viewed 3 December 2018, <http://www.neasc.org/downloads/aacu_high_impact_2008_final.pdf>

Leef, G 2018, *How could a professor make "the case against education?"* viewed 3 December 2018, <https://www.jamesgmartin.center/2018/01/professor-make-case-education/>

Lowney, C 2003, *Heroic leadership: Best practices from a 450-year-old company that changed the world*, Loyola Press, Chicago, IL.

Malone, TW, Laubacher, R & Johns, T 2011, *The big idea: The age of hyperspecialization*, Harvard Business Review, July-August 2011.

McCallum, DC, Connor, J & Horian, L 2012, *A leadership education model for Jesuit business schools*, viewed 3 December 2018, <https://www.stthomas.edu/media/catholicstudies/center/ryan/conferences/2012-dayton/McCallumFinalPaperRe.pdf>

Mulgan, G & Townsley, O 2016, *The challenge driven university: How real-life problems can fuel learning*, viewed 3 December 2018, <https://media.nesta.org.uk/documents/the_challenge-driven_university.pdf>

Palmer, PJ, Scribner, M & Zajonc, A 2010, *The heart of higher education: A call to renewal: Transforming the academy trough collegial conversations*, Jossey-Bass, San Francisco, CA.

Pelling, M 2011, *Adaptation to climate change: From resilience to transformation*, Routledge, New York.

Reed, B 2006, Shifting our mental model—"Sustainability to Regeneration", in *Proceedings of Rethinking Sustainable Construction 2006: Next Generation Green Buildings*, Sarasota, FL US.

Roberts, J 2013, "Experiencing sustainability: Thinking deeper about experiential education in higher education", *The Journal of Sustainability Education*, vol. 5, viewed 3 December 2018, <http://www.susted.com/wordpress/content/experiencing-sustainability-thinking-deeper-about-experiential-education-in-higher-education_2013_05/>

Robinson, J 2004, "Squaring the circle? Some thoughts on the idea of sustainable development", *Ecological Economics*, vol. 48, pp. 369–384.

Sanford, C 2017, *The regenerative business: Redesign work, cultivate human potential, and achieve extraordinary outcomes*, Nicholas Brealey Publishing, Boston, MA.

Scharmer, CO 2016, *Theory U: Leading from the future as it emerges*, Barrett-Koehler Publishers, Inc., Oakland, CA.

Scharmer, CO & Kaufer, K 2013, *Leading from the emerging future: From ego-system to eco-system economies*, Barrett-Koehler Publishers, Inc., Oakland, CA.

Senge, P, Scharmer, CO, Jaworski, J & Flowers, BS 2004, *Presence: Human purpose and the field of the future*, Doubleday, New York.

Squires, G 1992, "Interdisciplinarity in higher education in the united kingdom", *European Journal of Education*, vol. 27, no. 3, pp. 201–210.

Tbilisi declaration 1977, viewed 3 December 2018, <https://www.gdrc.org/uem/ee/tbilisi.html>

Williams, K 2012, "Regenerative design as a force for change: Thoughtful, optimistic and evolving ideas", *Building Research & Information*, vol. 40, no. 3, pp. 361–364.

14 Conclusion

Beth Schaefer Caniglia, John L. Knott, Jr., and Beatrice Frank

The focus of this book has been to bring together practitioners and social science scholars to examine the theory and practice of regenerative development. The challenges that emerge at the intersection of climate change, urbanization, and inequality necessitate a vision broader than sustainability. Practitioners and scholars alike argue that regenerative development provides such a vision. Before now, however, very few scholars have interrogated the concept of regenerative development through the lens of existing theories. Likewise, regenerative development practitioners had very little interaction with social science scholars to contextualize their work with existing sociocultural frameworks. Regenerative development purports to put people, prosperity, and planet into interaction with one another in ways that advance all three. To achieve such a lofty goal, scholars and practitioners alike need to build consensus definitions—not only of regenerative development itself, but also of significant framing concepts, such as systems and bioregions. Similarly, the scope and scale of regenerative development projects, as well as the types of indicators being used to measure progress toward regeneration, require considerable advancement beyond the current state of knowledge.

The contributors to this book have given us insights, theoretical and empirical, that provide significant direction for how to evolve current research and practice. Each chapter has provided a definition of regenerative development, allowing us to put forth a more detailed articulation of the concept. An experienced group of practitioners has shared the processes, skill sets, and frameworks that guide them in the field. Similarly, a diverse group of scholars from sociology, cultural geography, environmental studies, psychology, and philosophy/ethics examined the ways the concept of regenerative development resonates with their extensive scholarly literature. In this chapter, we will elaborate the themes of their contributions. Specifically, we will synthesize the key elements that characterize the most common definitions of regenerative development. We will also examine central considerations that we feel advance existing best practice in the field and articulate a research agenda to pursue as the social sciences incorporate regenerative development further into their frameworks.

Key themes

Defining regenerative development

It is critical to articulate an actionable definition of regenerative development that clearly distinguishes its dimensions from closely related concepts, such as sustainability and resilience. In Chapter 3, Caniglia argues that the following definitions should be used based upon the vast body of relevant literature:

- **Sustainable development:** The most common definition of sustainable development was put forth by the World Commission on Environment and Development in *Our Common Future*, and states that sustainable development is "development which meets the needs of the present without compromising the ability of future generations to meet their own needs."
- **Resilience:** The ability of a system to absorb external shocks without altering the existing relationships between species populations and other ecosystem characteristics. In this tradition, resilience can be measured by how long it takes to restore stability within the system or as the extent of disturbance a system can take before it crosses characteristic thresholds into a new stability regime.
- **Regenerative development:** The priority of regenerative development is to heal existing damages in communities and ecosystems, which are connected in bioregional contexts, in ways that create abundance for people, the economy, and the planet. The framework advocates applying holistic processes to create feedback loops between physical, natural, economic, and social capital that are mutually supportive and contain the capacity to restore healthy and prosperous relationships among these forms of capital.

Contributing authors presented a variety of principles that guide the practice of regenerative development, which we will quickly review here. For a detailed discussion of these, please see the referenced chapters.

Bioregional emphasis

Typically defined by watershed barriers, a bioregion is a geographic space wherein social and ecological systems are tightly coupled. Actions in one part of the bioregion have a direct effect on other parts of the system. Local practices related to waste management and litter will have a direct effect on water quality, the health of fish and wildlife, and in turn the quality of recreational experiences of those who visit local waterways. Proximity to green spaces in local neighborhoods will impact exposure to nature and wildlife and in turn shape citizens' values surrounding the outdoors. Regenerative development builds upon this tightly coupled human-natural system to leverage common interests in clean air, water, and overall quality of life. Our contributors consistently make the case that the relatively local scale (community, city, bioregion) is the most effective scale for building the vision, motivation, trust, and momentum needed to move the needle toward regenerative outcomes. Knott's chapter captures this sentiment when he argues that a scale is needed that leverages resources and drives efficiencies across sectors (public, private, and nonprofit). The pairing of appropriate scale and cross-sector collaboration attracts additional like-minded investment.

A systems perspective is required in order to see the connections between the human and natural system (see chapters by Caniglia, Cross and Plaut; Dietz; and Knott), and an understanding of the ways feedback loops in systems work is extraordinarily important. Cause and effect are not immediate in bioregions, and there are subsystems within overarching bioregions that exhibit patterns of interaction and feedback that impact outcomes in other areas. If we take an isolated look at the complex systems that make up a city, for example, we see that the placement of public transportation, even if it is meant to serve the poor or communities of color, can lead to increased housing costs that drive out the residents of origin. Dislocation of the poor and communities of color leads to homogenization of the urban core and leads us to lose what Pellow calls the "diversity advantage" that cities provide. Multiple subsystems within bioregions are impacted by economic pressures like these and can throw them out of whack over time. If, for example, we take Wilkerson and Dake's advice and create bioregional workforce development programs close to currently underserved communities, gentrification and displacement patterns will eventually move beneficiaries further away from those opportunities, resulting in program closures and potentially costly relocations of workforce development services. These feedback loops and time lags are complicated in cities and bioregions, but they are much easier to observe, acknowledge, and effectively manage at the bioregional level than at larger scales.

Strong consensus is found in the chapters regarding the need for regenerative development to have a positive impact on four kinds of capital: economic, physical, human, and ecological. Physical capital is the least commonly used and refers to infrastructure, such as power grids, roads, buildings, and bridges. Hunter Lovins points out that our current economy systematically depletes human and ecological capital, while bolstering economic and physical capital. Of course, the paradox is that economic and physical capital depends upon natural and human resources for their creation, operation, consumption, and overall valuation. As such, Lovins argues that we must build an economy in service to life. In bioregional systems, we can measure and evaluate the rise and fall of each form of capital over time at a scale that is more manageable to influence than at larger scales. These forms of capital have meaning in places. People have real faces; people see their neighbors displaced in ways that resonate much stronger than unemployment rates. No fishing or no swimming signs at a favorite recreation site spark much broader interest in environmental problems than discussions of waterborne diseases far away. Our contributors highlight the powerful impact of direct observation and personal experience of economic, physical, natural, and human capital.

Thus, the bioregional scale is also best for applying the broad principles of regenerative development, sustainability, and resilience, because it is the scale that allows the application in ways uniquely suited to local socio-ecological contexts.

There is debate among our contributors regarding how to capture and describe the "uniqueness" of places. For example, Carol Sandford refers to the principle of designing in ways that fulfill the "essence" of a place, organization, or community. Cross and Plaut describe the ways bioregional focus provides solutions that factor in the "uniqueness" of a place. However, consensus appears across these chapters that a critical benefit comes from working at the bioregional scale, because local cultural practices can be leveraged, intersectoral cooperation is easier to create, and impacts can be best observed at the bioregional and city scales. Dietz cautions that bioregions are not immune to actions at higher scales, nor are the actions of bioregions independent from those larger scales. Actions at the state, national, and global levels can and will have influence at bioregional levels despite efforts to curb the effects of outside forces; and actions at the bioregional level may have negative outcomes for areas outside of the local bioregion. The nested nature of coupled human and natural systems will always necessitate consideration of impacts at multiple scales.

Managing change in complex systems: Participatory processes

Strong consensus emerged from our contributors that participatory processes were essential to insure regenerative outcomes for everyone. However, even the best participatory processes fail to include all perspectives and prepare for all eventualities in complex and interdependent systems. As a result, Dietz elaborates the analytic-deliberative approach as a means of governing decision-making for regenerative development. This approach brings the best scientific consensus into dialogue with local priorities through a participatory process guided by eight key questions that require attention before making development decisions.

- How will we define the well-being of human beings, other species, and the environment?
- Are we allocating resources efficiently?
- Is our analysis taking adequate account of uncertainty about facts?
- Are we identifying and acknowledging value uncertainty and value conflict?
- Are we engaging the full range of useful experience?
- Is both the process of governance and the outcome fair to all?
- Does the process build on human cognitive strengths and compensate for weaknesses?
- Is the project designed to learn from experience?

Despite the advantages that stem from working at the bioregional scale compared to the national, state, or global levels, human beings are limited in the amount of information they have when making decisions in complex systems. Furthermore, Dietz highlights that conflicting values and the uneven distribution of voice and/or power in deliberative processes can lead to decisions that fail to deliver regenerative outcomes. Even well-designed deliberative processes can result in poor decision-making when important populations are left unrepresented.

How, for example, do we make decisions on behalf of the more-than-human residents of our bioregion, including plants, animals, and waterways (see chapters by Dietz and Pellow)? Similarly, the poor, single mothers, youth, and communities of color are often underrepresented in public hearings. Dietz provides us tools for keeping our deliberative processes focused on regenerative outcomes for everyone.

John Hardman—our good friend and collaborator in advancing regenerative development—describes regenerative leadership as the capacity to restore the damage caused by human activity on natural, social, and economic systems, while at the same time securing lasting, desirable futures for all living beings through the design of integrated approaches that lead to resilient, thriving, and life-affirming organizations, communities, and regions around the world (see chapter by Sagendorf and Jackson). Cross and Plaut articulate that critical shifts in frameworks are required to develop regenerative stakeholder participation processes. Specifically, to produce a regenerative future, a shift is required in six focal areas: whole systems, being of service, human interdependence with nature, accounting for uniqueness, focus on potential, and intentional network weaving. Regenerative practitioners and those organizing participatory processes for regenerative outcomes need to acknowledge and communicate these ground rules, establishing a baseline of shared understanding of the coupled human-natural system and the regenerative goals underpinning analytic-deliberative governance. These principles were echoed by multiple contributors to this book (see chapters by Dietz, Knott, Pellow, and Sanford).

The equity deficit

David Pellow's chapter highlights the "equity deficit" that exists in current elaborations of regenerative development—a theme that was echoed by several of our contributors. Similar to the literature and practice of sustainability, equity is often treated as an assumed outcome of regenerative development, rather than examined deeply as an explicit goal. Lovins highlights in her chapter the excessive inequality that currently exists around the world. Oxfam estimated in 2017 that the top 1% of wealth holders has more than the bottom 99% worldwide and that eight people have as much wealth as the poorest 3.5 billion people on the planet. She argues that the current economic system depletes human capital in devastating ways, restricting the ability of our citizens to contribute their talents to our highest aims. The injustices intrinsic in this system are multiplied by the vulnerabilities experienced by the poor, youth, single women, and communities of color around the world. Again, Lovins highlights what has been cited in multiple publications that the most vulnerable populations will be hit first and worst by climate change (see also Caniglia, Frank, and Vallee 2017). Inequality and power imbalances are also root drivers of ecological unsustainability (Pellow's chapter) and as such must be treated explicitly in regenerative development work.

Pellow highlights that where inequality is considered in the regenerative development literature, it fails to recognize the nuanced nature of inequality around the world. Social class is not the only predictor of oppression and marginalization.

Rather, Pellow argues that "racism, patriarchy, nativism, colonialism, and other systems of power" intersect. Just as any other relevant system must be examined in the bioregional contexts where regenerative development is the goal, systems of inequality and marginalization should be studied in ways that "build deeper analyses of the problems and stronger coalitions for charting solutions" (Pellow, p. 77–78). It is fair to say that every contributor to this book comprehends regenerative development to be a participatory processes designed to bring about shared prosperity on a healthy planet. However, chapters by Pellow, Sheehan, and Dietz argue clearly that even the best designed participatory processes will fail to be equitable, because existing power balances will underpin their implementation. All three authors suggest that practitioners and scholars can advance a more just form of regenerative development by explicitly considering the ways exclusion and dispossession play out in participatory processes. Pellow argues that advancement will come from the incorporation of environmental justice perspectives, while Sheehan and Dietz propose formal frameworks for insuring more equitable processes that account for diverse perspectives in ways that promote justice.

We stand in support of the need to promote a regenerative development research and practice agenda that systematically advances the extradition of those forms of inequality that stem from institutionalized oppression, greed, hate, and injustice. Regenerative development must examine systems of inequality with the goal of "healing the social fabric" of our communities (see chapter by John Knott) and improving human well-being (see chapter by Lovins). We can begin by examining the work of Agyeman, Bullard, and Evans (2003, p. 5) on just sustainability, which is "the need to ensure a better quality of life for all, now and into the future, in a just and equitable manner, while living within the limits of supporting ecosystems" (quoted in chapter by Pellow). Adding the components of environmental justice cited by Pellow is the next important step. These include the following:

- The right to accurate information from authorities concerning environmental risks (transparency).
- Public hearings (self-determination).
- Democratic participation in decision-making regarding the future of any threatened community.
- Compensation for injured parties from those who inflict harm on them (Polluter Pays Principle).
- Expressions of solidarity with survivors of environmental injustices.
- Call to abolish environmental racism/injustice.
- Procedural justice (Pellow; Schlosberg 2003, 2007).

Cross and Plaut argue that regenerative development is a processes that is "participatory, evolving, and constructivist", which calls upon practitioners and participatory design professionals to employ their considerable skills in ways that advance equitable and just outcomes for people and the biosystems where they reside.

Key skill sets

To advance the full realization of regenerative development potential, a variety of skill sets are needed. To begin with, integral thinking and systems thinking are the foundations of both research and the practice of regenerative development. As Caniglia argues in her chapter, systems thinking is very different from traditional ways of seeing the world. While traditional cause and effect models emphasize individual events, systems scholars focus on patterns of behavior over time. Systems thinking requires attention to feedback loops that cause particular patterns of behavior and recognizes that the effects of feedback loops can be delayed or nonlinear, depending on the ways they interact with the behaviors of subsystems within the overarching bioregion. Systems are structures with goals. Scholars and practitioners have to recognize that the goals of systems can be at odds with current values (see chapters by Burns, Boyd and Leslie and Caniglia) and develop tools for recognizing, measuring, and leveraging these system characteristics in ways that produce abundance for people, prosperity, and planet. Cross and Plaut call this skill "systems actualization." To leverage the complexity of systems, we need to develop our ability to observe system dynamics and structure, and we need to understand system behavior at multiple scales—including temporal delays and nonlinear interaction effects across subsystems within bioregions. We also need to measure the interaction effects that take place among the four forms of capital: human, natural, economic, and physical. Caniglia argues that this systems orientation is critical to the achievement of sustainability, resilience, and regenerative development. Cross and Plaut makes a very similar argument when they highlight framework thinking and living systems understanding as central skill sets needed by regenerative development practitioners.

Participatory processes are central to regenerative development practice. Aside from drawing upon existing best practices in stakeholder engagement, our contributors recommend a variety of additional skill sets to insure regenerative outcomes. Cross and Plaut suggest that facilitators "design and deliver group activities and processes that build capacity towards Systems Actualizing." In their framework, systems actualization is defined as "to awaken the regenerative capability embodied in all living systems to create increasing levels of vitality, viability, and capacity to evolve the systems of which they are a part." They draw upon positive psychology to illustrate the types of outcomes regenerative development practitioners are striving to produce: positive emotion, engagement, meaning, positive relationships, and accomplishment. In addition to developing these outcomes for individual participants, regenerative development practitioners also "attend to the creation of positive institutions and communities." The ultimate goals include building trust, collaborative learning, long-term engaged networks, and an orientation among participants of group efficacy in cocreating an alternative, more regenerative future. Dietz's chapter on applying an analytic-deliberative approach to making decisions in participatory processes is complementary in this regard. The process advocated in the analytic-deliberative model calls for the codesign of research and decision-making by all stakeholders, which builds transparency, shared understanding, and trust.

Chapters by Mang, Pellow, Sheehan, Dietz, and Burns, Boyd and Leslie highlight the need for facilitators and stakeholder processes to address ethics, morality, and social justice. This entails a shift away from stakeholder processes that develop consensus that serves those in attendance at stakeholder engagement meetings. Regenerative outcomes for all members of a given bioregion, along with the creation of increased vitality in multiple forms of capital need to remain central to the goals of regenerative participatory processes. Knott lists systems scale integration, capital mapping, capital integration, collective impact assessment, and measurement of desired outcomes as critical skill sets for the regenerative practitioner. Caniglia and Lovins highlight the principles of regenerative economies articulated by John Fullerton at the Capital Institute, which include the following:

- Right relationship among components in the system.
- Innovative, adaptive, and responsive to uncertainty and change.
- Capital integration.
- Empowered participation of all members of the bioregion.
- Robust circulatory flow of resources throughout the system.
- Attention to edge-effect abundance.
- Seeking balanced outcomes that attend to system dynamics at all levels of the bioregion.
- Honors community and place.

This set of skills—systems thinking, integrated capital mapping, systems actualizing, ability to understand coupled human and natural systems, building enduring networks, integrating contributions across sectors (nonprofit, industry, government, and educational)—is broad and will require new approaches to existing development practices. We elaborate these in the next section.

Lessons for research, practitioners, educators, and governance

Educational design

Based on these complex skill sets, educators will be on the spot to create even more opportunities for students to join transdisciplinary programs and develop curriculum that includes the frameworks needed by regenerative development practitioners. Sagendorf and Jackson describe current characteristics of contemporary educational paradigms that make regenerative education a challenge. First, subject silos lead to an emphasis on information delivery, rather than skill development. Binge and purge knowledge acquisition, which is encouraged by testing, leads students to see their coursework as a collection of unrelated requirements. Additionally, the skills required for regenerative work are difficult to teach, because they require students to develop tertiary analytical abilities, which many instructors also lack. The need for students to develop systems thinking *and* empathy *and* facilitation skills *and* basic knowledge of coupled human and natural systems pushes our educational systems to redesign curriculum in critically important ways. Educational curriculum is based in frameworks that form overarching narratives of how things came to be this way and what solutions might exist. As we described in the previous section on the bioregional and systems dimensions of regenerative development work, traditional analytics focus on particular problems at a given point in time. However, regenerative development practitioners need to develop the ability to analyze patterns of behavior over time, system characteristic feedback loops, temporal lags, and interaction effects that coexist within larger systems of oppression and exclusion and have subsystems with conflicting goals. Faculty need to develop coursework and cocurricular activities that (1) foster the ability to comprehend multilayered systems in this complex way, (2) illustrate the ways patterns of behavior in natural systems trigger patterns of behavior in social systems (and the other way around), (3) be a boundary spanning actor, who can speak the language of people from multiple sociocultural groups. These are just a few of the skill sets educators will need to understand when developing regenerative development curriculum.

John L. Knott, Jr describes one innovative attempt to create a regenerative educational collaboration in Denver, Colorado, called the CityCraft Integrated Research Center. The center brings together faculty from most of the major higher education institutions in Colorado to design and implement bioregional scale coupled human and natural systems research in the greater Denver metropolitan area and its related hinterlands. Several members of this collaboration are authors of chapters in this book, and their exchanges have built significant synergy in the articulation of regenerative development metrics, modeling, research, and curriculum development. The members of the center have developed a variety of projects that integrate educational sector members with government officials, nonprofits, local businesses, foundations, and neighborhood associations. The central organizing principle of CIRC is regenerative and encourages shared resources, cross-university program development, and applied insights in sites occupied by the area's most vulnerable citizens. While the ideal goal is to develop a cross-university curriculum accessible to students at each participating university, in reality it has been difficult to overcome university silo behaviors, resistance to new approaches within existing academic departments, and the perspective that universities are competitive with and independent of each other. We argue that to achieve bioregional regenerative development, systems of higher education within shared bioregions themselves must become regenerative, allowing resources to flow more freely in partnership with other sectors in their bioregions. Bringing together the intellectual and practical expertise from a range of higher learning institutions to conduct research and offer joint degrees in service to the goals of regenerative development can build local competencies. Land grant institutions and those with a mission to apply their skills and knowledge to improve the world around them are natural leaders in the creation of collaborative structures and joint degrees. We advocate specifically that consortium exchange systems be replicated, whereby students at one university or college in the consortium can take classes at other consortia universities for the same tuition.

Scholarly agenda

Before this book was published, the scholarship on regenerative development came predominately from the fields of architecture, planning, and design. As Pellow and others contributors point out, that literature has an "inequality deficit" and fails to examine either the natural or social science knowledge required to truly move the needle toward regenerative urban development. The range of fields where regenerative development is relevant is vast, as demonstrated by the diverse chapters in this book. Blind spots exist regarding how to measure regeneration, how to govern regenerative decision-making, how to design regenerative participatory processes, and how to develop regenerative leaders, regenerative workforces, and regenerative memorializing. The list of underexplored areas is long. We have benefitted greatly from bringing together the scholars and practitioners in this book to lend their insights, critiques, and recommendations as a starting point for clarifying the current state of knowledge and charting a rich set of areas poised to advance our understanding and practice of regenerative development. The conceptualization of regenerative development set forth by our contributors and articulated above clarifies some of the metrics that need to be included in studies designed to test regenerative development theories as they develop. Scholars in regenerative development studies need to measure baseline conditions, so that changes over time can be apprehended. And, because the goal of regenerative development is to put people, prosperity, and planet in interaction in ways that produce abundance across all three, scholars need to measure interaction effects and the feedback loops that exist between them.

The practice of regenerative development

Traditional development in cities is guided by existing structures, such as the availability of capital, preferential development zones, and presence of attractive amenities. Development projects within cities are perceived as independent from the bioregion, even if they are considered connected to the immediately surrounding areas. Regenerative development calls upon city officials, developers, financers, and those slotted to occupy newly developed places to approach their projects with broader impacts and potential in mind. John L. Knott, Jr, considered by many to be the father of sustainable urban development, shares the processes and principles that mark his company's regenerative development work. Rather than focused narrowly on developing buildings or campuses, CityCraft is focused on developing communities. Grounded in a master builder tradition and coupled with the Sanborn Principles, Knott lays out a thoroughly designed set of values, principles, and processes that can guide regenerative development projects from start to finish, resulting in projects that improve the long-term health and grows capacity across all related human, natural, infrastructure, and economic systems. The work is based on extensive research regarding the local bioregion, participatory processes that are both cross-sectoral and place based, and evolutionary in focus. We particularly encourage practitioners to thoroughly consider the culmination of Knott's life's work, because this chapter represents one of the most expertly designed set of regenerative development practices in the world.

We are also fortunate to have Carol Sanford, Josie Plaut, and Nicholas Mang as practitioner contributors to this book. Carol has published more than anyone else on the topic of regenerative business practices and provides a strong foundation for businesses and business schools to build upon as the regenerative framework is adopted more widely. Josie Plaut—an experienced stakeholder process practitioner lays out the characteristics and skill sets needed to serve as a system actualizer in service to regenerative development. These insights advance our current understanding of regenerative development in practice. After carefully evaluating the existing literature and the contributions of practitioners in this book, our editorial team feels that practitioners should increase their knowledge about and attention to the sociocultural dimensions of regenerative development. Knott's work at CityCraft Ventures serves as a shining example of a deeply integral approach, which includes capital mapping across all four forms of capital (human, natural, physical, and economic) and provides step-by-step guidelines for organizing participatory design processes. Cross and Plaut's chapter also provides resources for practitioners toward this end, including the skill sets needed by regenerative design practitioners and recommendations for best practices in participatory process design.

Why cities?

In very short time, urban areas will house three-quarters of the world's population. If we can get cities right, we can alleviate some of our greatest challenges, including climate change and extreme inequality. Our authors highlight several ways to transform the organizing principles of cities toward regenerative outcomes. The greatest hope lies in the energy system transformations that are already underway around the world, which Lovins elaborates in great detail. Other areas present challenges, however, including governance, community engagement, ethics, and building urban systems that coevolve in positive ways with their surrounding bioregions. As stated by Wilkerson and Dake, one of the key challenges faced by today's cities is to restore human capital in the context of social sustainability: Cities must pursue development practices that are "compatible with the harmonious evolution of civil society, fostering an environment conducive to the compatible cohabitation of culturally and socially diverse groups while at the same time encouraging social integration, with improvements in the quality of life for all segments of the population." Pellow highlights that cities have a "diversity advantage" that positions them to take advantage of the strengths that diversity provides, such as increased creativity. And the leadership role being played by cities in the climate arena, for example, lead many of our contributors to believe cities can extend their innovations toward challenges of diversity, equity, and inclusion faster and more effectively than states or nations can. Recent emphasis on local economic systems bodes well for workforce development that overcomes existing disenfranchisement and increased access to constantly improving educational systems.

Building a regenerative culture, both among residents of cities and their local authorities, will require a shift in many places, however. Community engagement processes, for example, need to offer genuine learning opportunities; too often these processes are used to placate the general population, rather than to gather insights that build on "diversity advantages." Community leaders across all sectors need to develop integral and systems thinking skills and to re-center their values around the creation of abundance for local human and natural systems. Current patterns of inequality in leadership lead the experiences of the poor, communities of color, the other abled, and the more than human populations outside of central governance agendas. Several of our contributors, including Knott, Dietz, Burns, Caniglia, Cross and Plaut, Wilkerson and Dake, and Pellow, argue that participatory processes designed around regenerative development principles provide the best pathway to create shared prosperity on a healthy planet. However, regenerative principles will need to be fostered and constantly kept at the forefront of governance processes and decision-making.

Conclusion

Hunter Lovins presents a powerful quote from Buckminister Fuller in her chapter: "You never change things by fighting the existing reality. To change something, build a new model that makes the existing model obsolete." Regenerative development is poised to provide such a model—one that sketches a path for cities to lead us toward shared prosperity on a healthy planet. According to Lovins, the world has to commit to three outcomes in order to create institutions that serve human well-being, rather than purely economic accumulation:

1 Enable all people to achieve a flourishing life within ecological limits.
2 Deliver universal well-being as we meet the basic needs of all humans.
3 Deliver sufficient equality to maintain social stability and provide the basis for genuine security.

Lovins rightly argues that our current economy is in service to the accumulation of wealth by the already wealthy, as proven by over a century of data highlighting massive disparities in wealth accumulation around the world. A new model is definitely needed, but the prospects for positive outcomes are strong. Solving the climate crisis will create millions of jobs. The New Climate Economy Report shows that decarbonization of the world economy will deliver a net economic gain of $26 trillion, and cities are leading the way to this new economy (Lovins).

The largest change needed is to transform our thinking away from sustainability toward regeneration. Our current system is not worth preserving, because it only serves the rich. We need a system that supports all life – human, flora, and fauna. Knott argues (p. 161–162):

> If we are to achieve a regenerative future, we must understand that as humans we are both biologic organisms and social beings with a spiritual dimension. As a biologic organism, we need adequate clean air and water, healthy soil to

provide healthy food and resilience shelter based on our unique bioregions. As a social being we require a network of others organized and connected to us supporting a healthy and thriving community each with access to the same resources. Any decision that threatens the capacity or health of these required resources is not regenerative.

As defined above, the priority of regenerative development is to heal existing damages in communities and ecosystems, which are connected in bioregional contexts, in ways that create abundance for people, the economy, and the planet. The framework advocates applying holistic processes to create feedback loops between physical, natural, economic, and social capital that are mutually supportive and contain the capacity to restore healthy and prosperous relationships among these forms of capital. Based on the insights of our experts—practitioners and scholars alike, a transition away from a sustainability approach toward regeneration holds the best promise for our common future.

References

Agyeman, J, Bullard, RD & Evans B 2003, *Just sustainabilities: Development in an unequal world*, The MIT Press.

Caniglia, BS, Vallee M & Frank B 2017, *Resilience, environmental justice and the city*, Routledge.

Schlosberg, D 2003, "The justice of environmental justice: Reconciling equity, recognition, and participation in a political movement", In A Light & A de Shalit (eds.), *Moral and political reasoning in environmental practice*, MIT Press, Cambridge, MA.

Index

Note: Page numbers in **bold** indicate tables; those in *italics* indicate figures.